Ecological Studies

Analysis and Synthesis

Edited by
W.D. Billings, Durham (USA) F. Golley, Athens (USA)
O.L. Lange, Würzburg (FRG) J.S. Olson, Oak Ridge (USA)
H. Remmert, Marburg (FRG)

Volume 66

Ecological Studies

Wayne T. Swank   D.A. Crossley, Jr.
Editors

Forest Hydrology and Ecology at Coweeta

With 151 Illustrations

Springer-Verlag
New York · Berlin Heidelberg
London Paris Tokyo

Wayne T. Swank
Southeastern Forest Experiment Station
Coweeta Hydrologic Laboratory
Otto, NC 28763
USA

D. A. Crossley, Jr.
Department of Entomology
University of Georgia
Athens, GA 30602
USA

Library of Congress Cataloging-in-Publication Data
Forest hydrology and ecology at Coweeta.
  (Ecological studies ; v. 66)
  "Based on papers presented during a 3-day
symposium held in Athens, Georgia, in October 1984
to commemorate 50 years of research at the Coweeta Hydrologic Laboratory" — Pref.
  Bibliography: p.
  Includes index.
  1. Hydrology, Forest — North Carolina — Congresses.
2. Forest ecology — North Carolina — Congresses.
3. Coweeta Hydrologic Laboratory (U.S.) — Congresses.
I. Swank, Wayne T. II. Crossley, D. A. III. Coweeta
Hydrologic Laboratory (U.S.) IV. Series.
GB705.N8F67   1987   551.48′0915′2   87-23431

Typeset by Publishers Service, Bozeman, Montana.
Printed and bound by Quinn-Woodbine.
Printed in the United States of America.

9 8 7 6 5 4 3 2 1

ISBN 0-387-96547-5 Springer-Verlag New York Berlin Heidelberg
ISBN 3-540-96547-5 Springer-Verlag Berlin Heidelberg New York

Foreword

As far as I have been able to ascertain, the 50 years of research on Coweeta watersheds represent the longest continuous environmental study on any landscape in North America. In a recent report prepared by the Institute of Ecosystem Studies (IES) at Millbrook, New York (Occasional Publication Number 2, 1986), the only longer study listed was the agricultural grassland research at Rothamsted in England, which has continued for over 100 years. Coweeta was also one of the first to adopt a large-scale experimental approach to the study of a natural landscape and one of the first to set up permanent water flow monitoring devices.

One of the major conclusions of the IES report is that dedicated leaders play an important role in the success of long-term studies. Cited as examples are the Likens-Bormann 25 year Hubbard Brook watershed study and Kendeigh's 35 year study of bird populations on an Illinois woodlot. Having one or two dominant leaders ensures that goals and methods will be comparable over an extensive period of time, thereby increasing the monitoring value of the project. But 20 to 30 years of effort is about the longest one could expect from any one individual. And even the most innovative leader can "get into a rut," so to speak, so there is merit in what I am choosing to call the "Evolving LTES (Long-Term Ecological Study)" with leadership changing as ecological theory and practice change, and as national needs, funding opportunities and government policies change.

Coweeta has been able to obtain financial support over these many years because it is a good example of an Evolving LTES that has taken advantage of new ideas, new personnel, and new funding opportunities as they have appeared. In its 50-year history there have never been any one or two persons who have dominated the research for very long; instead, leadership has shifted, and team work between governmental and academic researchers and between teachers and students has been promoted.

Although published papers from Coweeta are diverse, we can discern three major shifts in emphasis. The earliest studies focused on hydrology, especially water yield downstream as it is affected by different land uses and forestry practices. Then, logically, interest shifted from water quantity to water quality and the cycling of water and nutrients within the watersheds. In general, these studies showed that reducing the biomass of vegetative cover increased the water yield downstream but also altered its quality. The U.S. Forest Service provided major support for this early work and has continued to support the program throughout the 50 years. As University of Georgia ecologists and foresters became fully involved some 25 years ago, biotic components of the ecosystem—vegetation, insects, vertebrates, soil biota, stream life, and so on—began to receive increasing attention. The U.S. contribution to the International Biological Program (IBP), supported by the National Science Foundation (NSF), played an important role in this recent period, since Coweeta was selected as one of the major study sites within the Eastern Deciduous Forest Biome, and then later as one of NSF's first LTERs (Long-Term Ecological Research sites).

Over the years, Coweeta has become a major educational and training center. Some 1000 visitors—professionals, students, decisionmakers, and citizens—come each year. Coweeta alumni are currently found throughout the academic, private, and public sectors of science; in many cases they are leaders in their field. Lessons learned at Coweeta are especially applicable to land use planning.

Overall research at Coweeta has made major contributions to the understanding of the impact of man-made perturbations (such as clearcutting) on major inputs, outputs, and internal functions of forested watershed ecosystems, and—equally important—the effect of periodic natural stresses, such as droughts and caterpillar defoliations on such functions. The chapters in this volume summarize in considerable detail these findings.

Eugene P. Odum, Director Emeritus
Institute of Ecology
University of Georgia

Preface

This book is primarily based on papers presented during a 3-day Symposium held in Athens, Georgia, in October 1984 to commemorate 50 years of research at the Coweeta Hydrologic Laboratory. This undertaking was a challenge, since Coweeta is one of the oldest continuously operating facilities of its type in the world. Multiple authorships of numerous papers precluded approaching the topic from a tight, integrated synthesis of science. Our goal was to summarize and highlight major contributions from Coweeta to hydrologic and ecological understanding of Southern Appalachian forested lands. Moreover, it was our intent to provide a foundation of knowledge and baseline information essential to the future synthesis of more focused topics of ecosystem research at the site.

Over 550 published papers document the research findings at Coweeta and it was not possible to include all contributions of the program. Some major accomplishments omitted from this summary include research on hydrologic modeling, multiple use forest management, and several land use demonstrations. Similarly, some of the current efforts in process research are not reflected in this book. For example, more than 25 poster papers were presented at the Symposium, which included topics such as dynamics of dissolved organic carbon in streams, soil denitrification, leaf litter decomposition, stream meiofauna, forest stand dynamics, aluminum biogeochemistry, debris avalanches as land forming process, and sulfur gas emissions from forest floor and tree leaves. These studies and other research activities are derived from more than 30 cooperative research projects currently in effect at Coweeta involving collaborative studies with a variety of institutions and Federal and State agencies. Studies encompass systematic and population biology level investigations as well as more directly applied forest resource research.

Topics in the introductory and concluding sections, i.e., historical perspectives and relevance of ecosystem science to management needs, are frequently overlooked in documentation of long-term research programs. Yet these features provide evidence of why such programs have persisted, and because Coweeta has rich experience in both areas, we felt it was appropriate to address these topics. The level of detail varies substantially within the volume, since the contents range from updated syntheses of previously summarized subject matter to those containing previously unpublished analyses.

This volume is dedicated to the research support staff who have worked at Coweeta, including field and laboratory technicians, secretaries, and data processing personnel. These individuals are the fabric of and provide the continuity of any successful long-term field research program, but they seldom receive credit for their valuable contributions. Some of the staff served only a few years but contributed greatly to progress; others have diligently fulfilled their tasks for over 30 years. With apologies, we omit names of individuals for fear of excluding some, but we salute you as the unsung heroes who, collectively, are responsible for the stability, continuity, and productivity of the program. We also wish to recognize the contributions of more than 50 graduate students who have completed theses and dissertations at Coweeta and the cadre of 45 students currently conducting on-site studies. The originality and hard work invested by students have formed a consistent core of research effort in the interdisciplinary program.

Many individuals deserve acknowledgment for their roles in producing this volume. We are indebted to Frank Golley for his editorial advice and guidance during the preparation of the book. Manuscript typing was provided by Mary Lou Rollins and Jean Swafford and we are grateful to Polly Ann Casale, Salli Spaulding, and Kim Tidwell for word processing. Steve Waldroop drafted many of the figures including the cover illustration. Most authors of chapters in the volume provided substantive reviews of papers other than their own. We are also indebted to the following individuals for helpful technical reviews: Robert Berner, Peter Black, Donald Boelter, Parshall Bush, David Coleman, Scott Collins, Thomas Crow, Thomas Cuffney, Eric Davidson, Arthur Eschner, Sarah Green, Martin Gurtz, Howard Halverson, Mark Harman, Howard Halverson, Gray Henderson, James Hornbeck, Ginger Howell, Charles Hursh, Alex Huryn, Dale Johnson, Jeffery Lee, Donald Leopold, James Maxwell, Arthur McKee, Howard Neufeld, Joseph O'Hop, George Parker, David Perry, Michael Phillips, Jerry Qualls, Paul Risser, Thomas Rogerson, Fred Swanson, Robert Teskey, Peter White, Walter Whitford, Leonard Wiener, Stanley Ursic, and J. Zeller.

Support for research at Coweeta has been provided primarily by USDA, Forest Service and through numerous grants from the National Science Foundation. Other sources of support include: Georgia Research Foundation, Georgia Power Company, North Carolina Geological Survey, U.S. Department of Energy (Union Carbide and Martin Marietta Corporation, Oak Ridge National Laboratory), and U.S. Environmental Protection Agency.

Contents

Section 3   Forest Dynamics and Nutrient Cycling

Section 4   Canopy Arthropods and Herbivory

Contributors

ANDREW, T.L.
Department of Microbiology, University of Georgia, Athens, Georgia 30602, USA

ASH, J.T.
Department of Microbiology, University of Georgia, Athens, Georgia 30602, USA

BENFIELD, E.F.
Department of Biology, Virginia Polytechnic Institute and State University, Blacksburg, Virginia 24061, USA

BERISH, C.W.
Biology Department, Emory University, Atlanta, Georgia 30307, USA

BORING, L.R.
School of Forest Resources, University of Georgia, Athens, Georgia 30602, USA

CASKEY, W.H.
North Central Soil Conservation Research Laboratory, USDA-ARS, Morris, Minnesota 56267, USA

CROCKER, M.T.
Institute of Ecology and Zoology Department, University of Georgia, Athens, Georgia 30602, USA

CROSSLEY, D.A., JR.   Department of Entomology, University of Georgia, Athens, Georgia 30602, USA

CUNNINGHAM, G.B.   Southeastern Forest Experiment Station, Coweeta Hydrologic Lab., Otto, North Carolina 28763, USA

DAY, F.P., JR.   Department of Biological Sciences, Old Dominion University, Norfolk, Virginia 23508, USA

DOUGLASS, J.E.   Southeastern Forest Experiment Station Coweeta Hydrologic Lab., Otto, North Carolina 28763, USA

EDWARDS, R.T.   Institute of Ecology and Zoology Department, University of Georgia, Athens, Georgia 30602, USA

FITZGERALD, J.W.   Department of Microbiology, University of Georgia, Athens, Georgia 30602, USA

FRANKLIN, J.F.   Forestry Sciences Laboratory, 3200 Jefferson Way, Corvallis, Oregon 97331, USA

GIST, C.S.   Special Training Division, Oak Ridge Associate University, Oak Ridge, Tennessee 37830, USA

GOLLADAY, S.W.   Department of Biology, Virginia Polytechnic Institute and State University, Blacksburg, Virginia 24061, USA

GRANT, W.H.   Department of Geology, Emory University, Atlanta, Georgia 30322, USA

HAINES, B.L.   Botany Department, University of Georgia, Athens, Georgia 30602, USA

HALE, D.D.   Department of Microbiology, University of Georgia, Athens, Georgia 30602, USA

HARGROVE, W.W.   Department of Entomology, University of Georgia, Athens, Georgia 30602, USA

HATCHER, R.D. JR.*   Department of Geology, University of South Carolina, Columbia, South Carolina 29208, USA

Present Address:
Department of Geological Science, University of Tennessee, Knoxville, Tennessee 37966-0156, USA

HELVEY, J.D.   Timber and Watershed Laboratory, P.O. Box 445, Parsons, West Virginia 26287, USA

HIBBERT A.R.   Forestry Sciences Laboratory, Arizona State University, Tempe, Arizona 85281, USA

HOOVER, M.D.   1012 West Park, Wenatchee, Washington 98801, USA

KAZMIERCZAK, R.F., JR.   Department of Biology, Virginia Polytechnic Institute and State University, Blacksburg, Virginia 24061, USA

KELLER, H.M.   Swiss Federal Institute of Forestry Research, CH-8903 Birmensdorf, Switzerland

MEYER, J.L.   Institute of Ecology and Zoology Department, University of Georgia, Athens, Georgia 30602, USA

MONK, C.D.   Botany Department and Institute of Ecology, University of Georgia, Athens, Georgia 30602, USA

NEARY, D.G.   Southeastern Forest Experiment Station, School of Forest Resources and Conservation, University of Florida, Gainesville, Florida 32611, USA

PATRIC, J.H.   P.O. Box 2154, Alexandria, Virginia 22301, USA

PERRY, W.B.   Department of Biology, Virginia Polytechnic Institute and State University, Blacksburg, Virginia 24061, USA

PETERS, G.T. — Department of Biology, Virginia Polytechnic Institute and State University, Blacksburg, Virginia 24061, USA

PHILLIPS, D.L. — Department of Biology, Emory University, Atlanta, Georgia 30322, USA

RAGSDALE, H.L. — Department of Biology, Emory University, Atlanta, Georgia 30307, USA

RISLEY, L.S. — Department of Entomology, University of Georgia, Athens, Georgia 30602, USA

SCHOWALTER, T.D. — Department of Entomology, Oregon State University, Corvallis, Oregon 97331, USA

SEASTEDT, T.R. — Division of Biology, Kansas State University, Manhattan, Kansas 66506, USA

STRICKLAND, T.C. — Department of Microbiology, University of Georgia, Athens, Georgia 30602, USA

SWANK, W.T. — Southeastern Forest Experiment Station, Coweeta Hydrologic Lab., Otto, North Carolina 28763, USA

SWIFT, L.W., JR. — Southeastern Forest Experiment Station, Coweeta Hydrologic Lab., Otto, North Carolina 28763, USA

TATE, C.M. — Division of Biology, Kansas State University, Manhattan, Kansas 66506, USA

TODD, R.L. — Department of Microbiology, South Dakota State University, Brookings, South Dakota 57007, USA

TROENDLE, C.A. — USDA, Forest Service, 240 West Prospect, Fort Collins, Colorado 80526, USA

VELBEL, M.A. — Department of Geological Sciences, Michigan State University, East Lansing, Michigan 48824-1115, USA

WAIDE, J.B.   Southeastern Forest Experiment Station,
                              Coweeta Hydrologic Laboratory, USDA-FS,
                              Otto, North Carolina 28763, USA

WALLACE, J.B.   Department of Entomology, University of
                              Georgia, Athens, Georgia 30602, USA

WALLACE, L.L.   Department of Botany and Microbiology,
                              University of Oklahoma, Norman, Oklahoma
                              73109, USA

WATWOOD, M.E.   Department of Microbiology, University of
                              Georgia, Athens, Georgia 30602, USA

WEBSTER, J.R.   Department of Biology, Virginia Polytechnic
                              Institute and State University, Blacksburg,
                              Virginia 24061, USA

1. The Coweeta Hydrologic Laboratory

1. Introduction and Site Description

W.T. Swank and D.A. Crossley, Jr.

This volume is a collection of chapters describing results of a variety of research efforts with forested catchments as the common basis of ecosystem interpretation. The Coweeta Hydrologic Laboratory was established 50 years ago as a testing ground for certain theories in forest hydrology. That research required development of a firm data base describing the hydrologic cycle in watersheds (see Chapter 2). Later, in 1968, efforts began to establish an extensive data base on nutrient cycling phenomena in Coweeta watersheds with joint USDA Forest Service-National Science Foundation funding. This research was a logical extension of the research on watershed hydrology. Both types of research have been based on an ecosystem concept, explicitly for nutrient cycling studies and implicitly even for early studies in forest hydrology (see Chapter 30).

The research program at Coweeta represents a continuum of theory, experimentation, and application using watersheds as landscape units. Two underlying philosophies have guided the research approach at Coweeta as revealed by the contents of this volume; i.e. (1) that the quantity, timing, and quality of streamflow provide an integrated measure of the success or failure of land management practices, and (2) good resource management is synonymous with good ecosystem management. Response to disturbance has frequently been used as a research tool for interpreting ecosystem behavior. The use of perturbation or disturbance has allowed specific hypotheses to be tested with subsequent revision and development of theories and application of results when appropriate. We have continuously attempted to integrate individual research efforts into a holistic concept of watershed response.

Understanding hydrologic processes and responses of the various watersheds was an enormous advantage to research in nutrient cycling (see Chapters 22 and 25). Water is the vehicle by which nutrients are moved: inputs into ecosystems, movement through the soil, uptake by vegetation, and exit via streamflow are water mediated; their interpretation requires extensive knowledge of site-specific hydrology. Construction of the basic input-output budgets for nutrients in watersheds (Chapter 4) requires a very detailed knowledge of hydrology, as well as biotic transfer and storage.

The ecosystem concept was rescued from relative obscurity by E. P. Odum (1953). Ecology in the early 1950s was dominated by autecological approaches, but questions on the horizon could only be approached from an ecosystem basis, i.e., worldwide radioactive fallout, insecticides in the environment, and chemical pollution of water-ways, etc. Initially, testing of ecosystem theory was largely confined to aquatic systems; particularly to ponds and lakes. A grasp of the forest watershed as exemplifying ecosystems, and the experimental analysis of forested ecosystems using nutrient

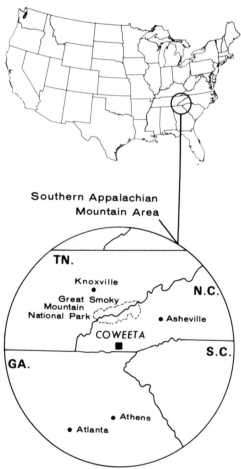

Figure 1.1. The Coweeta Hydrologic Laboratory is located in southwestern North Carolina, about a 2-hr drive north from the University of Georgia in Athens. The Laboratory is also in close proximity to other universities and centers of ecosystem research.

dynamics led to singular advances in the understanding of how natural ecosystems function (Bormann and Likens 1967; Franklin and colleagues, Chapter 30; others). Forested ecosystems are open, as are all units of the biosphere, but the amounts of many nutrients cycling internally within watersheds are much larger than the input-output flows. Observations about the responses of these cycles to disturbance have led to further theoretical development concerning ecosystem function (Odum 1969; Vitousek and Reiners 1975; Monk et al. 1977). Hypotheses concerning mechanisms of resistance to disturbance and resilience following disturbance, closing of nutrient cycles, important shifts in successional species accompanying changes in substrate quality, alterations of nitrogen cycles, shifts in herbivory, and changes in stream communities have all been tested experimentally. Results of studies with forested catchments have continued to make important contributions to understanding the nature of ecosystems.

Finally, we believe that the only level of ecological theory that will provide the necessary guidelines for proper resource management is the ecosystem level. The evaluation of landscape management practices, in the context of basic scientific inquiry into ecosystem structure and function, is one of the major benefits to be derived from cooperative work such as that described in this volume. Chapters 22 through 27 provide summaries of some of the management interpretations and implications of the research.

We begin the volume with a description of the Coweeta Basin and a historical account of the establishment and development of the program and facilities. The subsequent section focuses on climatology, geology and hydrology of the Laboratory. Then, five chapters address forest dynamics and nutrient cycling. Groups of chapters then describe herbivory, forest floor processes, and within-stream processes, followed by reviews of management implications of the research. The final section encompasses broader interpretations of the research program. It begins with a revision of an important component of the underlying theory which has guided the ecosystem approach at Coweeta, including illustrations of principles which utilize data from other chapters. The last two chapters consider the research within the context of national and international perspectives of forest hydrology and ecology programs.

Site Description

General Features

The Coweeta Hydrologic Laboratory is located in the Nantahala Mountain Range of western North Carolina within the Blue Ridge Physiographic Province, latitude 35°/03′N, longitude 83°25′W (Figure 1.1). The 2185 ha Laboratory is comprised of two adjacent, east-facing, bowl-shaped basins. Coweeta Basin encompasses 1626 ha (Figure 1.2) and has been the primary site for watershed experimentation, while the 559 ha Dryman Fork Basin has been held in reserve for future studies. Ball Creek and Shope Fork are fourth-order streams draining the Coweeta Basin and they join within the Laboratory boundary to form Coweeta Creek, a tributary that flows 7 km east to the Little Tennessee River. Elevations range from 675 m in the administrative area to 1592 m at Albert Mountain. The relief has a major influence on hydrologic, climatic, and vegetation characteristics (Chapters 3 and 10).

Figure 1.2. Aerial view of the Coweeta Basin taken looking west toward the main ridge of the Nantahala Mountains. The bowl-shaped physiography is typical of the region. The arrow in the photograph indicates the location of the Laboratory administrative facilities.

Access within the Laboratory is facilitated by two main graveled roads which are open throughout most of the year, and by service roads and trails which are closed to public vehicles. The Laboratory is open to the public, including hunting and fishing activities as regulated by state laws. No camping or fires are permitted in the area. Responsibility for fire protection, road maintenance, experimental timber harvest, and law enforcement is provided by and coordinated with appropriate National Forest System personnel. The administrative area currently consists of three office buildings totaling 600 m² of floor space, a complete analytical laboratory (370 m²), residence 165 m²), storage building (230 m²), maintenance shop and instrument repair building (245 m²), record storage vaults (100 m²), and several house trailers. The remaining facilities such as weirs and instrument houses are located throughout the experimental area.

Watershed Description and Treatments

Since the establishment of Coweeta, 32 weirs have been installed on streams in the Laboratory. Locations of individual watersheds and other laboratory features are shown in Figure 1.3. Many weirs are no longer operational due to termination of studies or lack of resources for maintenance. Currently, 16 streams are gaged; a summary of physical characteristics for these watersheds, along with information for other catchments discussed in subsequent chapters, is given in Table 1.1. Watersheds range in size from 3 ha to 760 ha. Weir structures vary from the 90° V-notch type used on first order streams draining watersheds less than about 28 ha to the 3.7 m Cipolletti type

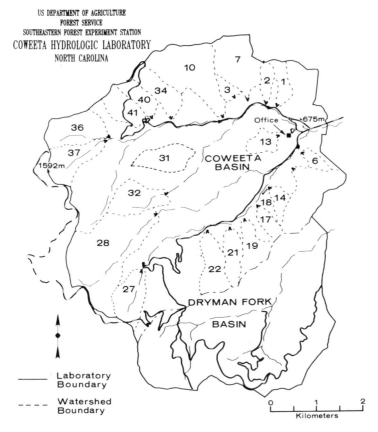

Figure 1.3. The 2185 ha Coweeta Hydrologic Laboratory is comprised of 2 main basins—Coweeta and Dryman Fork. Experimentation has focused on the Coweeta Basin and the numbers indicate individual watersheds referred to in this book.

used on fourth order streams. A 120° V-notch is typically used on intermediate sized streams. Stream gaging was initiated on most watersheds (Table 1.1) between 1934 and 1938. The most recent weir was constructed in 1981 on Watershed 31 (WS 31). Relief (from weir to top of the watershed) averages 300 m on small catchments and 550 m on large catchments. Side slopes average about 50% at low and mid elevations and increase to 60% or more on some high elevation watersheds. A variety of aspects are present in the Coweeta Basin, but most unit watersheds are predominately either north-or south-facing. For watersheds with long-term flow records, the maximum instantaneous discharge occurred either in May 1973 or May 1976, and both events were associated with intense orographic thunderstorms. Eight watersheds have remained relatively undisturbed since the establishment of the Laboratory and serve as controls in paired watershed experiments (Hewlett 1971). They range from 12 ha to 49 ha in size and are distributed throughout the Coweeta Basin. The experimental treatments at Coweeta

Table 1.1. Summary of Physical Characteristics of Watersheds at Coweeta Hydrologic Laboratory

Watershed Number	Name of Stream	Area (ha)	Date of First Record	Type of Notch	Elevation at Weir (m)	Maximum Elevation (m)	Maximum Discharge ($m^3sec^{-1}km^{-2}$)	Aspect
1	Copper Branch	16	6-13-34	90° V-notch	705	988	2.29	S
2	Shope Branch	12	6-22-34	90° V-notch	709	1004	2.41	SSE
3[a]	Little Hurricane	9	7-5-34	CIA deep notch	739	931	20.27	E
6	Sawmill Branch	9	7-10-34	90° V-notch	696	905	3.76	NW
7	Big Hurricane	59	7-31-34	90° V-notch	722	1077	2.18	S
8	Shope Fork No. 1	760	10-6-34	3.66 m Cipolletti	702	1600	1.68	—[b]
9	Ball Creek No. 1	724	10-12-34	3.66 m Cipolletti	687	1554	2.46	—[b]
10[a]	Camprock Creek	86	3-7-36	120° V-notch	742	1159	0.85	SSE
13	Carpenter Branch	16	3-12-36	120° V-notch	725	912	1.09	ENE
14	Hugh White Branch	61	5-26-36	120° V-notch	707	992	1.11	NW
16[a]	Shope Fork No. 2	382	6-4-36	1.83 m Rectangle	739	1600	1.88	—[b]
17	Hertzler Branch	13	6-6-36	90° V-notch	760	1021	1.06	NW
18	Grady Branch	13	7-3-36	120° V-notch	726	993	1.36	NW
19[a]	Snake Den Branch	28	5-16-41	120° V-notch	796	1119	1.28	NW
21[a]	Sheep Rock Branch	24	7-22-38	120° V-notch	823	1174	1.79	N
22[a]	Lick Branch	34	2-18-37	120° V-notch	847	1244	1.75	N
27	Hard Luck Creek	39	11-2-46	120° V-notch	1061	1454	5.65	NNE
28	Henson Creek No. 2	144	5-31-37	1.83 m Rectangle	964	1551	2.03	E
31	Mill Branch	34	10-1-81	120° V-notch	869	1146	NA	ENE
32	Cunningham Creek No. 2	41	10-25-41	120° V-notch	920	1236	1.35	ESE
34	Bee Branch	33	10-31-38	120° V-notch	866	1184	1.00	SE
36	Pinnacle Branch	49	4-29-43	120° V-notch	1021	1542	4.36	ESE
37	Albert Branch	44	4-15-42	120° V-notch	1033	1592	5.37	ENE
40[a]	Wolf Rock Branch	20	12-4-38	90° V-notch	872	1219	1.00	SSE
41[a]	Bates Branch	29	8-23-40	120° V-notch	893	1298	1.18	ESE
49[a]	Barker's Cove	3	3-14-38	90° V-notch	922	971	0.69	E

[a] Weirs are inactive or have been removed from service.
[b] Aspect not calculated for large watersheds.

Table 1.2. Summary Descriptions of Coweeta Watershed Treatments

Watershed No.	Treatment Description
1	Entire watershed prescribed burned in April, 1942. All trees and shrubs within the cove-hardwood type (areas adjacent to stream) deadened with chemicals in 1954. This treatment represented 25% of both land area and total watershed basal area. Retreated as necessary for three consecutive growing seasons. All trees and shrubs cut and burned in 1956–57, no products removed; white pine planted in 1957. In subsequent years, pine released from hardwood competition by cutting and chemicals as necessary.
3	All vegetation cut and burned or removed from the watershed in 1940. Unregulated agriculture (farming and grazing) on 6 ha for a 12-year period, followed by planting yellow poplar and white pine.
6	All woody vegetation cut and scattered in the zone 5 m vertically above the stream; reduced total watershed basal area 12%. Clearcut in 1958, products removed and remaining residue piled and burned. Surface soil scarified, watershed planted to grass, limed and fertilized in 1959; fertilized again in 1965. Grass herbicided in 1966 and 1967; watershed subsequently reverted to successional vegetation.
7	Lower portion of watershed grazed by an average of six cattle during a 5-month period each year from 1941 to 1952. Commercially clearcut and cable logged in 1977.
8, 9, 16	Combination watersheds containing both control and treated watersheds.
10	Exploitive selective logging during the period 1942–1956 with a 30% reduction in total watershed basal area.
13	All woody vegetation cut in 1939 and allowed to regrow until 1962 when the watershed was again clearcut; no products removed in either treatment.
17	All woody vegetation cut in 1940 and regrowth cut annually thereafter in most years until 1955; no products removed. White pine planted in 1956 and released from hardwood competition as required with cutting or chemicals.
19	Laurel and rhododendron understory cut in 1948–1949; comprised 22% of total watershed basal area.
22	All woody vegetation within alternate 10 m strips deadened by chemicals in 1955; reduced total watershed basal area 50%. Treatment repeated from 1956 to 1960 as required to maintain conditions.
28	Multiple use demonstration comprised of commercial harvest with clearcutting on 77 ha, thinning on 39 ha of the cove forest and no cutting on 28 ha; products removed.
37	All woody vegetation cut in 1963; no products removed.
40	Commercial selection cut with 22% of basal area removed in 1955.
41	Commercial selection cut with 35% of basal area removed in 1955.
2, 14, 18, 21, 32, 34	Controls with mixed hardwoods stands remaining undisturbed since 1927.
27	Control, but partially defoliated by fall cankerworm infestation from 1972 to 1979.
36	Control, but partially defoliated by fall cankerworm infestation from 1975 to 1979.

have produced a diverse array of ecosystems with respect to vegetation composition and structure. An abbreviated treatment history is provided in Table 1.2. More detailed descriptions can be found in previously published papers which focus on each watershed. The earlier mountain farming and exploitive logging demonstration studies on WS 3 and 10, respectively, have been among the most severe watershed disturbances (Figures 1.4 and 1.5). The main focus of these studies was to demonstrate the effects of land use on stream turbidity; experiments were terminated following fulfillment of objectives. The most recent severely disturbed ecosystem is WS 6, which was converted from hardwoods to grass (with applications of lime and fertilizer) and maintained in a grass cover for 5 years (Figure 1.6). The grass subsequently was herbicided for 2 consecutive years, and the watershed was then allowed to regrow with no additional manipulation (Figure 1.7). A variety of cutting prescriptions have been applied, including light selection cutting, clearcutting without roads and no products removed (Figure 1.8), commercial clearcutting (Figure 1.9), and a combination of thinning and clearcutting (Figure 1.10). Another manipulation included hardwood to white pine conversion on WS 1 and WS 17 (Figure 1.11). Vegetation on two control watersheds (27 and 36) was partially defoliated by insects each spring for several years. Watershed experiments were originally designed to test specific hypotheses or meet objectives related to assessing the effects of disturbances on the quantity, quality, or timing of streamflow. Subsequently, many of these ecosystems have provided a unique opportunity to examine ecological processes related to biogeochemical cycles.

Figure 1.4. One of the earlier land use demonstration experiments at Coweeta was conversion of WS 3 from a hardwood forest cover to mountain farming, a prevalent practice in the region until the late 1940s. This view shows 3 ha in pasture (*foreground*) 3 ha in corn (*center*), and 3 ha of hardwood regrowth (*background*).

Figure 1.5. The effects of unregulated logging on water quality was demonstrated on WS 10. This 1952 view of a logging skid trail crossing a perennial stream is representative of practices in the region during this time period.

Figure 1.6. Conversion of hardwoods to grass on WS 6 (9 ha) was designed to examine changes in the quantity and timing of streamflow due to changing the vegetal cover.

Figure 1.7. Luxuriant herbaceous vegetation covered WS 6 in the summer of 1968, two years after the grass cover was herbicided.

Figure 1.8. The first experiment conducted in the eastern United States to study the influence of forests on streamflow was initiated in 1939 on WS 13. All woody vegetation was clear-felled and no products removed; the treatment was repeated in 1962. This aerial view of the coppice regrowth was taken in 1967, five years after the second cut.

Figure 1.9. WS 7 (*left center*) was clearcut in 1977 as shown in this view and logged with a mobile cable yarding system. WS 2 (*center*) is the hardwood covered control for both WS 7 and for WS 1 (*right center*), a white pine plantation.

Figure 1.10. The concept of multiple use forest management was demonstrated on WS 28, a 144 ha catchment. The treatment included commercial harvest with clearcutting on 77 ha and thinning on 39 ha within the cove forest. There was no cutting on 28 ha of the steeper upper slopes. Improved design and proper location of logging roads were major features of study. The resources assessed included timber, water, wildlife, and recreation.

Figure 1.11. All woody vegetation was clear-felled on WS 17 in 1940 and regrowth was cut back annually for a period of years (*upper view* in 1951). White pine was planted on the watershed in 1956 and the lower view was taken when the plantation was 14 years old. WS 18 (*below*) is the adjacent hardwood covered control for WS 17.

Resource Inventories

In addition to the climatic, hydrologic, vegetation, and water chemistry assessments presented in later chapters, a variety of other baseline data are available for the conduct of research. A third order survey in the Laboratory forms the basis for two topographic maps: one at a scale of 1:7200 with 3 m (10 ft) contour intervals and one at a scale of 1:14,400 with 15 m (50 ft) contour intervals. Streams, watershed boundaries, roads, trails, control elevations, and other topographic features are accurately displayed on the maps. Vegetation analyses are aided by a variety of permanent plots, including 408 (0.08 ha) plots established in 1934 and distributed over the entire basin (supplemented by 50 plots (0.08 ha) located on specific control watersheds), two 1 ha reference stands with detailed records, and numerous plots in disturbed watersheds. A herbarium comprised of 604 taxa and represented by 327 genera of 97 families has been established at Coweeta. Documentation includes efforts to fill life-stage gaps, accurate mapping of collection sites, and duplicate specimens for many species (Pittillo and Lee 1984).

The soils within the Laboratory have been mapped several times. The most recent and detailed effort was a first order survey completed in 1985 by the Soil Conservation Service. This survey was greatly facilitated by Hatcher's (1980) map of the basin bedrock geology. Soils are mapped at the phase level (10 series) on the small scale topo-

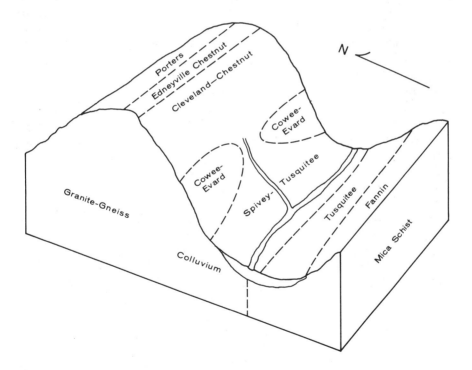

Figure 1.12. Block diagram of interrelationships between selected soil series and bedrock, aspect, and relative elevation at Coweeta Hydrologic Laboratory. (Prepared by Douglas Thomas, SCS, N.C., and Scott Keenan, N.C. Department of Natural Resources and Community Development).

Table 1.3. Selected Characteristics of Coweeta Soils

Depth (cm)	Bulk density (gm cm⁻³)	Organic matter (%)	pH	CEC[a] (meq 100 g⁻¹)	B.S.[b] (%)	K	Ca	Mg	P
							(mg kg⁻¹)		
0–10	1.24	15.9	4.74	11.6	17.2	22	71	18	6.2
10–20	1.30	10.0	4.89	9.4	18.8	21	21	10	5.0
20–30	1.39	7.7	5.02	6.8	19.1	15	16	13	2.5
30–40	1.42	5.8	5.11	7.2	27.7	22	18	20	3.7
40–50	1.47	5.8	5.14	6.2	16.1	19	12	18	1.3
50–60	1.52	5.4	5.13	6.0	11.7	15	14	18	1.3

[a] Cation exchange capacity.
[b] Base saturation.
Source: After McGinty, 1976.

graphic map. More refined soil boundaries are available for some individual experimental watersheds. The relative topographic position of selected soil series is given in Figure 1.12. Soils within the Laboratory fall within two orders: immature Inceptisols and older developed Ultisols. The Inceptisols found at Coweeta fall within seven "Great Groups." Umbric Dystrochrepts of the Porters series are found on steep, rocky faces at high elevations on the north- and south-facing aspect of the Laboratory. The Typic Dystrochrepts, as represented by the Chandler gravelly loam series, are found on south-facing slopes underlain by the Tallulah Falls formation. The Typic Haplumbrepts as represented by the Spivey-Tusquitee complex, are formed in colluvium of long narrow areas associated with watershed hollows and coves.

The Ultisols are represented by Typic Hapludults and Humic Hapludults. The Typic Hapludults are the largest soil group in areal extent at Coweeta and fall into two geomorphic settings: the Cowee-Evard gravelly loam series is found on strongly sloping to very steep ridges and sideslopes, and the Fannin sandy loam series is found on gently sloping sideslopes. Both soil series are formed in residuum weathered from schist, gneiss, or granite. Humic Hapludults include the Trimont gravelly loams, which are found on cool, steep north-facing slopes.

Chemical and physical analyses have been conducted for many of Coweeta soils as part of studies on specific watersheds. Soil properties vary substantially over the Laboratory as implied from the variety of soil series. However, McGinty's (1976) assessment is the most generalized soils data available, and is based on samples taken on north-facing control watersheds at eight locations over the elevational range of the Ball Creek drainage. Taken collectively, selected features in Table 1.3 generally show that soils are relatively high in organic matter and moderately acid with both low cation exchange capacity and percent base saturation, characteristics that are typical for highly weathered Ultisols.

More detailed descriptions for some of the watersheds and their ecosystem compartments are contained in subsequent chapters including the history of early land use in the Coweeta Basin (see Chapter 2).

2. History of Coweeta

J.E. Douglass and M.D. Hoover

The Early Years

Coweeta, according to some, means little star. Actually, the meaning of the Cherokee word is unknown, but Coweeta was certainly a lucky star for forest hydrology. The hydrological laboratory was born at a favorable place at a fortunate time and despite trials and tribulations, the lucky star still shines brightly. Coweeta may actually be a Creek rather than Cherokee word meaning "War Town."[1] Certainly a lot of luck was involved, and sometimes battles were fought to keep the little star shining.

To understand the origin and development of 50 years of research at Coweeta, one must understand something of the social, economic, and political circumstances as these changed through time. It was concern about soil erosion, flood control, and sustained flow of streams as well as future timber supplies that led to establishment of the first forest reserves (soon to become national forests) from the public domain lands of the West. The role of the forest in regulating the flow of navigable streams was the constitutional basis for the Weeks Act of 1911. This act allowed the Federal government to purchase private lands for national forest in the East. At the time there was considerable debate about the influence of forests upon regulation of streamflow and flooding. The need for quantitative information led the Forest Service, with help from the Weather Bureau, in 1909 to begin the first experiment to measure streamflow before

---

[1]Personal communication, R. E. McArdle.

and after removal of trees at Wagon Wheel Gap in Colorado (Bates and Henry 1928). Controversy over the role of forests in regulating streamflow peaked after the disastrous 1927 flood of the Mississippi River, and an extensive effort to gather information on topography, soils, climate, forest cover, and runoff was made. Reconnaissance surveys throughout the country revealed serious land use, erosion, and flood problems and accentuated the need for scientific studies of factors controlling these problems.

At the time, E. N. Munns was in the Forest Service Branch of Research in Washington. Munns was convinced of the need for sound watershed research from his previous work in California. At the Appalachian Forest Experiment Station, Dr. Charles R. Hursh, an ecologist, was hired in 1926 to begin research in forest influences there. With the strong support of Munns, Dr. Hursh was the guiding genius behind the selection of Coweeta as a research site and the development of the research program. His earliest work dealt with erosion control, stabilization of roadbanks, runoff as influenced by land use, accumulation of organic matter in soils, and infiltration and percolation of water through soils. He also studied the role the Ohio River watersheds played in the 1927 flood. By 1932, he had studied the various needs of the mountains and Piedmont, and had prepared a comprehensive analysis of forest influences problems and an approach for solving them (Hursh 1932a).

Discussing streamflow and erosion studies, Hursh stated that "the purpose of the streamflow and erosion study in its broader sense is to determine the principles underlying the relation of forest and vegetative cover to the supply and distribution of meteorological water." His study of available literature led him to two major conclusions. The first was the need to establish the relation of forest and other plant cover to streamflow where erosion was of minor importance. The second was the need to establish the relationship of agricultural land use to erosion and streamflow. In the latter conclusion, forests were important chiefly as corrective cover. Hursh observed that forest influences investigations required study of individual components of the problem to develop practical conservation measures. He recognized that several major fields of science, namely meteorology, plant physiology, soil science, geology, and hydrology would be involved. He felt that this complexity required division of research into components for critical study, and he stated "not only is it impractical to attack the entire problem as a whole, but at the same time . . . it is impractical to attack all of the many factors at one time. Hence, it becomes necessary to determine as a factor for study one that may be considered as fundamental and basic . . ." To better understand the relation of forest cover to flood, the fundamental question was which cover conditions would contribute to maximum water yield and still ensure the most effective use of the watershed for power, industrial, and municipal water supplies.

Hursh visited lysimeter installations in 1931, but was not convinced of their usefulness in forest influences research. He returned to search out suitable areas in the Appalachian Mountains where comprehensive forest influences and forest management investigations could be made. His site criteria were a complete headwater drainage basin with well developed stream systems, perennial flow, high rainfall, deep soils, and complete forest cover. John Byrne, supervisor of the Nantahala National Forest, suggested several possible research sites and Hursh and Herbert Stone (deputy supervisor of the Nantahala National Forest) finally selected the Coweeta basin in 1931.[2]

Research Philosophy

Although we like to believe that models and studies of basic processes are recent ideas, Hursh's (1932a) model was the hydrologic cycle which dealt with individual studies of the cycle. Originally, he visualized Coweeta as an area in which the impacts of forest management practices on soil and water characteristics could be studied. Three phases were visualized; the first involving study of processes on individual watersheds with the objective of quantifying individual components of the hydrologic cycle. He planned for a research forest made up of unit watersheds, multiple watersheds, and finally the entire basin. Specific studies were to be established on individual watersheds in a process of watershed standardization (or calibration of a watershed on itself), as opposed to the paired watershed approach. Standardization required detailed information on climate, soils, vegetation, ground water and runoff from each catchment. An early conceptual picture of this phase of the research program is portrayed in Figure 2.1.

Once watersheds were standardized, watershed experimentation would begin. In this second phase, the changes in the water balance produced by treatment would be measured. This phase logically would lead to the third phase of study, the objective being to develop and test methods of managing forests for water and other resources.[3]

Some of the researchers at the Station wanted to install management studies early in the development of Coweeta, something Hursh opposed. He appealed to C. L. Forsling, then station director, who had experience in western forest influences studies. Forsling decreed that no manipulation of vegetation would take place until watersheds were standardized. Thereafter, interest in Coweeta by other scientific disciplines waned, and administrative responsibility fell entirely to Hursh.[4]

Research Begins

In 1932, the nation was in the depths of the depression, with millions unemployed and hungry. Emergency relief and recovery programs to provide work for the unemployed and to stimulate industrial recovery were on the horizon. Hursh foresaw in them the

---

[2]No history of the research program would be complete without consideration of land use prior to establishment of the experimental forest. To prevent interruption of chronology, this information is summarized as Appendix I to this chapter.

[3]Research followed these phases generally, but after experimentation began, all three phases progressed concurrently as the need for supporting information dictated.

[4]Forsling's decision had a far-reaching influence on continuity of research. It put responsibility for Coweeta entirely in the hands of one person, and indirectly provided a clear objective or mission which persists to this day. Many other watershed experiments failed because research priorities and/or interest in the research changed with changes in administrators.

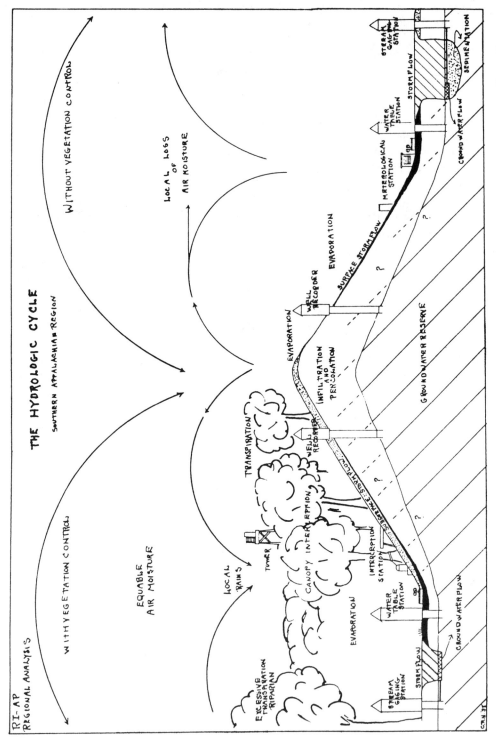

Figure 2.1. Conceptual diagram developed by C. R. Hursh in 1938 of the hydrologic cycle and instrumentation of water balance studies at Coweeta.

possibility of securing manpower and funds to facilitate the work. He had already developed the research strategy and selected the appropriate research site. He secured permission from the National Forest to use the basin for research even though formal approval of transfer of the land to Research was months in the future. In early 1933, the Public Works Administration (PWA) established in the last days of the Hoover Administration began providing money to put the unemployed to work.[5] When the word went out that work was available paying $26 to $28 per month, unemployed local residents turned out at dawn by the hundreds, bringing their own hand tools and ready to work.

The people of the lower Coweeta Valley and surrounding areas were living a true pioneer life, dependent on their own intelligence, skills, and available resources for survival. Many were skilled builders; others were blacksmiths, rock masons, and loggers. What they didn't know, they could learn. They appreciated the opportunity to work close to their homes, were hard workers, and were interested in the investigations. Part of Coweeta's luck was having these resourceful and hardworking men available.

Roosevelt greatly expanded the PWA in 1933 and established the Civilian Conservation Corps (CCC), the Federal Relief Administration (ERA), and Civil Works Administration (CWA) (reorganized and consolidated into the Works Program Administration, WPA, in 1935). Each of these programs played important roles in developing facilities at Coweeta.

Manpower for the first work came from the CCC camp at Franklin, North Carolina. CCC camps were operated by the Army and work of the Corpsmen was directed by the Forest Service. The CCC first built 5 km of road from Otto into the basin to provide access. They provided labor for establishing and inventorying the permanent vegetative plots at Coweeta and for boundary and topographic surveys. The CWA built three chestnut log cabins for a residence, bunkhouse, office-shop and later a superintendence residence, machine shop, and a hydroelectric plant were built (Figure 2.2). Other developments included three trails along which 60 rain gages were located, a nursery for growing planting stock to stabilize roadbanks, and the first 9 weirs. The order establishing the Coweeta Experimental Forest was signed in June 1933 and finally approved by the chief on March 28, 1934.

Coweeta's luck held. To stimulate private industry, Government agencies put all types of serviceable property on surplus and issued orders for new property and equipment. Hursh's men became expert scroungers—a machine shop was outfitted from a surplus World War I Army mobile ordnance shop, and office furniture, tools, pipes, and valves for water system and weirs, pumps, cement mixers, and vehicles were obtained at every opportunity. Surplus property equipped the CCC Camp (NC-23) that was built in the Coweeta basin in 1935 (Figure 2.2). This camp supplied CCC manpower for both Coweeta and the surrounding Nantahala National Forest.

Still, Coweeta's luck held. When leaders were needed to plan and direct construction, permanent positions were frozen, but individuals with necessary skills could be employed on 3-month temporary appointments which could be extended indefinitely. Dean Dana at the University of Michigan was contacted for foresters, but all graduate

---

[5]At the time, the Appalachian Forest Experiment Station consisted of a director and five or six scientists. Almost overnight, the station budget for manpower and operations increased about 60-fold (Hursh, personal communication.)

Figure 2.2. *a.* CCC Camp NC-23 built in the administrative area in 1935 provided manpower for Coweeta and the surrounding Nantahala National Forest until 1942. *b.* A machine shop was built by WPA personnel and equipped with WWI surplus.

c

d

Figure 2.2. *c.* The log residence, bunkhouse and office was built in 1934 with Civil Works Administration personnel. *d.* The CCC, PWA, and other relief programs of the New Deal Administration supplied manpower for the weir construction.

foresters were already working for the CCC program (R. A. Hertzler, personal communication). He referred the inquiry to the School of Engineering, and in 1934, Hursh employed five graduate engineers, each having unique personalities and talents. One was an office man; he stayed in Asheville and, with a staff of 16 to 18 clerical workers, kept records up to date, did drafting, and performed similar duties for field operations at Coweeta, Bent Creek, and Copper Basin.[6]

Another was assigned primarily research responsibilities and the others oversaw construction of the facilities. Richard A. Hertzler had a direct pipeline to surplus property and, more than any other, served as resident superintendent during those early years.[7] As luck would have it, when men with hydrology training and construction skills were needed, engineering graduates were the very ones available.

But one can make his luck. The Forest Service had no funds for research equipment, but $6 per man per month was available for tools, supplies and materials. When one finance official told Hursh he could not use these funds for instruments and equipment, Hursh solved the problem by finding another who said he could. If equipment or facilities were not available commercially, Hursh and his people built their own. When money for supplies was short and charts for recording instruments expensive, they drafted and printed their own (C. R. Hursh, personal communication). When rating a larger capacity weir was necessary, the testing station was built (Schaill 1936), and Hertzler (1938) designed and calibrated the 120° V-notch weir which was used on many streams at Coweeta and subsequently throughout the world.

By June 1936, 19 km of roads and 38 km of trails, 16 stream gages, the hydraulic testing station, a 60-gage network for measuring precipitation, 4 ground water wells, 3 log cabins, a shop, a nursery, 10 weather stations, and a 1.4 kw powerplant had been built (Hursh 1936). Manpower from the various relief projects increased from 74 man-months in 1934 to 732 in 1935, to almost 2000 in 1936. By 1939, 25 weirs were operating, 18 on unit watersheds up to 140 ha in size and 7 on larger streams containing one or more unit watersheds.

The First Watershed Treatments

Enough records were available by the fall of 1939 to begin tests of the effects of changing watershed cover on three watersheds. During the winter of 1939 to 1940, a 9-ha watershed (WS 3) was cleared to be used for a cornfield and pasture; on a 16 ha watershed (WS 13) all trees were felled but soil was not disturbed in order to determine the amount of water used in transpiration; and a 58-ha watershed (WS 7) was fenced to measure the consequences of grazing cattle in woodlands.

---

[6]Hursh had major research underway at each of these locations. Work at Copper Basin and Bent Creek preceded that at Coweeta and experience gained at these locations played an important part in developments at Coweeta.

[7]Because work was underway concurrently at Coweeta, Bent Creek, and Copper Basin, Hursh directed activities by giving broad assignments—design and build a shop at Coweeta, build a weir on Shope Fork, etc.—and his staff of engineers and technicians traveled from place to place to accomplish their assignments (personal communication from C. R. Hursh).

This bold beginning caused concern among watershed researchers in Washington and National Forest Administration.[8] A conference was set for April 1940 and attended by specialists from all regions of the country and by Marvin Hoover, the newly appointed resident superintendent. The purpose of the conference was to review the entire Coweeta program, including methods of standardization of watershed experiments and the proposed cover manipulations (Hursh 1942a). Hursh planned to evaluate effects of treatments by changes in the water balance (Figure 2.1) – the balance between precipitation and runoff (Hursh, Hoover, and Fletcher 1942). This would allow estimates of changes in interception, transpiration, soil moisture, and ground water storage, and is similar to the present day hydrological models for partitioning precipitation on a watershed. The paired watershed method, pioneered at Wagon Wheel Gap, was to serve as a check on the total change in streamflow. The consensus was that the planned progress was daring, but that early results were needed. The overall methods and objectives were endorsed, and many valuable suggestions were made for improving gaging installations, especially for protection against flooding.

The years 1940 and 1941 were the peak of pre-World War II activity (USDA 1940; Hursh 1943). As recommended at the Coweeta conference, stream gages were "flood proofed" by raising instrument shelters, which also created better hydraulic conditions in weir ponds and approach channels, and cutoff walls were strengthened. A large measuring section was built to contain sediment from the mountain farm watershed (Watershed 3). A new office and fireproof vault for record storage were built. A total of 130 man-months of WPA and 865 man-months of CCC labor were used in Calender Year 1940 (USDA 1940).

Research personnel were added to begin studies in litter accumulation and decomposition, soil water movement, ground water fluctuation, water temperature, and water quality. In the winter of 1941, all vegetation was cut on a 13-ha watershed (WS 17) with no disturbance to soil or litter. The sprout growth was cut back to the ground in the summer of 1942. Riparian vegetation on a 9-ha watershed (WS 6) was cut in July 1941 and the understory of a 16-ha watershed (WS 1) was burned to simulate a spring wildfire in April 1941. Logging began in 1942 on a 81-ha watershed (WS 10). Five new stream gages were built between 1940 and 1943 to provide more watersheds for experiments.

World War II and the Research Program

The War rapidly reduced the manpower and supplies available to Coweeta. The Army reclaimed most of the equipment made available to the CCC. By 1942, the CCC labor force had shrunk to 33 men and the young mens' CCC company was replaced in March 1942 by a CCC company of World War I veterans until July 13, 1942, when it too closed. The WPA workers were terminated at the end of July 1942. Three of the four technicians who serviced recording instruments and made preliminary computations

---

[8]To prevent damage to soil, logs were not harvested on watershed 13, but were allowed to decompose in place. Cutting good timber and leaving it to rot greatly disturbed administrators of National Forests.

were in service by the end of 1942, and no labor was available until September, when funds became available to employ a crew of two to six local men.

When the labor supply dried up, Hoover was directed to terminate all but the most important watershed studies. Nevertheless, he was determined to maintain the essential records of streamflow, ground water, rainfall, and climate so that watershed treatment results could be evaluated and more treatments could be made without loss of time when the war ended. This was an almost unbelievable feat because of the sheer number of instruments and gages to be read. For example, all nonrecording rain gages were read after each storm (it rains one day in three) along 50 km of foot trails. By learning some shortcuts and strengthening their legs, three men read all rain gages in less than 6 hr; a job that once took 12 CCC corpsmen! The loyalty and hard work of local workers kept Coweeta alive during the war.

Not only were essential operations maintained during the war, but Hursh's dream of establishing the credibility of Forest Service research in the eyes of the world was also fulfilled by publishing papers on the water balance and on results from early treatment (Hoover 1944; Hoover and Hursh 1943; Hursh et al. 1942; Hursh and Brater 1941; Hursh and Fletcher 1943).

Post-War Years

When the war ended, Coweeta was famous. This did not bring much more funding, but it did bring a multitude of visitors. Coweeta provided conclusive demonstrations, with measured results in terms of water and sediment, and of watershed changes caused by land use. Many training sessions to teach fundamentals of watershed management were held, and Coweeta staff was involved in putting better land use into practice throughout the Appalachian region.

Dryman's Fork, a 600-ha area lying south of Coweeta was added to the experimental forest in 1946 to provide additional area, especially for study of timber harvesting.

The original watershed treatments were continued and new ones started. These included removal of understory shrubs below the forest canopy (WS 19), a test of narrow strip clearcuts (WS 22), conversions from hardwood to white pine (WS 1 and 17), and from forest to grass (WS 6). Tests were also started of integrated management for timber and water production (WS 40 and 41). In cooperation with the State of North Carolina, a fisheries biologist was stationed at Coweeta for a few years to study stream environment for trout (Hassler and Tebo 1958).

To create more awareness of the possibilities and importance of watershed management, the Forest Service produced the award winning film "The Waters of Coweeta" in 1955. This film was widely shown to introduce people to watershed management research. It added to the international prestige of the Lab and brought many visitors to Coweeta.

Back to Basics

The next stage of the program was to develop and test methods of managing watersheds for water and other uses. Despite the considerable effort that went into watershed experimentation, quantitative prediction of watershed response still eluded the scientists, responses to manipulations being highly variable and unpredictable within reasonable limits. In the 1950s, Dr. John D. Hewlett concluded that it would be impractical to replicate watershed experiments in sufficient quantities to obtain needed accuracy in predicting effects on water yield and quality. He felt that if the prediction goal was to be achieved, it would be through mathematical modeling of watershed processes. He reorganized scientists at Coweeta into four teams to conduct basic research in plant water, soil water, streamflow, and atmospheric relations (Meginnis 1964; Hewlett 1964a, 1964b). Timing seemed perfect—computers were coming into general use, new experimental techniques and equipment were being developed and refined, and across the nation, Forest Service research staffs and research facilities were being expanded. Plans called for a major lab to be built at Coweeta and a substantial increase in the scientific and technical staff. Although these plans were well founded, resources were not available to carry them out. Nonetheless, the new research organization was partly successful and many scientific advancements were made. Building on Hursh's original ideas of subsurface flow, Hewlett was able to demonstrate with soil models that slow, unsaturated flow of water through steep soil was the source of water feeding Coweeta's streams (Hewlett 1961; Hewlett and Hibbert 1963). Definitive work was done on interception of water by pine and hardwood (Helvey 1964, 1967, 1970, 1971; Helvey and Patric 1965, 1966) and the functional relationship between vegetative type, evapotranspiration and streamflow (Hibbert 1966, 1969; Swank and Helvey 1970; Swank and Miner 1968). Automatic data processing techniques were applied, which streamlined data handling and analysis and allowed the large backlog of data from the war years to be put into usable form (Hibbert and Cunningham 1966). In 1962, field operations of the Piedmont Research Center were consolidated with those at Coweeta. During 1963 through 1965, the Accelerated Public Works (APW) program of the Kennedy Administration (Southeastern Forest Experiment Station 1964), which was designed to stimulate the economy by putting the unemployed back to work infused the first new money into Coweeta in 20 years. A "Wet Lab," water system, residence, and an extension on the old office were built, all directed toward supporting the expected laboratory and expanded research program. Three major watershed experiments were initiated. The multiple use study (WS 28) demonstrated conclusively that management of forests for timber, water, and other resources was economically feasible and environmentally defensible when proper logging and road design and construction techniques were used. This study was an example of the third phase of Coweeta research as originally envisioned by Hursh. The first replication of a watershed experiment in time was achieved when WS 13 was clearcut a second time in 1962. The third experiment was conducted when a high elevation watershed (WS 37) was clearcut to determine whether or not high elevation watersheds with steeper, thinner soils and higher rainfall responded as did lower elevation watersheds. Nevertheless, the era of rapid expansion in research programs and facilities passed without Coweeta receiving either new laboratory facilities or an expanded program. Indeed, through the 1960s, with the exception of the APW period, Coweeta lost ground in both professional and nonprofessional support as inflation ate into a constant budget.

University Cooperation and Ecological Research

In 1967, Dr. Eugene Odum asked the Forest Service if a student, whose dissertation study area had been rendered unusable by flooding, might study breeding bird populations on experimental watersheds at Coweeta. This study was followed immediately by a small mammal study on the same watersheds, and in 1968, a cooperative study of mineral cycling on four of Coweeta's watersheds began with the Institute of Ecology, University of Georgia. From this small beginning grew one of the largest and most successful cooperative research efforts in Forest Service history.[9]

The National Science Foundation (NSF) funded the original mineral cycling investigations by the University of Georgia at a level approximately equal to the annual Forest Service budget at Coweeta. These investigations later became part of the Eastern Deciduous Forest Biome studies of the United States International Biological Program (IBP) funded by NSF. The objective of the mineral cycling studies was to investigate effects of experimental manipulations on nutrient cycling and productivity of forested watersheds as revealed by nutrient and water budgets. The research approach was to examine the emergence of higher level ecosystem properties through the examination of lower level component ecological processes (Johnson and Swank 1973; Monk et al. 1977). Small watersheds were used as a means of describing basic ecosystem phenomena much as earlier, these same watersheds had been experimental units for studying basic hydrologic processes. Coweeta, with its 35-year climatological and hydrological data base, was an ideal location for such investigations. Because of the large increase in research use of the experimental area, principal investigators were appointed to oversee the cooperative program—W. T. Swank for the Forest Service, and for the University, P. L. Johnson, later C. D. Monk, and finally D. A. Crossley, Jr. Thus, as the Forest Service during the preceding decade had sought out the university environment for their laboratories, university scientists were moving to the woods, where the effects of forest management on the environment could be studied.

Forest Service research was changing in other ways, as efforts to predict how forest management practices would affect water yield were drawing to a close as empirical and process models were developed (Douglass and Swank 1972, 1975; Swift et al. 1975; Douglass 1983). Research emphasis shifted from water yield to water quality investigations, which had previously received less attention. The early 1970s were a troubling period for Coweeta, however. Research administrators were uncertain of the proper direction for future research at Coweeta. They were unwilling to support the research emerging from scientists at Coweeta and, at the same time, were unwilling to define a new research direction for the unit. In 1971, a wildlife biologist was assigned to the Coweeta staff, but persistent rumors circulated that the unit was to be closed. The decision was finally reached that a major laboratory would not be built at Coweeta. Although not supported in a positive way, neither was the research direction emerging

---

[9]There had previously been cooperation with universities, industry, and other state and Federal agencies, but nothing even approached the scale which was now unfolding.

from Coweeta opposed, and Coweeta's research program gained momentum and support from outside the Forest Service.

In 1972, the amendments to the Federal Water Pollution Control Act, [which contained the now famous nonpoint source pollution provisions (NPSP)] were passed. This legislation stimulated a national shift in watershed research emphasis to nonpoint source pollution and, in 1974, the station director, J. B. Hilmon, directed that Coweeta make a major thrust in the NPSP area. He also endorsed research emphasis on water quality and supported the cooperative mineral cycling investigations with the University of Georgia's Institute of Ecology. The NPSP work led to expanded studies outside the Coweeta basin on effects of mechanical site preparation, herbicides, and fire on water quality. Because plans for a new lab had been scrapped and modern research facilities were needed, staged renovation of existing and construction of new facilities were carried out over the next 5 years. Support came from Coweeta's research appropriations, the Lyndon B. Johnson Civilian Conservation Center (operated by the National Forests in North Carolina), the Young Adult Conservation Corps (YACC), and one small construction appropriation. The wet lab was completely renovated into a first rate soils and water chemistry laboratory, a conference room, and a sample preparation room. The old shop was refurbished into a modern shop and an instrument repair and storage facility. The LBJ corpsmen built a petroleum storage house and gas island nearby, and completed a 2000 ft² office in 1980. The YACCs also repaired buildings, and, with partial funding from NSF, constructed the first permanent weir built since the 1940s. As IBP wound down, NSF began funding detailed studies of ecological impacts of cable logging on WS 7 (treated in 1976 to 1977) in 1975. Coweeta became an experimental ecological reserve and was also designated a biosphere reserve by UNESCO and the State Department in 1978. In 1980, Coweeta became a Long-Term Ecological Research (LTER) site of the NSF. Through a number of competitive grants to several universities, NSF is putting about $2 into University research at Coweeta today for each dollar of Forest Service funding.

The last major event of significance, and therefore deserving of some mention, is the National Acid Precipitation Assessment Program. Acid rain became an issue of national and international concern in the 1970s and, in 1980, Congress passed the NAPAP Act (Title VII of PL-294). Of concern was the effect of increased rain acidity due to air pollution on fish, forest, manmade objects, and human health. Twelve federal agencies and four national laboratories are cooperating in a 10-year program to recommend solutions to or solve this problem. Coweeta is a part of this program, and the Forest Service funds both cooperative and inhouse research dealing with the acid deposition questions. This program perhaps as well as any illustrates the ultimate value of the 50 years of hydrologic, climatic, and vegetative data and 16 years of ecological data gathered at Coweeta. Atmospheric and stream chemistry monitoring began at Coweeta 10 years before there was a National Acid Precipitation Assessment Program, and provides some of the best baseline data in the world for assessing the impact on the forest ecosystem of increasing rainfall acidity. The long term data base and the types of studies underway at Coweeta offer the best hope for evaluating impacts of acid rain or other perturbations on the health and productivity of our forests. In the words of John Hewlett (1964a): "The watershed is still the ultimate testing ground." This will always be so.

Appendix I: History of the Basin Before Establishment of the Experimental Forest (Hertzler 1936)

The basin was part of the Cherokee Indian Nation until the Indians were moved out in 1837. The Indians practiced spring and fall burning of the woods to control understory and weeds, to expose nuts and other mast in the fall, and to control milk-sick, a disease which killed cattle and infected humans as well. Apparently the burning controlled understory vegetation, because early white settlers describe riding horseback through the forest and driving wagons through the woodland and over the mountains.

In 1835, a major hurricane struck Jones Creek to the north and blew down much of the timber in the Coweeta basin. Little and Big Hurricane drainages and Hurricane Flats derive their name from this storm. Two years later the Indians were removed, and in 1842, Charles Dryman built the first house on the present administrative site. He was followed in 1847 by John Shope, who built in Hurricane Flats at the mouth of Hurricane Creek. At least three other families established permanent residences on the experimental area, although a total of 32 land grants were issued. Some families from further down the valley cleared and farmed fields for brief periods, but never established residences on Coweeta. Grants were limited in acreage and cost 5 cents per acre, but early settlers got around the acreage restriction by entering land in the names of children and other relatives, one living as far away as Texas. Holdings ranged in size from 26 to about 640 acres in size and were inexactly described, which inevitably led to boundary disputes and unclaimed land. When the area was surveyed for the Nantahala Company in 1901, 1,267 acres were unclaimed and were granted to the surveyors for 12.5 cents per acre.

Estimates are that in 1902, when the land was bought by the Nantahala Company, less than 100 acres were under cultivation. Because most of the land was rough, steep mountain land, only the bottoms or "flats" were initially cultivated. Predominant crops were corn, rye, some wheat and oats, beans, and pumpkins. Crops were good, because soils were rich and rainfall was plentiful except in drought years. When productivity dropped, new land was cleared. In this way small fields were established on Watershed 2, near Dyke Gap (Garland Patch) in WS 28, at Thomas Fields, and elsewhere high up in the basin.

Grazing of livestock was the more important land use and early settlers continued the Indian practice of semiannual burning of woodlands. Estimates of grazing pressure were given as 1500 sheep, 3500 hogs, 600 cattle, 30 horses, and 50 mules annually. Although this is undoubtedly an overestimate, it illustrates the relative importance of woodland grazing to the early economy. Other land uses included: one iron mine, which was developed and worked briefly in 1872; three mica mines located along the ridges, which produced good mica; and one copper deposit, which was opened but never mined commercially. Markets were too far removed and prices too low to make mining important at that period in history.

In 1902, most of the basin was bought by the Nantahala Company for $1.00 per acre. Tenants were moved from the land and in 1906, land was sold to the W. M. Ritter Lumber Company as part of a 11,500 ha, $175,000 land sale. Three logging operations were conducted.

In 1909, a Mr. Bingham purchased timber rights from the east boundary to Ball Flats on the south and to Shope Branch on the north fork of the basin. Good oak and poplar were cut and skidded with cattle to the present administrative site. The Gennette Brothers bought the logs, sawed them and hauled the lumber to the railroad at Otto. (This was not part of the tract purchased by the Ritter Lumber Company.) In 1914, the Gennette Brothers bought the timber lying between Shope Branch and Camprock Creek from the Ritter Company. They cut about 200,000 board feet of logs and skidded these with cattle and horses to the sawmill located about 100 yards northeast of the present administrative site.

The Ritter Company surveyed for a railroad from Otto to Mooney Gap and planned to log the entire basin. However, in 1918, the Forest Service proposed a purchase of all the Ritter holdings. Because of the impending logging, the Forest Service surveyed only the standing timber under 15 inches at the stump (that which would remain after logging) in order to establish a price. The timber stand in the Coweeta basin was among the best of the Ritter Company holdings. Mr. C. E. Marshall wrote that the basin "... contains the heaviest stand of hardwood timber found on any lands in the Southern Appalachians ... For reproduction of desirable hardwoods, there are no better lands in Western North Carolina." R. C. Hall, timber examiner, wrote, "On the whole, I believe that there is not a more desirable property than this in the Southern Appalachians ... It is superior in both timber and soil to the Vanderbilt Pisgah Forest ... It does not contain as much old, overmature timber in immediate need of cutting as do most virgin tracts, but has a great stand of thrifty young and middle-aged timber ..." due to "cyclonic storms blowing down old timber in bolts and patches in the past."

Thus, the Forest Service purchased the basin with rights to timber larger than 15 inches at the stump reserved (except for construction purposes, but in no case was poplar, basswood, cherry, red or white oak under 15 inches to be cut), and the basin was logged by J. A. Porter in 1919. Shope Fork was cut first and Ball Creek last. Mr. Porter moved two mills into the basin and with a crew of about 40 men, logged the basin moving the mills three times in about 3 ½ years. A railroad spur was built from Otto into the basin, of which one fork ended at Ball Flat and the other stopped at Chestnut Flats on Shope Fork. Tram roads were built through Dyke Gap to the mid-elevations of WS 28 on the south and to the lower reaches of WS 36 and 37 on the north. A flume was constructed on the headwaters of Ball Creek to carry boards to the rail head at Ball Flats, but was reportedly never used. All skidding was done by horses and after about 8 million board feet of timber had been removed, in 1923 the Forest Service took over administration of the desolate, recently logged, uninhabited tract of land that was soon to become the Coweeta Experimental Forest. Thereafter, management consisted mostly of protection of the land against fire. The only major occurrence of significance, until 1933, was the intensification of the chestnut blight, which, by 1940, had killed virtually all chestnut trees over 10 cm dbh. This was a major impact, because chestnut made up an estimated 40% of the forest stand.

2. Hydrology, Geology, Climate and Water Chemistry

3. Climatology and Hydrology

L.W. Swift, Jr., G.B. Cunningham and J.E. Douglass

In a problem statement that preceded establishment of Coweeta Experimental Forest, Hursh (1932a), recognized two problems for forestry research in the southern Appalachian Mountains: (1) how is streamflow regulated by vegetative cover, and (2) how can erosion be controlled by vegetative cover? In early reports on the new research program, Hursh champions studies to describe and develop understanding of climate, precipitation, soils, and topography of these mountains and how they interact with forest vegetation to produce streamflow. The state of our knowledge of climate and streamflow in the Coweeta Basin and our understanding of interactions with topography and vegetation are subjects of this chapter.

History of Climatic and Hydrologic Measurements

Climatic Stations

During the last 50 years, 30 climatic stations were operated at Coweeta. Most were active for only a few years to measure the influence of various watershed treatments. Six stations were sited around the valley from 1936 to 1939 to establish the climatology of the basin. Only the main Climatic Station 01 (CS01) has been continuously active since the initiation of precipitation measurements there in August 1934. In April 1936, air and soil temperature, relative humidity, wind travel, evaporation, and cloud cover observations were added. Some early temperature and precipitation measurements were taken near the CCC Camp but the major site has always been in the Laboratory administrative area at elevation 686 m. Initially, CS01 was on the south side of Shope Fork at the approximate location of a parking lot below the present office. In 1963, the site was moved 85 m NE across the stream. Solar radiation measurements were added in 1965 and conversion from manual and strip chart observations to automatic data loggers began in 1974.

Although many Coweeta streamflow and precipitation stations have long and continuous records, climatic stations outside the administrative area do not. A goal to establish continuing climatic measurements above the valley bottom on north-, south-, and east-facing slopes is being realized. In 1969 and 1974, records began at two low-elevation sites (Table 3.1) and in 1985, we initiated climatic measurements high on the east-facing slope. An additional station is registering the climatic changes associated with clearcutting and deciduous forest regrowth on WS 7 and another began in the adjacent *Pinus strobus* plantation in 1986. All of these sites now use battery-powered electronic data loggers.

Precipitation Stations

Over 135 precipitation sites have been gaged at various times in the last 50 years. Density of gaging in the main Coweeta Basin was greater than is reported for any other mountain area of similar size. On several experimental watersheds, density was as high as 1 gage for every 4 to 5 ha while the base network of 50 gages covered the entire basin with a density of 1 per 32.5 ha. This network was maintained for 20 years and, through much of this period, the 8-inch standard gages were measured after each storm. It was reduced in 1958 to a long-term network of 9 sites, each with a pair of gages. The amount in the 8-inch standard gage is the most accurate measure of the week's total precipitation and observations from the recording gage are adjusted to this amount. The recording gage trace for each storm is broken into unequal-duration intensity periods and maximum 5-, 15-, 30-, and 60-min intensities are interpolated for all storms over 6 mm total precipitation. Daily (midnight to midnight) totals are also calculated from the intensity period data.

Table 3.1. Active Climatic Stations at Coweeta Hydrologic Laboratory

Station Name	Elevation (m)	Slope	Type of Site	Record Begins
		Permanent Stations		
CS01	685	Valley floor	Grassed field	August 1934
CS17	887	North-facing	Forest opening	October 1969
CS21	817	South-facing	Forest opening	July 1974
CS28	1189	East-facing	Forest opening	May 1985
		Study-Related Stations		
CS20	866	South-facing	Regrowing forest after logging	August 1973
CS24	740	South-facing	White pine canopy	April 1986

Streamflow Stations

Permanent weirs were constructed on 32 Coweeta streams with records initiated on 9 streams by 1934. By 1941, 28 streams were gaged on watersheds ranging from 2.8 to 760 ha. Presently, 16 streams are being gaged. Head or flow depth was recorded on strip charts until 1964. Now, head is recorded on 16-channel punched tape every 5 min. Recorders are checked by hookgage weekly. Chart records were manually integrated until 1958, when computer processing of flow records began. Methods of data extraction, flow calculation, and storm analysis developed at Coweeta have been utilized by hydrologists throughout the world. Details of climatic and hydrologic measurements and analyses are given by Hibbert and Cunningham (1966) and Swift and Cunningham (1986).

Climate

The climate of the Appalachian Mountains is distinguished from that of surrounding lowlands by characteristics of precipitation and temperature or evaporation. Under Köppen's system, Coweeta's climate is classed as Marine, Humid Temperature (Cfb) because of high moisture and mild temperatures (Critchfield 1966; Trewartha 1954). The lower elevations of Coweeta are borderline, because the basis for separating the Marine climate from the adjacent Humid Subtropical is the mean monthly air temperature and whether it is below or above 22 °C in the warmest month (see Table 3.2). For Thornthwaite's original scheme, Coweeta is in the wet, mesothermal, adequate rainfall (AB'r) climate. Thornthwaite's modified classification (1948), based on water surplus or deficiency calculated from precipitation and potential evaporation, gives essentially the same result—perhumid, mesothermal with water surplus in all seasons.

Precipitation

Precipitation, as modified by topography, is a dominant factor of the Coweeta climate. At the basin scale, precipitation consists of frequent, small, low-intensity rains in all seasons with little snow. Occasional large storms occur at shorter return intervals than experienced by adjacent, less mountainous areas.

In general, precipitation increases with elevation (about 5% per 100 m) along the east-west axis of the Coweeta valley but changes little with elevation over north- and south-facing side slopes (Figure 3.1). The pattern is fairly regular, except near ridgelines where gage catch is consistently lower than at sites immediately downslope. Hibbert (1966) found reductions of 30% in a 30 m change of elevation extending about 100 m either side of a ridge. A compact array of 18 gages installed near the top of WS 1 after it was cleared showed 50% less rainfall in a cove facing the prevailing wind where a strong updraft was observed. Ninety-three percent of Coweeta's precipitation falls when winds are calm or blowing less than 2.2 m/s from southwest, south or southeast (Swift 1964). High precipitation points occur on the lee side of major gaps in the mountains that rim the south and west boundary of Ball Creek. We hypothesize that air moving across the ridges is accelerated, particularly through the gaps, and points of

Table 3.2. Averages and Ranges of Monthly Means for Climatic Variables at Coweeta CS01

Climatic Variable	JAN	FEB	MAR	APR	MAY	JUN	JUL	AUG	SEP	OCT	NOV	DEC
Precipitation (mm/month)												
Average	174	181	203	156	140	130	144	139	125	112	139	178
Minimum	50	28	72	13	32	32	9	41	6	1	24	23
Maximum	371	423	433	287	477	318	300	340	333	291	478	388
Evaporation (mm/month)												
Average	34	43	77	97	104	107	108	98	81	67	46	30
Minimum	16	11	50	57	83	79	88	77	60	48	34	14
Maximum	60	76	132	137	133	157	135	120	100	83	57	57
Air temperature (°C/day)												
Average	3.3	4.4	8.0	12.6	16.4	20.0	21.6	21.3	18.4	13.0	7.8	4.1
Minimum	-4.0	-0.4	1.7	10.0	13.4	17.4	19.6	19.5	16.5	10.8	4.8	-0.6
Maximum	10.9	8.8	12.8	15.0	18.8	22.9	23.1	22.8	20.4	16.0	10.9	8.5
Solar radiation (MJ/m^2/day)												
Average	8.0	11.1	14.1	17.8	18.2	19.2	17.9	16.2	14.0	12.5	9.0	7.0
Low	6.0	9.0	10.3	12.9	15.4	14.5	13.9	12.5	12.3	9.2	7.2	5.2
High	10.8	13.6	17.6	21.3	21.4	22.5	24.3	19.8	16.9	17.7	11.1	8.7

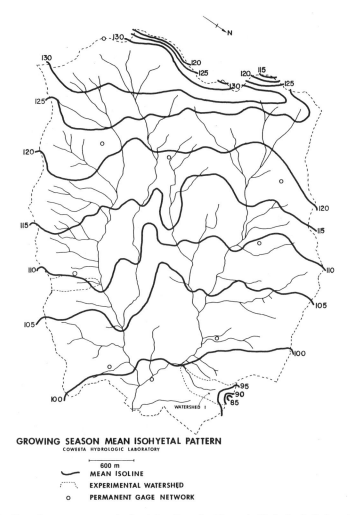

GROWING SEASON MEAN ISOHYETAL PATTERN
COWEETA HYDROLOGIC LABORATORY

600 m
⌣ MEAN ISOLINE
⌐⌐⌐ EXPERIMENTAL WATERSHED
o   PERMANENT GAGE NETWORK

Figure 3.1. Growing season mean isohyetal pattern for Coweeta Hydrologic Laboratory. Values on mean isolines represent 100 times the ratio between season total precipitation at each measured point in the basin and the catch at Standard Gage 15.

high precipitation represent dump zones for rain transported away from the ridges. Based on long-term data and these studies of precipitation distribution, Swift (1968) developed mean isohyetal maps and from these derived weighting factors for estimating watershed precipitation. Unlike the previously used mean polygon weighting system, isohyetal weighting uses ridgeline and dump zone patterns. Isohyetal lines were based on ratios of precipitation at each site to precipitation at a base or normalizing gage. March et al (1979) extended the normalizing procedure while relating precipitation pattern for six storm types to topography.

Monthly mean precipitation amounts for the upper and lower elevations (Figure 3.2) show a consistent difference in all seasons. Precipitation is greater in late winter and

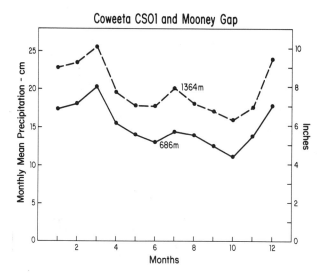

Figure 3.2. Long-term monthly mean precipitation at two gages near the elevational extremes at Coweeta. Mooney Gap (Gage 31) is near the Appalachian Trail at 1364 m above mean sea level (47-year record). Climatic Station 01 (Gage 6) is at 686 m (50-year record).

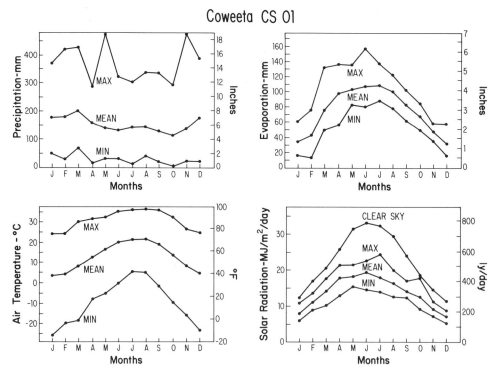

Figure 3.3. Mean monthly precipitation, evaporation, air temperature, and solar radiation at Climatic Station 01 compared with maximum and minimum monthly values.

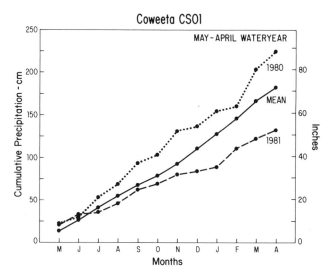

Figure 3.4. Cumulative precipitation at Climatic Station 01 for 2 extreme years compared with cumulative monthly means. The May 1979 to April 1980 water year had the largest total precipitation out of 50 years (224.7 cm). The May 1980 to April 1981 year had the third lowest total (132.5 cm). The minimum year at Coweeta was 1986 with 122.9 cm.

spring with March the wettest month. Although fall months are the driest, half of the record high rain days have been in the fall, typically caused by tropical storms. But, amounts of 50 to 100 mm/day have been measured in every season. Rain falls in all months. Snow comprises only 2 to 10% of annual precipitation at lower elevation CS01. Precipitation averages 152 mm/month at CS01 and has ranged from 0.8 mm in October 1963 to over 475 mm in November 1948 (Table 3.2). Figure 3.3 shows the means and extremes for monthly precipitation totals at this site, whereas Figure 3.4 compares the cumulative precipitation for our wettest (1980) and third driest (1981) years with accumulated mean precipitation. Except for the dry year, accumulation was most rapid in the dormant season (November through April). High precipitation input in this low evapotranspiration period provides fairly consistent soil moisture recharge by early spring.

The probability of measurable precipitation (>0.254 mm) on any date ranges from 5% on November 30 to 73% on February 6 but a 5-day moving average puts the probability at around 40% for over 6 months of the year (Figure 3.5). On the average, 133 storms are recorded annually. A storm is defined as a period of precipitation separated by rainless periods of at least 4 hr. Many storms are short and small (Figure 3.6). Consequently, interception loss accounts for over 20% of rainfall in half the storms each year. For discussion on interception losses see Helvey and Patric (Chapter 9) and for interception processes see Murphy (1970). Daily precipitation amounts are skewed toward 0 and fit the S_B, a three-parameter log-normal distribution (Swift and Schreuder 1981), whereas groups of consecutive rain days best fit the gamma distribution.

Precipitation intensity is often used in simulation models to estimate streamflow and soil erosion, although Hewlett et al (1984) strongly question the value of intensity as

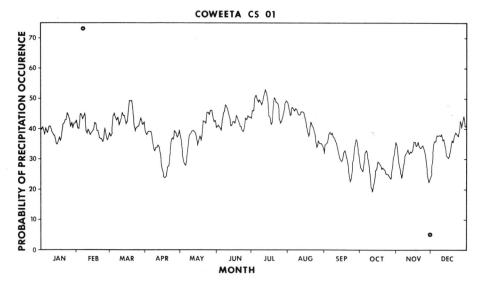

Figure 3.5. Five-day moving average of probability of precipitation occurrence for each date of the year. Extremes are 73% on February 6, and 5% on November 30.

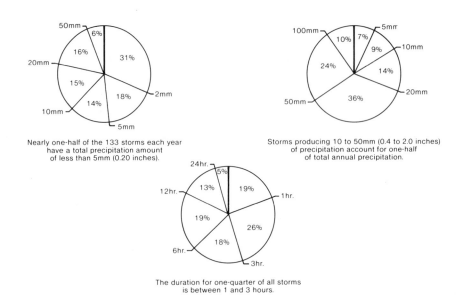

Figure 3.6. Storm size and duration relationships at Coweeta Climatic Station 01. Diagrams are based on 40 years' data from Recording Gage 6.

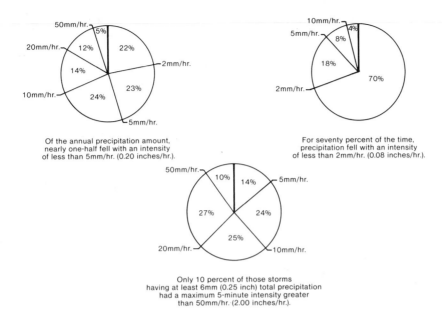

Of the annual precipitation amount, nearly one-half fell with an intensity of less than 5mm/hr. (0.20 inches/hr.).

For seventy percent of the time, precipitation fell with an intensity of less than 2mm/hr. (0.08 inches/hr.).

Only 10 percent of those storms having at least 6mm (0.25 inch) total precipitation had a maximum 5-minute intensity greater than 50mm/hr. (2.00 inches/hr.).

Figure 3.7. Precipitation intensity relationships at Coweeta Climatic Station 01. Diagrams are based on 40 years' data from Recording Gage 6.

a predictor of stormflow. Figure 3.7 suggests a contributing cause for their conclusion −nearly three-quarters of Coweeta's precipitation falls with an intensity of less than 10 mm/hr, and only 10% of storms have maximum intensities over 50 mm/hr. However, intensity and raindrop size are useful for estimating erosion rates. Drop size distributions were measured at Mooney Gap (Gage 31) for 102 storm periods during 15 months. Drop sizes ranged from 0.4 to 5.4 mm with the peak frequency at 1 to 2 mm in diameter (Cataneo and Stout 1968). Drops of 1.5 mm have a terminal velocity of about 650 cm/sec.

Because of point-to-point variability in mountain topography, standard tables and maps seriously underestimate rainfall intensities and derived parameters for Coweeta and possibly other upper elevation sites in the southern Appalachian Mountains. The erosivity index for the Universal Soil Loss Equation is based on both the intensity and amount of rain. The index for Coweeta is given as 300 in a frequently referenced map in Wischmeier and Smith (1978). However, annual erosivity indexes for Coweeta rain Gage 6 give typical values of 400 to 500. Extreme rainfall frequency maps (Weather Bureau 1961) predict intensities of annual maximum storms for various return periods that are uniform over all the Coweeta Basin and surrounding terrain. These closely match the storm history for the low-elevation gage at CS01. In contrast, Bradford's (1977) depth–frequency–duration curves for high-elevation Gage 31 at Mooney Gap (Figure 3.8) are more similar to the Weather Bureau prediction for a high-rainfall zone . southeast of Coweeta. Earlier publications contain less detail and are not suitable for use in the mountains because they were based on limited data from first-order stations such as Asheville, where precipitation is 53% of that at Coweeta. Greater precipitation

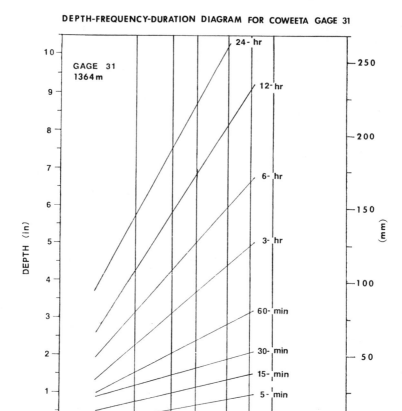

Figure 3.8. Precipitation depth-frequency-duration diagram for Rain Gage 31 at Mooney Gap. (Bradford 1977.)

at higher elevation sites along the southeastern edge of the Appalachian Mountains is attributed to uplift and frontal contact of moist air masses moving inland. For 11 of 17 years (1963 through 1979) Coweeta Gage 31 (elevation 1364 m) caught more rain than any other gage in the Tennessee Valley Authority system and was in the top five for most of the other years. Thus, rainfall amount and intensity at the upper elevations on Coweeta are probably similar to those at Rosman, Highlands, and several mountaintop gages along the southeastern front of the Appalachian Mountains which regularly report high annual precipitation totals.

Solar Radiation, Temperature, Wind, and Evaporation

Coweeta's temperate climate is due to the combined influences of elevation (678 to 1592 m) and latitude (35°N). The low latitude increases the potential solar radiation available for sensible heat, plant growth, and evaporation, but the actual energy at any point is influenced by the inclination and aspect of mountain slopes, shading by adjacent hills, and cloud cover. The seasonal difference between the mean monthly global solar radiation and the curve for clear skies (Figure 3.3) is due to frequent afternoon clouds in the summer months. Only 10 to 15% of the summer solar radiation is transmitted through the fully leafed canopy, but in the dormant season about half penetrates to the forest floor (Swift 1972). The record at Climatic Station 01 is slightly faulty, because the site is shaded by adjacent mountains from direct beam radiation approximately 60 and 90 min at the beginning and end of a midwinter day when radiation is least. Nevertheless, radiation on mountain slopes can be estimated from the CS01 record (Swift and Knoerr 1973) using an algorithm for calculating potential radiation (Swift 1976). Mean monthly recorded solar radiation is approximately 45% of potential radiation. The algorithm has been useful in modeling evapotranspiration (Huff and Swank 1985; Swift et al. 1975), moisture content of leaf litter (Moore and Swank 1974), and streamflow increases due to forest cutting (Douglass and Swank 1975).

The majority of solar radiation retained on site is converted to heat or evapotranspiration. Figure 3.3 and Table 3.2 show the annual cycles of air temperature and evaporation at CS01. Warmest months are June through August, while December through February are the coldest. Pan evaporation follows the same trend except that it peaks a month earlier in the summer. The 50-year record temperature extremes were −27.8°C in January 1985 and 36.7°C in July 1952. Growing season degree days accumulate, on the average, from early May through early October (Figure 3.9). However, frost has been recorded at CS01 as late as June 2 and as early as September 14.

Relative humidity in the forest drops to an average 50% during the day and rises to 75% on winter or spring nights and over 90% most summer and fall nights.

Little is known about wind speed and direction around the Coweeta Basin. Long-term records show that wind speeds are low at CS01, except during some storm periods, probably because of the sheltered position of this valley bottom site. More

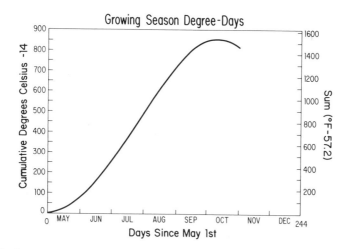

Figure 3.9. Growing season degree-days at Climatic Station 01, based on temperature mean for each date over the period of record.

complete data are forthcoming from upper elevation stations CS 17, 21, and 28, now equipped with moderately sensitive anemometers and vanes. In a detailed study, Mowry (1980) consistently found linear wind profiles over the pine forest on WS 1 proving that the theoretical logarithmic profile does not universally apply.

Annual evapotranspiration can be estimated several ways: from the evaporation pan at CS01, empirical and energy balance equations, and the water balance of control watersheds. None of these methods are error free (Lee 1970), but several approximate 900 mm at the lower elevations. The mean annual total for the evaporation pan is 892 mm. With an admittedly low estimate for winter, Thornthwaite's PE equals 696 mm while the PROSPER simulation (Swift et al. 1975) estimated ET, using a Penman-Montieth equation, to be 890 and 910 mm for 2 specific years. Average precipitation minus runoff values (P − RO) approximate 90 and 50 cm per year respectively for low-elevation WS 2 and high-elevation WS 36, 2 undisturbed watersheds at Coweeta (Figure 3.10). Huff and Swank (1985) present an application of long-term Coweeta climatic data to simulate evapotranspiration and streamflow.

Hydrology

Hydrologic behavior of a watershed is determined by precipitation input, as modified by gradients of soil moisture storage and by evapotranspiration use. These combine to define the annual hydrograph (cycle of monthly flow totals), the storm hydrograph, and characteristics for flow frequency and peak discharge for each stream.

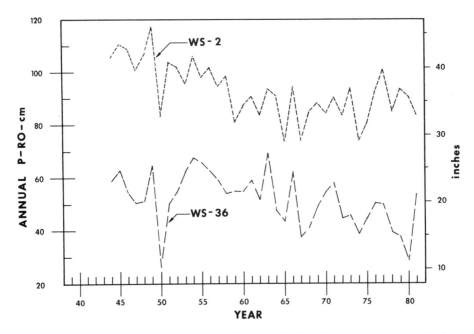

Figure 3.10. Range of annual precipitation minus runoff values for two control watersheds at Coweeta. Midelevations are 857 m for WS 2 and 1281 m for WS 36.

Hydrograph Analysis

Hursh (1932b) concluded that Horton's concept of rainfall excess and overland flow was not applicable to most forest land because virtually all rain infiltrates into the soil. Thus, his early work was in the area of infiltration, storage, and movement of water in forest soils. Hursh (1942a) felt that runoff coefficients (such as total runoff/precipitation, storm runoff/precipitation, and storm runoff/total runoff) could be used to characterize watersheds. Hursh and Brater (1941) discussed the need for and methods of separating hydrographs into direct and delayed (base) flow components and the relative importance of the different sources of water that contribute to runoff from Appalachian forests. With computers, consistent, systematic separation of the hydrograph into components became a practical reality (Hibbert and Cunningham 1966) and runoff responses to rainfall could be examined in detail. Hewlett and Hibbert (1966) found that many traditional runoff coefficients gave highly variable results between watersheds, even within the comparatively small Coweeta Basin. Hewlett finally selected the ratio of annual mean quickflow (storm runoff) to mean precipitation as the most useful response characteristic and used it to develop a response map, first of Georgia (Hewlett 1967) and then of the entire eastern United States (Woodruff and Hewlett 1971). Such maps are useful for land management planning and for estimating the quickflow response to precipitation on ungaged watersheds.

Hoover and Hursh (1943) noted that large differences in runoff occur between watersheds in the Coweeta Basin. They attribute runoff differences primarily to differences in rainfall amount, soil depth (i.e., soil water storage and release characteristics), and topography. Although Hursh and Brater (1941) describe the components of soil water flow that contribute to the storm hydrograph and Hursh and Hoover (1942) discussed the importance of water holding and transmitting characteristics of soil profiles, proof of the importance of unsaturated flow through soils on steep slopes as a source of baseflow was obtained only after a series of soil model studies (Hewlett 1961; Hewlett and Hibbert 1963; Scholl and Hibbert 1973). Hydrologic models had been using the Hortonian concept of effective rainfall, a fixed percentage of precipitation, as input, but Hewlett (1961) and Hewlett and Hibbert (1966) postulated the now generally accepted concept of a dynamic, variable sized area adjacent to stream channels and drainage ways as being the source of quickflow during rainfall events (see Chapter 8).

The historical records of control watersheds at Coweeta allow examination of a considerable range of hydrologic responses from relatively undisturbed forest systems within the 1625-ha basin. Some physiographic and hydrologic characteristics of six control watersheds are presented in Table 3.3. The hydrograph data for each watershed are averages of 1440 to 3555 separate storm events. No more detailed data base for forested watersheds exists anywhere in the world, either in length or in quality of records.

Hydrologic Response

Figure 3.11 shows the average hydrograph for four of the six watersheds presented in Table 3.3 (the other two watersheds fall within the extremes set by these four). This figure illustrates the large range in response to rainfall that occurs at Coweeta. WS 2 and 18 typify low elevation, deeper soiled and gentler sloped watersheds. WS 27 and 36 are the contrast—high elevation, steeply sloping watersheds with shallow soils and high rainfall that yield a greater percentage of their precipitation as streamflow. Low-

Table 3.3. Some Physiographic and Mean Hydrologic Characteristics of Coweeta Conrol Watersheds

Characteristic		Watershed					
		2	14	18	34	27	36
(1) Area	ha	12.26	61.03	12.46	32.70	39.05	48.60
(2) Maximum elevation	m	1004	992	993	1184	1455	1542
(3) Minimum elevation	m	709	707	726	852	1061	1021
(4) Land slope	%	60	49	52	52	55	65
(5) Record length	yr	37	44	45	18	35	39
(6) Mean annual precipitation	cm	177.17	187.55	193.90	200.94	245.08	222.25
(7) Mean annual runoff	cm	85.39	98.81	103.42	117.47	173.74	167.51
(8) Precipitation-runoff	cm	91.78	88.74	90.48	83.47	71.34	54.74
(9) Hursh's runoff coefficient = $\frac{(7)}{(6)}$		0.482	0.527	0.536	0.585	0.709	0.754
(10) Initial flow rate	l/s/km²	24.52	29.46	29.54	36.57	36.30	40.81
(11) Quickflow before peak	cm	2.22	3.30	2.73	1.83	17.65	11.33
(12) Quickflow after peak	cm	5.93	6.62	7.00	3.74	34.15	25.86
(13) Delayed flow	cm	77.20	88.90	93.69	111.91	121.94	130.33
(14) Storm duration	hr	13.78	13.45	13.50	11.10	29.93	26.75
(15) Time to peak	hr	4.20	4.65	4.11	3.95	9.31	8.02
(16) Absolute peak	l/s/km²	68.78	81.12	79.97	75.65	242.60	154.05
(17) Recession time	hr	9.57	8.81	9.39	7.15	20.62	18.73
(18) Mean runoff events/year		66	80	79	80	68	68
(19) Response factor = $\frac{(11)+(12)}{(6)}$		0.046	0.053	0.050	0.028	0.211	0.167
(20) Hursh's storm runoff ratio = $\frac{(11)+(12)}{(7)}$		0.095	0.100	0.094	0.047	0.298	0.222

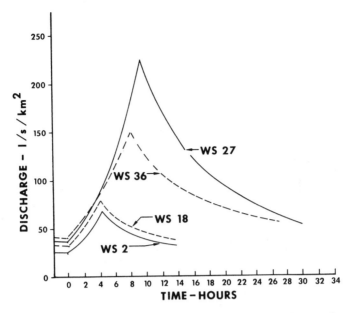

Figure 3.11. Average hydrograph for each of four control watersheds at Coweeta.

elevation watersheds release about half of their rainfall as runoff (Table 3.3, line 9), and 5% as quickflow (or storm runoff) (line 19). High-elevation watersheds with steep slopes and shallow soils yield 75% of their rainfall as runoff with 20% of rainfall leaving as quickflow. Thus, over a horizontal distance of 4.5 km and a vertical distance of 915 m, average annual runoff response to precipitation (Hursh's runoff coefficient) increases by a factor of 1.5. Over this same distance, the time span from the beginning of storm runoff to peak runoff rate doubles from about 4 to nearly 9 hr; storm duration doubles from about 14 to 28 hr and peak discharge increases by a factor of 2.6. Volume of quickflow, which is the primary factor influencing downstream flooding, increases about fivefold.

The large differences in response to rainfall shown in Table 3.3 and Figure 3.11 reflect physical differences between watersheds. Figure 3.1 shows that rainfall increases with elevation. Drilling to determine depth of regolith indicated a general decrease in soil depth with elevation. Although weakly correlated, slope steepness increases with elevation; thus, the hydrologic head driving water through soil increases. Shallow soil at upper elevations provides a smaller cross-sectional area through which subsurface water moves. The combined result is that rate of discharge and volume of storm runoff both increase with elevation. Hence, elevation can represent the integrated effects of physiographic and hydrologic variables which determine rates and volumes of discharge.

Hewlett (1967) showed that runoff response was a hydrologic characteristic which could be mapped. His quickflow map (Figure 3.12) indicates a 10-fold increase in potential flood waters between low and high elevations in the basin, an astounding 5 to 50 cm range.

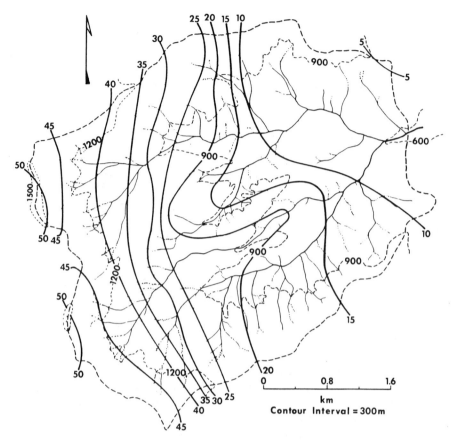

Figure 3.12. Runoff map of Coweeta Basin (Hewlett 1967). Units are cm of mean annual quick-flow.

Peak Discharge

Although average response factor is a useful watershed characteristic, the range of streamflow response is also needed to define the expected minimum and maximum storm events. The minimum stream response to a small storm may appear to be zero, but some minor response due to channel interception does occur. The maximum response is of more interest because it is related to maximum peaks and storm runoff volumes. The maximum storm response for WS 2 occurred in a frontal storm on May 27–29, 1976. This storm of 30 hr duration came 7 days after a 15-cm storm; thus, soil moisture was recharged and the second rain produced approximately a 100-year runoff event (Douglass 1974). Table 3.4 shows the quickflow reponse of 4 control watersheds to this storm. Twenty-six percent of the precipitation left each low-elevation watershed as quickflow compared with 50 to 54% of rainfall from the high-elevation watersheds.

Table 3.4. Maximum Quickflow Responses of Four Control Watersheds to the May 27–29, 1976, Frontal Storm at Coweeta

Watershed	Midslope Elevation (m)	Precipitation (cm)	Quickflow (cm)	Quickflow / Precipitation
2	857	20.02	5.21	0.26
18	860	20.81	5.41	.26
27	1257	25.86	13.92	.54
36	1281	23.80	11.99	.50

Although precipitation was 3.4 to 5.4 cm greater on the upper watersheds, quickflow was 6.6 to 8.6 cm greater. Again, upper elevation watersheds yield a greater percentage of rainfall, but the range in response between watersheds for this large event is less than the 10:1 ratio shown for average annual response.

Douglass (1974) used elevation as an integrator of climatic and physiographic factors to develop equations for predicting peak discharge from forested watersheds of the Appalachian Mountains of North Carolina. These equations were derived from both Coweeta's experimental watersheds and from larger, predominately forested watersheds (up to 13,400 ha in size) elsewhere in the mountains of North Carolina. The correlation indices (r^2) of equations for peak discharge versus watershed area and maximum watershed elevation all exceeded 0.98 for the 2.33- to 50-year recurrence intervals. Figure 3.13 illustrates a set of curves for the 20-year equation. While eleva-

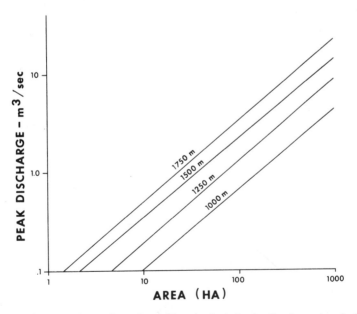

Figure 3.13. Peak discharge from forested watersheds in the Southern Appalachians for the 20-year return interval as a function of area and maximum elevation of watershed.

tion is a convenient integrator of several effects, the individual contribution of variables such as slope, precipitation, and soil moisture storage have yet to be quantified.

Another flow characteristic of watersheds useful in engineering applications is the distribution of flow. Flow frequency or the percent of time a given discharge rate is equalled or exceeded is one measure of flow distribution. Figure 3.14 is the average flow frequency distribution for the same four control watersheds at Coweeta. The pattern for flow frequency follows the same general order shown by Figure 3.11 and Tables 3.3 and 3.4. The smallest rates of discharge for all exceedence percentages is observed on low elevation WS 2 and 18, which have less rainfall, gentle slopes, and deeper soils.

Annual Hydrograph

The annual hydrograph tracks monthly total streamflows. Figure 3.15 shows maximum, mean, and minimum monthly total flows for WS 2. The minimum trace shows

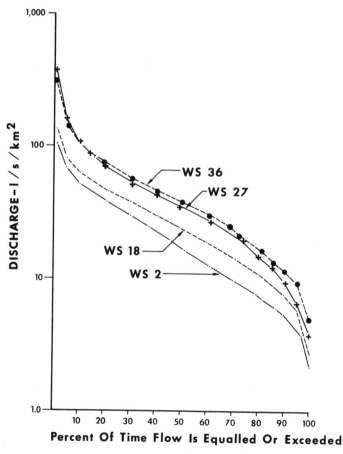

Figure 3.14. Flow frequency distribution for 4 control watersheds at Coweeta showing increasing flow rates with increasing elevation.

Figure 3.15. Mean monthly total
streamflow from control WS 2
compared with the maximum and
minimum monthly observed.

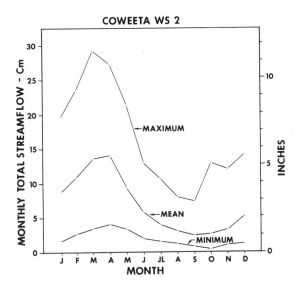

that in any month of the year, the total flow can be less than 4 cm, but flow does not
cease. Comparing the minimum and mean, low flows in late summer and fall have been
half the mean flows for those months. The maximum monthly runoff for WS 2 ranges
from 7.5 cm to over 25 cm/month during the March to April recharge period, which
ends the dormant season. Rainfall for WS 2 during the peak flow month of April averages 16 cm; thus, 87% of the rain falling on this watershed during April appears as
streamflow during that month. As a watershed becomes recharged, rainfall added to the
soil surface quickly displaces into the stream an equivalent amount of water from
storage in the soil profile.

The long-term P − RO value for each control watershed is given in Table 3.3. It
varies from about 55 cm for more responsive high elevation watersheds, which have a
shorter growing season and generally cooler and wetter climate, to 90 cm for low elevation watersheds with their longer growing season. P − RO values for 39 years for WS 2
and 36 (Figure 3.10) illustrate the very large year-to-year variation, typical also for
other watersheds at Coweeta. One might expect that the year-to-year variation in P −
RO for a given watershed could be related easily to amount and timing of precipitation.
However, fluctuations in successive years sometimes may be positively and other times
negatively correlated with precipitation while adjacent watersheds may not respond
alike in the same year. Nevertheless, the P − RO value is a summary of the watershed
response to climate and hydrologic processes and with the length of records at
Coweeta, detailed investigation of the causes for P − RO variation over time and
between watersheds could be a fruitful area of study.

Summary

Streamflow from an undisturbed forested watershed is the net result of the physiography of the catchment and its climate. Various hydrologic response characteristics
correlate well with watershed elevation and it has been used as a predictor of stream-

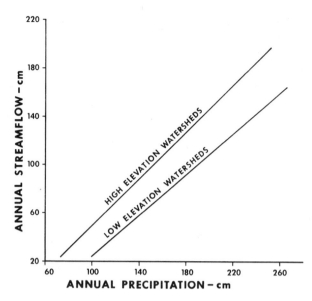

Figure 3.16. A streamflow versus precipitation relationship for forested watersheds over the range of elevations at Coweeta.

flow response. Elevation, however, is a surrogate variable representing several other variables whose exact physical relationships to streamflow are incompletely defined at this time.

Precipitation amount and timing have the greatest influence upon streamflow. Precipitation increases with elevation along the east-west axis of the Basin, but is not closely correlated with elevation along the slope for most of the north- and south-facing watersheds.

At all elevations, precipitation is distributed fairly evenly throughout the year with large individual storms occurring in nearly every month. Generally, monthly total precipitation is less in April, late summer, and fall. Lagged streamflow response with snowmelt is a minor factor, because heavy snows and long lasting snowpacks are rare. The majority of rains have short durations and low intensities, but when large, high intensity storms do occur, they occur with shorter recurrence intervals than predicted from standard reference works.

On an annual basis, precipitation exceeds evapotranspiration demand and streams flow yearround. Solar radiation is the primary source of energy for evapotranspiration. This energy input is influenced by land slope, aspect, topographic shading, and ground fog. Thus, radiation on various watersheds is only partly correlated with elevation. However, solar energy is converted to heat and in this way influences temperature, humidity, and wind, variables which do exhibit elevation effects. Because evapotranspiration rates are sensitive to temperature, humidity, and wind, evapotranspiration also can be correlated with elevation.

Soil depth decreases and slope steepness increases with elevation. Both factors reduce the ability of the watershed to retain precipitation and thus increase the percentage that appears as streamflow. Upper elevation watersheds have lower precipitation-runoff (P − RO) factors because they have less soil moisture storage capacity, return a higher percentage of precipitation as quickflow, and have less evapotranspiration demand to create soil moisture storage opportunity before a rain. Figure 3.16 proposes a relationship between annual streamflow and precipitation for forested watesheds at the high and low elevation extremes in the Coweeta Basin. Here, elevation again represents the net effect of several watershed physical factors, precipitation input, and evapotranspiration demand. Other, fully forested deciduous forests in the eastern United States, for which we have data, generally fall within the bounds of the two lines in Figure 3.16.

4. Characterization of Baseline Precipitation and Stream Chemistry and Nutrient Budgets for Control Watersheds

W.T. Swank and J.B. Waide

Budgets of solute inputs and outputs for undisturbed watershed ecosystems provide a conceptual and empirical framework both for examining ecosystem function in diverse geographic regions and for evaluating man's impact on the natural landscape (Bormann and Likens 1967; Henderson et al. 1978; Swank and Waide 1980). Thus, the primary objective for initiating cooperative research between the Institute of Ecology, University of Georgia, and Coweeta Hydrologic Laboratory, USDA Forest Service in 1968 was the measurement of annual and seasonal fluxes of select nutrients for several forested watershed ecosystems in the Coweeta Basin (Johnson and Swank 1973). As the research program developed and studies of nutrient recycling processes within experimental watersheds were initiated, the baseline network for precipitation and stream chemistry measurements was expanded to the entire Basin and additional inorganic constituents were added to the routine analyses. By 1972, most of the basic system for long-term water chemistry measurements was established (Swank and Douglass 1977).

In this chapter our objectives are to summarize the long-term record of Basin precipitation and stream chemistry for control watersheds. Specifically, we will (1) evaluate the sampling network, (2) characterize the average solute composition of precipitation and stream water for select watersheds, (3) describe long-term annual and seasonal trends of specific solutes, and (4) describe average annual nutrient budgets for control watersheds. The Coweeta precipitation and stream chemistry record is among the most extensive, long-term data bases available for any single location. Thus, this record provides an excellent basis for characterizing the solute composition of precipitation and streamflow in Southern Appalachian forests, as well as for detecting trends in precipitation and stream chemistry unaffected by local site manipulation.

The Sampling Network

Bulk precipitation samples are collected weekly using plastic funnels at eight sites located over the 1626 ha Coweeta Basin, at stations previously established to quantify precipitation. The representative nature of these stations for estimating precipitation is based on more than 2,000 gage years of record (Chapter 3). To ensure that adequate amounts of sample are available for all chemical analyses, samples are not collected unless weekly precipitation exceeds 1.3 cm. When this is not the case, samples remain in the field for additional weekly periods until accumulated precipitation is at least 1.3 cm. Because of frequent precipitation events, this situation rarely occurs. However, preservative is added to samples to inhibit degradation (Swank and Henderson, 1976).

The arithmetic mean solute concentrations of weekly samples for the eight bulk precipitation gages over a 13-year period of record are shown in Table 4.1. One-way analyses of variance were performed to examine homogeneity of gage means for all ions. Individual gage means were not statistically different (5% level of significance) for H^+, NO_3^--N, NH_4^+-N, Cl^-, or SO_4^{2-}-S. One gage mean was statistically different from the remaining seven for PO_4^{3-}-P. Some significant differences were observed for K^+, Na^+, Ca^{2+}, and Mg^{2+}, but these differences were not consistent across gages or ions. (Ion charge notation is assumed hereafter.) Further attempts to rank gages failed to reveal any consistent patterns; therefore, it appears that only small increases in the precision of estimating nutrient inputs to individual watersheds would be gained by gage stratification. Based on this analysis, all eight gages are used to calculate a weekly arithmetic mean ion concentration for the Basin, and that weekly mean is used along with precipi-

Table 4.1. Mean Weekly Concentrations of Dissolved Chemical Constituents in Bulk Precipitation Within the Coweeta Basin[a]

Ion	Units	$\bar{x}$	$s_{\bar{x}}$	CV
NO_3^--N	mg L^{-1}	0.174	0.002	0.032
NH_4^+-N	mg L^{-1}	0.117	0.003	0.073
PO_4^{3-}-P	mg L^{-1}	0.00564	0.00039	0.190
Cl^-	mg L^{-1}	0.283	0.007	0.066
K^+	mg L^{-1}	0.124	0.006	0.145
Na^+	mg L^{-1}	0.172	0.009	0.146
Ca^{2+}	mg L^{-1}	0.236	0.008	0.098
Mg^{2+}	mg L^{-1}	0.0471	0.0017	0.104
SO_4^{2-}-S	mg L^{-1}	0.675	0.008	0.033
H^+	μeq L^{-1}	21.01 (pH = 4.68)	0.08	0.109

[a] Data shown as the arithmetic mean ($\bar{x}$), standard error ($s_{\bar{x}}$), and coefficient of variation (CV), calculated over the mean concentrations for eight rain gages located throughout the Coweeta Basin. Individual gage means were calculated as arithmetic averages of weekly samples collected over all water years (June–May) of record (1971–1983). Sample sizes for specifications range between 285 and 415 for each gage.

tation amounts to estimate watershed nutrient inputs. The standard errors of estimate for weekly mean ion concentrations were generally less than 5%, with lowest variability for NO_3-N and SO_4-S and highest variability for PO_4-P, K, and Na (Table 4.1).

Samples for estimating dry deposition inputs of nutrients to the Coweeta Basin have been collected since 1971 in dryfall samplers located at two low-elevation sites within the Basin. Both collectors close automatically to exclude rain or snow. Samples are recovered weekly from each collector in 500 ml of deionized water and analyzed for inorganic constituents. This collection method provides reasonable estimates for inputs of such ions as Ca, Cl, K, and Na, which are deposited via sedimentation and as large (>1 mm) aerosols, but substantially underestimates inputs of ions such as NO_3 and SO_4, which may be deposited primarily as small aerosol particles or via gaseous

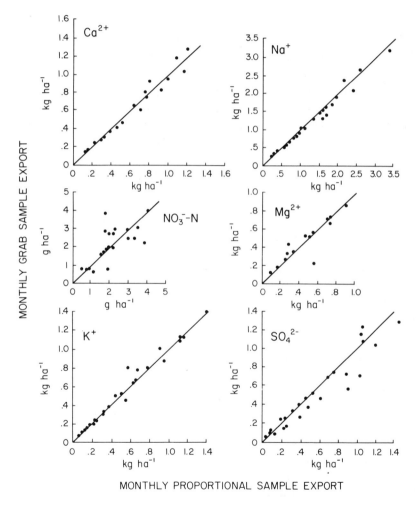

Figure 4.1. The relationship between monthly export of selected ions estimated from weekly grab samples and monthly export estimated from samples collected with a flow-proportional water sampler (Coweeta WS 2).

Table 4.2. Comparison of 3-Year Average Annual Export of Dissolved Constituents in Stream-flow for Coweeta WS 2 Calculated from Weekly Grab Samples Versus Flow-Proportional Samples

	Dissolved Export (kg ha^{-1} yr^{-1})		$(P - G)/G$
Constituent	Grab Sample (G)	Proportional Sample (P)	(%)
NO_3^--N	0.022	0.020	-9
NH_4^+-N	0.021	0.031	+48
PO_4^{3-}-P	0.020	0.025	+23
Cl^-	6.93	7.04	+2
K^+	5.28	5.33	+1
Na^+	12.75	12.93	+1
Ca^{2+}	5.98	6.12	+2
Mg^{2+}	3.36	3.52	+5
SO_4^{2-}-S	1.59	1.89	+19
SiO_2	93.64	93.66	<1

absorption. Moreover, lead burden data in Coweeta litter layers (Chapter 27) suggest that dry deposition inputs to high-elevation watersheds at Coweeta may be underestimated by the dryfall collected at low-elevation sites.

Stream water is routinely sampled at the same time each week by collecting a 250 ml sample in the flowing stream above the weir. For some studies, proportional water samplers have been used to collect and composite samples in proportion to discharge volumes. The latter procedure more adequately integrates changes in ion concentrations during storm events and provides the most precise estimate of periodic solute export. The adequacy of grab samples compared to proportional samples to estimate monthly export for some ions is illustrated in Figure 4.1 for a 12-ha control watershed [Watershed 2 (WS 2)] during a 3-year period. For low values, grab sample exports show good agreement with proportional sample exports for most ions. As export increases, grab samples show greater deviation from the expected 1:1 relationship with proportional samples, but there is no consistent positive or negative bias in the relationship except for SO_4. For this solute, monthly exports calculated from grab samples are frequently less than those calculated from proportional samples. This indicates that SO_4 concentrations increase substantially with increases in discharge, a response substantiated by preliminary storm event analyses at Coweeta.

The reliability of grab samples to estimate annual exports of dissolved ionic constituents in the stream draining control WS 2 is shown in Table 4.2. Grab sample estimates for most constituents are within 1 to 5% of values calculated from proportional samples. The largest percentage differences occur for NH_4-N and PO_4-P, but export of both ions is very low and the quantitative bias from grab samples is minor for many interpretations. Sulfate shows a large discrepancy between grab and proportional samples with about a 20% underestimate; this difference should be recognized in subsequent budget analyses. Furthermore, these comparisons are for a low-elevation watershed where quickflow accounts for only about 10% of total annual discharge. As shown by Swift et al. (Chapter 3), quickflow accounts for a much greater proportion of annual discharge for high-elevation Coweeta watersheds. Thus, weekly grab samples may provide larger underestimates of SO_4-S export for such watersheds. In general,

Table 4.3. Volume-Weighted Mean Annual Concentrations of Dissolved Inorganic Constituents in Bulk Precipitation and Stream Water for Coweeta WS 2[a]

Constituent	Precipitation		Stream Water	
	mg L^{-1}	µeq L^{-1}	mg L^{-1}	µeq L^{-1}
H^+	0.027	26.64	<0.000	0.2
Ca^{2+}	0.194	9.70	0.583	29.1
Na^+	0.170	7.38	1.22	53.2
NH_4^+	0.095	6.80	0.002	0.2
Mg^{2+}	0.041	3.35	0.326	26.9
K^+	0.094	2.41	0.499	12.8
SO_4^{2-}	1.59	33.1	0.450	9.4
NO_3^-	0.143	10.2	.003	0.2
Cl^-	0.271	7.65	0.662	18.7
HCO_3^-	0.074	1.21	4.97	81.5
PO_4^{3-}	0.013	0.42	0.006	0.2
Dissolved silica	0.030	–	8.80	–
		$\Sigma+$ 56.3		$\Sigma+$ 122.3
		$\Sigma-$ 52.6		$\Sigma-$ 109.9

[a] Data span the period 1973–1983 (June–May water year) for all ions except SO_4^{2-} (1974–1983), dissolved silica (1975–1983), and HCO_3^- (1977–1982).

weekly grab samples are sufficiently frequent to provide confidence in export estimates for most constituents.

Characterization of Precipitation and Stream Water Chemistry

Precipitation and Stream Water Concentrations for Representative Watersheds

The volume-weighted mean annual concentrations of dissolved constituents in precipitation and stream water over an 11-year period for control WS 2 are summarized in Table 4.3. On a charge equivalent basis, the chemical composition of bulk precipitation is dominated by H and SO_4 ions. Hydrogen comprises 47% of total cation equivalents, followed by about equal amounts of Ca, Na, and NH_4. Sulfate comprises 63% of total anions while NO_3 contributes about 20%. Thus, bulk precipitation at Coweeta is characterized as a dilute solution of sulfuric and nitric acids that is somewhat buffered by base cations to produce a mean pH of 4.6. There is only a 7% difference between the long-term equivalent balance of cations and anions (Table 4.3).

The relative contribution of different ionic constituents to the chemical composition of stream water for WS 2 is representative of other streams draining low-elevation watersheds in the Coweeta Basin. Sodium accounts for 43% of total cations, followed by nearly equal amounts of Ca and Mg, each contributing about 23% to total cations. Bicarbonate is clearly the dominant anion, comprising 74% of total anion equivalents, followed by Cl with 17%. Thus, stream water is characterized as a cation-bicarbonate solution with a mean pH of 6.7. There is a 12% difference between the long-term equivalent balance of negative and positive ions; this anion deficit may be partially due to unmeasured organic acids in stream water. Based on a reasonable organic acid: DOC charge equivalence of about 10 µeq L^{-1} per mg L^{-1}, this assumption is consistent with

Table 4.4. Volume-Weighted Mean Annual Concentrations of Dissolved Inorganic Constituents in Stream Water for Coweeta WS 27[a]

Constituent	Concentration (μeq L^{-1})
H^+	0.3
Ca^{2+}	18.2
Na^+	21.1
NH_4^+	0.3
Mg^{2+}	17.1
K^+	5.9
SO_4^{2-}	24.2
NO_3^-	1.3
Cl^-	13.9
HCO_3^-	22.5
PO_4^{3-}	0.1
$\Sigma+$	62.9
$\Sigma-$	62.0

[a] Data span the period 1973–1983 (June–May water year) for all ions except SO_4^{2-} (1974–1983) and HCO_3^- (1982–1983).

measured DOC levels in streams draining control watersheds (Chapter 20). The ionic strength of stream water is about double that of incident bulk precipitation. As precipitation leaches through the canopy, forest floor, and soil to the stream, H, SO_4, NO_3, and NH_4 ions show large depletions with concurrent enrichment of cations and HCO_3. A variety of physical, geochemical, and biological processes interact to produce this response; these topics are covered elsewhere in this volume.

The chemical composition of streams draining steep, high-elevation watersheds at Coweeta (Table 4.4) provides an interesting contrast with the chemistry of low-elevation streams (Table 4.3). Partly due to higher precipitation and streamflow at higher elevations, total ionic strength is about 50% of lower elevation values, and some interesting shifts in relative proportions of different ions are evident. In particular, HCO_3 concentrations are much lower for WS 27 as compared with WS 2, while SO_4 concentrations are much higher. Thus, SO_4 is the dominant anion in stream water for

Table 4.5. Summary of Factors Postulated to Affect SO_4^{2-} and HCO_3^- Availability and Mobility in Forested Watershed Ecosystems at Coweeta

Factors	High-Elevation Relative to Low-Elevation Watershed
Total SO_4^{2-} deposition	Greater
Potential for soil SO_4^{2-} adsorption	Lesser
Rates of SO_4^{2-} incorporation and immobilization by soil microbes	Lesser
Ecosystem carbon flow[a]	Lesser
Precipitation and streamflow	Greater
Quickflow[b]	Greater
Mean soil–water contact time	Lesser

[a] Primary production, decomposition and root respiration.
[b] Total amount and as percentage of total discharge.

WS 27. These comparisons indicate major differences in processes regulating solution chemistry of low- and high-elevation watersheds at Coweeta.

Factors postulated to regulate the differences in SO_4 and HCO_3 availability and mobility are summarized in Table 4.5. Briefly, higher elevation watersheds receive greater precipitation and SO_4 deposition, and exhibit higher streamflow. Soils are quite shallow at these elevations, quickflow volumes are higher (Chapter 3) and the potential for SO_4 adsorption is reduced. Also, air and soil temperatures are lower, leading to reduced rates of SO_4 incorporation into organic S forms by litter-soil microbes (Chapter 18) and to reduced carbon flow through these high-elevation forests. Thus, high-elevation watersheds are characterized by an increased flux of SO_4 but a reduced flux of HCO_3 as compared to low-elevation catchments. These differences imply that high-elevation forests have an enhanced susceptibility to acidic precipitation.

Stream Water Concentrations for All Control Watersheds

Mean annual flow-weighted concentrations of dissolved constituents for all seven Coweeta hardwood control watersheds are summarized in Table 4.6. These catchments are located throughout the Coweeta Basin and thus span a wide range of environmental conditions and overlay different bedrock formations.

All controls exhibit low stream water concentrations of NO_3-N, NH_4-N, and PO_4-P. The higher mean NO_3-N values for WS 27 and WS 36 resulted from the partial defoliation of vegetation on these catchments by the fall cankerworm, *Alsophila pometaria*, during the years 1974–79 (Swank et al. 1981; Chapter 25). Phosphorus is known to be immobile and strongly conserved by forest ecosystems; NH_4-N and NO_3-N are immobilized by soil biota or taken up by the root mycorrhizal complex.

Differences among watersheds are apparent for Cl, SO_4-S, SiO_2, and cation concentrations. Chloride and Na concentrations are considerably higher on WS 2; WS 27 has much lower concentrations for K, Na, Ca, Mg, and SiO_2 compared to all other catchments; SO_4-S values are twofold to threefold greater on high-elevation WS 27 and WS 36. Low stream water concentrations of cations and SiO_2 on WS 27 might reflect dilution due to higher streamflow (see Table 4.5). However, flow on WS 36 is also quite high, whereas cation and SiO_2 concentrations do not differ substantially from values for other controls. Another hydrologic variable which may explain these differences concerns the amount of quickflow or direct storm runoff. A much higher fraction of total flow on WS 27 appears as quickflow than on other control watersheds, including WS 36. Perhaps the major source of variation among control catchments in stream water composition of cations and SiO_2 concerns differences in bedrock mineralogy and weathering rates, as discussed by both Hatcher (Chapter 5) and Velbel (Chapter 6). This is especially true since the cation and SiO_2 data summarized in Table 4.6 reflect differences among watersheds in the relative proportions of the ions present in stream water. Other possible reasons for differences in SO_4-S concentrations between low-elevation (WS 2, 14, 18, 32, 34) and high-elevation (WS 27, 36) catchments were summarized earlier (Table 4.5).

Sources of Ions in Bulk Precipitation

Atmospheric deposition occurs through both wet and dry processes which entail complex physical, chemical, and aerodynamic interactions. Precipitation scavenging is the

Table 4.6. Volume-Weighted Mean Annual Concentrations (mg L^{-1}) of Dissolved Inorganic Constituents and Mean Annual Flow (cm) for Streams Draining Control, Hardwood Covered Watersheds at Coweeta[a]

Watershed Number	Flow	Ion Concentration									
		NO_3^--N	NH_4^+-N	PO_4^{3-}-P	Cl^-	K^+	Na^+	Ca^{2+}	Mg^{2+}	SO_4^{2-}-S	SiO_2
2	93	0.002	0.002	0.002	0.66	0.50	1.22	0.58	0.33	0.15	8.27
14	119	0.004	0.003	0.002	0.54	0.34	0.74	0.45	0.27	0.13	4.98
18	119	0.004	0.003	0.001	0.53	0.41	0.86	0.59	0.29	0.14	7.01
27	187	0.018	0.004	0.001	0.49	0.23	0.49	0.36	0.21	0.38	3.51
32	171	0.002	0.004	0.001	0.50	0.30	0.65	0.48	0.28	0.15	4.77
34	128	0.003	0.003	0.002	0.57	0.39	0.90	0.69	0.34	0.14	6.21
36	183	0.008	0.003	0.001	0.54	0.30	0.75	0.59	0.26	0.30	5.80

[a] Summarized by ion, data span the following water years: 1974–83 for SO_4^{2-}, 1975–83 for SiO_2, 1973–83 for other ions. Summarized by watershed data span these water years: 1973–83 for WS 2, 18, 27, 36; 1973–74 for WS 14, 34; 1973–74 and 1982–83 for WS 32. Specific ion-watershed pairs may be determined from these separate lists.

Table 4.7. Summary of Dryfall Contributions to Nutrient Inputs in Bulk Precipitation for Coweeta WS 18[a]

Ion	Water Years of Record	Bulk Precipitation (kg ha⁻¹ yr⁻¹)		Dryfall (kg ha⁻¹ yr⁻¹)		Dryfall
		$\bar{x}$	CV	$\bar{x}$	CV	%
NO_3^--N[b]	1973–82	2.90	0.183	0.22	0.271	7.6
NH_4^+-N	1973–82	1.95	0.289	0.30	0.637	15.2
PO_4^{3-}-P[c]	1973–82	0.09	0.307	0.07	0.687	70.6
Cl^-[b]	1973–82	5.56	0.316	0.78	0.430	13.9
K^+	1971–82	3.56	1.23	0.70	0.305	19.6
Na^+[b]	1971–82	3.29	0.459	0.55	0.339	16.7
Ca^{2+}[b]	1971–82	4.14	0.313	0.91	0.314	21.9
Mg^{2+}	1971–82	2.16	1.58	0.23	0.480	10.4
SO_4^{2-}-S	1974–82	10.69	0.126	1.14	0.307	10.7

[a] Data shown as means ($\bar{x}$) and coefficients of variation (CV) over the indicated water years (June–May) of record.
[b] For the indicated ion, annual bulk precipitation and dryfall inputs are positively correlated at the 5% level of significance.
[c] For PO_4^{3-}-P, a significant (1% level) negative correlation exists between annual precipitation and the % annual dryfall input, suggesting washout and dilution at higher precipitation amounts.

primary mechanism of the wet process, while dry deposition includes sedimentation, aerosol impaction, and gaseous absorption processes (Swank 1984). The mean annual dryfall inputs of ions to WS 18 and the percent contributions to bulk precipitation chemistry indicate the relative importance of wet and dry sources of solutes at Coweeta (Table 4.7). Dry fallout contributes between 8 and 14% of anion inputs except for PO_4-P, where dry deposition accounts for an average of 70% of the input. Dryfall contributions to cation inputs range from 10 to 22% with the highest values for Ca and K. Due to measurement techniques, dry depositions of NO_3-N and SO_4-S are substantially underestimated. However, these results show that dryfall can be a major source of chemical loading to forest ecosystems for some ions and should be quantified in deposition studies. Annual contributions of dryfall total inputs exhibit substantial year-to-year variability, with the highest coefficient of variation for PO_4-P and the lowest for NO_3-N (Table 4.7). Previous observations at Coweeta have shown large, episodic loadings of PO_4-P and other ions associated with dust particles originating in the southwestern United States (Swank 1984).

Plots of annual dryfall inputs (kg ha⁻¹ or percent of bulk precipitation input) over time also reveal substantial year-to-year variability (Figure 4.2). There are no apparent temporal trends in annual dryfall inputs except for SO_4-S and Ca. In the 9-year period from 1974 to 1982, SO_4-S dryfall inputs appeared to decrease about 50% (4.5 kg ha⁻¹ yr⁻¹ to 2.7 kg ha⁻¹ yr⁻¹). The correlation of dryfall SO_4-S input with time is statistically significant ($r = -0.83$, $p < 0.001$). There is also some tendency for a decline in Ca inputs over time, with an r value of -0.47 ($p < 0.10$).

In addition to wet versus dry sources of ion inputs to the Coweeta Basin, the relative importance of marine versus terrestrial ion sources can be estimated. Based strictly upon stoichiometric considerations and average element ratios in sea water, total bulk precipitation inputs can be partitioned into marine and terrestrial sources. The analysis

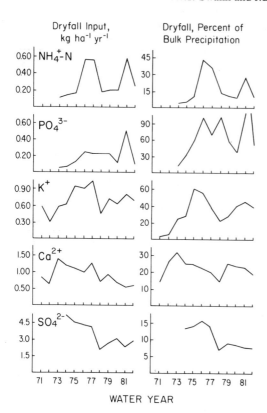

Figure 4.2. Long-term trends of annual dryfall inputs of select ions and percent contributions to bulk precipitation at Coweeta.

also yields an estimate of the major sources of precipitation acidity at Coweeta. Marine inputs of a specific ion are calculated as the product of ion:Cl or ion:Na ratios times the measured Cl or Na concentration. If the Cl:Na ratio is greater than that of sea water, then all Na is assumed to be of marine origin and the calculations are based on measured Na concentrations. Otherwise, Cl is assumed to be entirely of marine origin and forms the basis for the calculations. The use of this approach and the assumptions underlying it are discussed by Granat (1972), Cogbill and Likens (1974), and Galloway et al. (1982). Basically, the approach assumes that no ion fractionation occurs in the formation of sea salt aerosols, that anions calculated to be of terrestrial origin are equally likely to be neutralized by terrestrial cation sources, and that remaining anions not so neutralized are equally likely to produce H ions. The validity of this approach for examining ion sources in Coweeta precipitation is generally supported by detailed correlation analyses (not presented here) of individual rain gage samples over 11 years of collection.

Results of the source calculations, averaged (volume-weighted basis) over more than 2500 individual rain gage collections for 10 years of record, show that about 90% of the Cl, 80% of the Na, 40% of the Mg, and less than 5% of the K, Ca, and SO_4 in Coweeta bulk precipitation are of marine origin (Table 4.8). Remaining ion sources, including all NO_3 and NH_4, are terrestrial in origin. Of the terrestrial anions, approximately 51% are neutralized by terrestrial cation sources (sum of terrestrial cations divided by sum

Table 4.8. Summary of Calculations Indicating Marine vs. Terrestrial Sources of Ions in Bulk Precipitation Inputs to the Coweeta Basin[a]

Ion	Mean Concentration μeq L^{-1}	Sea Salt Contribution μeq L^{-1}	%	Terrestrial Contribution μeq L^{-1}	%	Neutralized μeq L^{-1}	%	Acid-Forming μeq L^{-1}	%
NO_3^-	10.26	0	0	10.26	100	5.21	51	5.06	49
NH_4^+	6.96	0	0	6.96	100	6.96	100	–	–
PO_4^{3-}	0.41	0	0	0.41	100	0.21	51	0.20	49
Cl^-	7.61	6.86	90	0.75	10	0.38	5	0.37	5
K^+	2.45	0.12	5	2.32	95	2.32	95	–	–
Na^+	7.27	5.86	81	1.41	19	1.41	19	–	–
Ca^{2+}	9.73	0.25	3	9.48	97	9.48	97	–	–
Mg^{2+}	3.36	1.33	40	2.03	60	2.03	60	–	–
SO_4^{2-}	33.05	0.70	2	32.34	98	16.40	50	15.94	48

Sum = 21.57
Predicted pH = 4.67
Measured pH = 4.58
(volume-weighted mean)

[a] Calculations based on volume-weighted mean ion concentrations over all individual rain gage collections for water years (June–May) 1974–1983.

of terrestrial anions equals 0.507). The remaining anions (approximately 22 μeq L^{-1}) are assumed to contribute to precipitation acidity. Of these acid-forming equivalents, SO_4 contributes about 74%, and NO_3 about 23%. The pH predicted by these analyses is 4.67, which compares favorably (i.e., within measurement error) with the volume-weighted mean measured pH 4.58, a result which supports the method used to estimate ion sources.

These calculations show clearly that precipitation acidity results from the availability of both (i) potentially acid-forming anions and (ii) neutralizing cations. The apparent trends of declining dryfall inputs of Ca and SO_4 (Figure 4.2) are evidence for reduced inputs of particulate neutralizing ions to Coweeta, perhaps due to reduced fly ash outputs from tall smokestacks or to relative conversion from coal to petroleum burning in the contributing airshed (e.g., Likens and Bormann 1974). If these trends continue, then precipitation inputs to Coweeta may become more acid with time, even in the absence of increased SO_4 or NO_3 concentrations in bulk precipitation.

Long-Term Trends in Precipitation and Stream Chemistry

Annual Trends

With the current interest in acid deposition we will focus our attention on long-term trends of stream chemistry in relation to precipitation chemistry, with an emphasis on SO_4 and H ions. It is important to recognize that this period of chemistry record includes the second highest annual precipitation amount recorded at Coweeta in 1980, followed in 1981 by the driest year on record.

Analyses of volume-weighted mean annual H ion concentrations in precipitation and streamflow for WS 2 from 1973 to 1983 show no significant trends of increasing or decreasing acidity over time ($p > 0.10$) (Figure 4.3). Correlation coefficients for stream and precipitation H ion concentrations versus time were -0.43 and -0.24, respectively. However, mean annual stream water H ion concentrations are partially related to precipitation concentrations, with an r of 0.66 ($p < 0.05$). Thus, precipitation at Coweeta is currently acidic, but shows no tendency for becoming more acidic over time.

Volume-weighted mean annual precipitation SO_4 concentrations fluctuate widely from year to year, with a peak concentration of 46 µeq L^{-1} coinciding with the year of lowest precipitation (Figure 4.3). Over the entire period, concentration shows a weak positive correlation with time ($r = 0.39$, $p > 0.10$). The most definitive annual trend observed is the increase in mean annual SO_4 concentrations in stream water ($r = 0.83$, $p < 0.005$). Similar time trends of increasing SO_4 concentrations are apparent for all control watersheds as illustrated by high-elevation WS 36 (Figure 4.4). Although sulfate levels are significantly higher on WS 36, the slope of the relationship with time is not different from that on WS 2 and averages about $+0.7$ µeq L^{-1} yr^{-1}. Regression

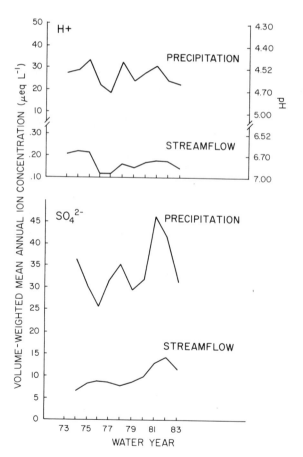

Figure 4.3. Long-term trends of volume-weighted mean annual H$^+$ and SO_4^{2-} concentrations in precipitation and stream water of Coweeta WS 2.

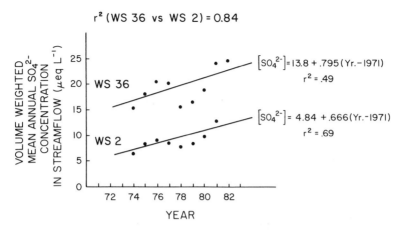

Figure 4.4. Time trends of volume-weighted mean annual SO_4^{2-} concentrations in streams draining low (WS 2) and high (WS 36) elevation Coweeta control watersheds. Analysis of covariance showed that residual variances around the regression lines are not heterogeneous ($p > 0.05$), and that regression slopes do not differ ($p > 0.250$) and have a common value of 0.731 µeq L^{-1} yr^{-1}. However, the analysis also revealed that regression elevations or intercepts do differ significantly ($p < 0.001$).

analysis of SO_4 concentrations showed that precipitation concentrations accounted for only about 35% of the variation in stream water concentrations on WS 2. The fact that the two regression intercepts differ significantly again reflects differences in SO_4 (and HCO_3) flux through high- and low-elevation watersheds as discussed earlier (Tables 4.3 through 4.5).

Alternative hypotheses could be proposed to account for increased SO_4 release from undisturbed forested watersheds over the past decade. However, even a qualitative evaluation of these hypotheses is beyond the scope of this paper. Nonetheless, it is apparent that Southern Appalachian forests are gradually releasing more SO_4, possibly in response to atmospheric inputs of H and SO_4, and it is appropriate to consider other trends in stream chemistry which accompany increasing SO_4 concentrations. Briefly, the following statistically significant trends in ion concentrations over time are evident in the data record: the anion deficit (= organic acids?) has increased, whereas concentrations of HCO_3, Ca, and the sum of the four base cations (Na, K, Ca, Mg) have declined. These trends suggest that forested watershed ecosystems in the Southern Appalachians may be in the initial phase of responding to atmospheric inputs of air pollutants.

In contrast to these trends in steam chemistry, very few ions in bulk precipitation show any evidence of long-term trends, especially when the confounding influences of time and precipitation amount on ion concentration are separated with partial correlation analyses. Concentrations of NO_3 show a weak tendency to increase, but this is due to a slight negative correlation of precipitation amount with time rather than to any significant time trend of NO_3 concentrations. The only significant relationships with time are reductions in concentrations of Ca and the sum of the four base cations.

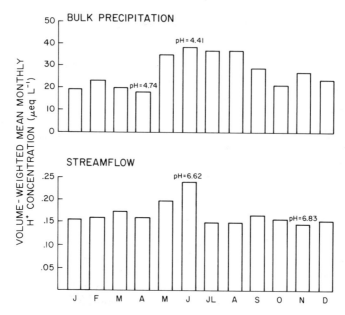

Figure 4.5. Volume-weighted mean monthly H⁺ concentrations in bulk precipitation and stream water of Coweeta control WS 2 (1973–1983).

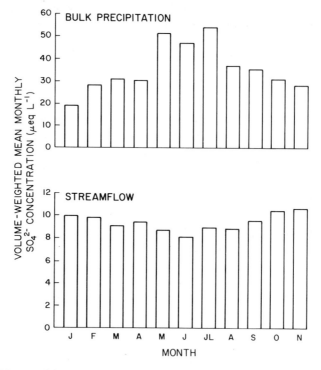

Figure 4.6. Volume-weighted mean monthly SO_4^{2-} concentrations in bulk precipitation and stream water of Coweeta control WS 2 (1974–1983).

Monthly and Seasonal Trends: H and SO$_4$

The climatology of Coweeta is distinctly seasonal, and the influence of climatic variables and biological activity on biogeochemistry is evident in the long-term mean
monthly chemistry record. Concentrations of H ions in bulk precipitation are seasonal,
with peak values in May through August and lowest concentrations in late winter and
spring (Figure 4.5). Monthly extremes range from pH 4.4 to pH 4.7. Peak concentrations of H ions in stream water also occur in May and June (pH 6.6), but remain rather
constant throughout the remainder of the year (Figure 4.5). The relationship between
mean monthly stream and precipitation H ion concentrations is weak with an r value
of 0.42 ($p > 0.10$). As would be expected, monthly SO$_4$ concentrations in bulk precipitation follow H concentrations ($r = 0.82$, $p < 0.005$), but there is a negative relationship of stream SO$_4$ concentrations and precipitation SO$_4$ values with an r of -0.74
($p < 0.01$) (Figure 4.6). Because of the deep soils at Coweeta and the resultant effect
of storage on transmission lags of atmospheric inputs, the lack of stronger agreement
between precipitation and stream water concentrations for H and SO$_4$ is not surprising.
Furthermore, biological nitrogen and sulfur transformations exert a dominant
influence on the timing and magnitude of H and SO$_4$ export.

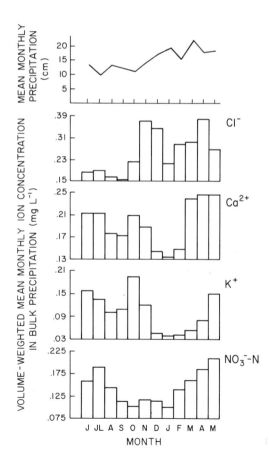

Figure 4.7. Mean monthly precipitation and volume-weighted mean
monthly concentrations of select ions
for Coweeta control WS 2. Data for all
ions span the period 1973–1983.

Monthly and Seasonal Trends: Precipitation Ions

Chloride concentrations in precipitation begin to increase in late fall and remain high throughout the winter and into the spring, with very low summer values (Figure 4.7). Sea salt aerosols are the major source of Cl (Table 4.8), and this pattern reflects the dominance of frontal storms with precipitation originating in the Gulf of Mexico and subsequent transport of marine aerosols into the region. In contrast, concentrations of Ca and K are minimum in winter and maximum in early spring, remain high through the summer months, and exhibit a second peak in October. These ions are derived from terrestrial sources (Table 4.8); the seasonal patterns shown in Figure 4.7 are related to the timing of local agricultural activities such as spring plowing, liming, and fertilizing followed by summer cultivation (Swank and Henderson 1976). The second peak in October coincides with traditional burning of leaves and woody debris accumulated from summer clearing operations. Nitrate-nitrogen also has a distinctly seasonal pattern, with minimum concentrations in the fall and midwinter months, increasing concentrations in late winter and spring, peak concentrations in May, and relatively high values in summer.

Correlations between mean monthly ion concentrations and bulk inputs versus precipitation amounts are summarized in Table 4.9. Although some trends of both positive and negative correlations of concentrations versus precipitation amount exist, coefficients of determination (r^2) are low for most ions. Three ions (H, SO_4, K) show a negative relationship between ion concentration and precipitation amount, reflecting the washout of atmospheric ion sources by high rainfall storms. Concentrations of NO_3 and Ca are unrelated to precipitation amount. Because higher Cl concentrations occur during high rainfall winter months (Figure 4.7), this ion exhibits a weak positive correlation between concentration and precipitation amount. For the three ions that show no relationship or a weak positive relation between concentration and precipitation

Table 4.9. Summary of Correlation Analyses of Mean Monthly Ion Concentrations and Mean Monthly Inputs vs. Mean Monthly Precipitation Amounts[a]

Ion	Concentration vs. Precipitation Amount r^2	Input vs. Precipitation Amount r^2
H^+	$(-)\,0.31*^{b}$	0.10
SO_4^{2-}	$(-)\,0.31*$	0.06
NO_3^-	0.005	$0.51***$
K^+	$(-)\,0.43**$	$(-)\,0.05$
Ca^{2+}	0.004	$0.56***$
Cl^-	$0.28*$	$0.69****$

[a] Correlations based on mean monthly inputs (g ha^{-1} mo^{-1}), volume-weighted mean monthly concentrations (mg L^{-1}), and mean monthly precipitation amounts (cm), calculated over the period of record (water years 1974–83 for SO_4^{2-}, 1973–83 for other ions).
[b] Negative sign in parentheses indicates a negative correlation coefficient (i.e., $r < 0$). Asterisks indicate level of significance of the correlation coefficient, r, as follows: $*p < 0.10$; $**p < 0.05$; $***p < 0.01$; $****p < 0.001$.

Figure 4.8. Mean monthly discharge and volume-weighted mean monthly concentrations of select ions in stream water of low elevation Coweeta WS 18 (HCO_3^- is for WS 2). Sulfate is for the period 1974–1983; HCO_3^-, 1977–1982; and all other ions, 1973–1983.

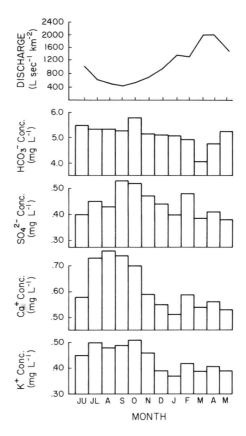

amount (i.e., NO_3, Ca, Cl), precipitation amount accounts for a significant fraction of the variation in mean monthly bulk precipitation inputs, with r^2 values ranging from 0.51 to 0.69. For the remaining three ions (H, SO_4, K), which tend to become diluted at higher rainfall amounts, mean monthly inputs are essentially independent of monthly precipitation amounts.

Monthly and Seasonal Trends: Stream Water Ions

Streams draining undisturbed watersheds show distinct seasonal trends in concentrations as illustrated by monthly weighted concentrations of K, Ca, SO_4, and HCO_3 for low-elevation watersheds (Figure 4.8). Concentrations of most constituents peak during mid-autumn when flow is lowest, and then decline over the winter months as precipitation and flow increase, with minimum values occurring in January. Low concentrations persist until early summer and then begin to increase again. High-elevation streams show similar seasonal patterns of concentrations, but the amplitude of monthly changes appears to be greater (Figure 4.9). This behavior is probably due to overall higher and more seasonably variable hydrologic response factors for these steeper watersheds having shallower soils. Concentrations of NO_3, NH_4, and PO_4 are very low in all streams draining control watersheds and show little seasonal change.

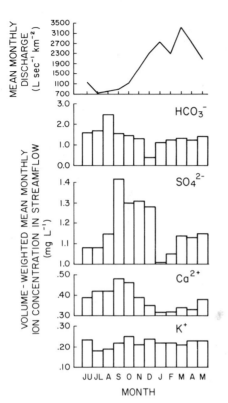

Figure 4.9. Mean monthly discharge and volume-weighted mean monthly concentrations of select ions in stream water of high-elevation Coweeta WS 36. Sulfate is for the period 1974–1983; HCO_3^-, 1981–1983; and all other ions, 1973–1983.

The quantity and timing of streamflow are dominant factors regulating both monthly ion concentrations and export (Table 4.10). Monthly concentrations of K, Ca, SO_4, and HCO_3 show strong negative relationships with flow amounts; r^2 values for three of these ions exceed 0.60. Sulfate concentration has the lowest correlation with flow. With regard to ion flux, flow accounts for at least 96% of the variation in monthly export of the four ions shown.

The extensive data base available on stream solute concentrations for all Coweeta watersheds (Table 4.6), and the high predictability of mass export from continuous flow measurements (Table 4.10), provide an excellent baseline for detecting and quantifying forest ecosystem changes resulting from natural disturbance, management practices, and other anthropogenic effects. In many cases, the impacts of human activities on forest ecosystem structure and function, as reflected by changes in stream chemistry, are quite subtle and difficult to elucidate in any cause-effect fashion. Only by integrating well-documented, long-term, landscape-scale studies, such as those discussed in this chapter, with detailed research on those processes which regulate forest response to the stress of interest, can such complex environmental questions be addressed meaningfully. The impact of atmospheric deposition on Coweeta forests, and its relation to changes in stream chemistry summarized herein, represents an excellent example of such complex relationships. In contrast, we have successfully detected impacts of such natural system dynamics as insect defoliation on forest biogeochemistry through changes in stream chemistry (Swank et al. 1981). In fact, changes in stream chemistry were observed in some cases prior to detection in the field of an insect outbreak. Changes in stream chemistry in response to such natural events and to forest management will be considered in Chapter 25.

Average Annual Nutrient Budgets

Long-term measurements of net nutrient budgets (bulk precipitation inputs minus stream exports) for experimental watersheds provide a quantitative means of characterizing integrated biogeochemical processes at the ecosystem level and of quantifying system-level changes due to human disturbance. Long-term mean annual budgets of nutrient inputs and outputs smooth out year to year variability in precipitation, evapotranspiration, nutrient deposition, and biological activity and thus reflect the integrated behavior of natural landscape units. At Coweeta, continuous chemistry data are available for at least 12 years for two low-elevation watersheds (2 and 18) and two high-elevation watersheds (27 and 36); hydrologic data are available for nearly 50 years. For the remaining three control watersheds (14, 32, and 34), stream chemistry for most ions spans water years 1973 and 1974 and 1982 to present. Based on this extensive, long-term data base, mean annual inputs, outputs, and net differences for these seven control catchments are summarized in Table 4.11.

Both precipitation and runoff tend to increase with watershed elevation over the Basin whereas evapotranspiration generally decreases, thus resulting in lower net water budgets at high elevations. The long-term record shows that more NO_3-N, NH_4-N, and PO_4-P is added annually than is lost in streamflow for all control ecosystems. Apparent accumulations of NO_3-N range from 2.6 to 4.0 kg ha^{-1} yr^{-1}; NH_4-N, from 1.8 to 2.5

Table 4.10. Summary of Correlation Analyses of Mean Monthly Ion Concentrations and Mean Monthly Export vs. Mean Monthly Flow for Control Hardwood Forests at Coweeta[a]

Ion	Concentration vs. Flow[b] r^2	Export vs. Flow[c] r^2
K^+	$(-)\,0.70^d$	0.99
Ca^{2+}	$(-)\,0.63$	0.98
SO_4^{2-}	$(-)\,0.45$	0.97
HCO_3^-	$(-)\,0.72$	0.96

[a] Correlations based on mean monthly exports (g ha^{-1} mo^{-1}), volume-weighted mean monthly concentrations (mg L^{-1}), and mean monthly streamflow (L s^{-1} km^{-2}) calculated over the period of record (water years 1977–82 for HCO_3^-, 1974–83 for SO_4^{2-}, 1973–83 for other ions). WS 18 data used for K^+, Ca^{2+}, and SO_4^{2-}; WS 2 data used for HCO_3^-.
[b] Levels of significance for the correlation coefficient, r, as follows: SO_4^{2-}, $p < 0.01$; Ca^{2+}, $p < 0.005$; K^+ and HCO_3^-, $p < 0.001$.
[c] All correlation coefficients significant at $p < 0.001$.
[d] Negative sign in parentheses indicates a negative correlation coefficient (i.e., $r < 0$).

Table 4.11. Average Annual Nutrient (kg ha⁻¹) and Water (cm) Budgets for Control Watersheds at Coweeta Hydrologic Laboratory

Watershed Number	Water			NO_3-N			NH_4-N			PO_4-P		
	Input	Output	Net Difference	Input	Output	Net Difference	Input	Output	Net Difference	Input	Output	Net Difference
2	187	93	+94	2.67	0.02	+2.65	1.78	0.02	+1.76	0.08	0.02	+0.06
14	214	119	+95	3.15	0.05	+3.10	2.14	0.04	+2.10	0.09	0.02	+0.07
18	206	119	+87	2.97	0.04	+2.93	1.97	0.03	+1.94	0.09	0.02	+0.07
27	265	187	+78	4.38	0.34	+4.04	2.54	0.07	+2.47	0.12	0.03	+0.09
32	240	171	+69	3.77	0.04	+3.73	2.52	0.06	+2.46	0.11	0.02	+0.09
34	215	128	+87	3.22	0.04	+3.18	2.24	0.04	+2.20	0.09	0.02	+0.07
36	225	183	+42	3.25	0.14	+3.11	2.15	0.06	+2.09	0.10	0.02	+0.08

Watershed Number	SO_4-S			Cl			K			Na		
	Input	Output	Net Difference	Input	Output	Net Difference	Input	Output	Net Difference	Input	Output	Net Difference
2	9.69	1.37	+8.32	5.07	6.18	-1.11	1.76	4.66	-2.90	3.17	11.43	-8.26
14	11.34	1.89	+9.45	6.98	6.44	+0.54	1.80	4.08	-2.28	4.51	8.73	-4.22
18	10.72	1.64	+9.08	5.72	6.30	-0.58	1.98	4.93	-2.95	3.56	10.31	-6.75
27	13.17	7.06	+6.11	9.44	9.21	+0.23	1.86	4.31	-2.45	5.16	9.07	-3.91
32	13.56	2.58	+10.98	8.35	8.48	-0.13	2.20	5.08	-2.88	5.24	11.18	-5.94
34	11.55	1.82	+9.73	6.60	7.37	-0.77	1.91	4.99	-3.08	4.14	11.58	-7.44
36	11.77	5.53	+6.24	6.21	9.94	-3.73	2.18	5.49	-3.31	3.89	13.79	-9.90

Watershed Number	Ca			Mg			SiO_2			HCO_3		
	Input	Output	Net Difference	Input	Output	Net Difference	Input	Output	Net Difference	Input	Output	Net Difference
2	3.63	5.45	-1.82	0.76	3.05	-2.29	0.55	77.25	-76.70	1.27	40.44	-39.16
14	3.83	5.28	-1.45	0.88	3.25	-2.37	0.48	59.09	-58.61	—	—	—
18	4.00	7.03	-3.03	0.85	3.49	-2.64	0.62	83.69	-83.07	—	—	—
27	3.31	6.80	-3.49	1.17	3.87	-2.70	0.46	65.50	-65.04	—	—	—
32	4.25	8.25	-4.00	1.04	4.82	-3.78	—	81.43	—	—	—	—
34	3.90	8.82	-4.92	0.88	4.40	-3.52	0.44	79.81	-79.37	—	—	—
36	4.40	10.80	-6.40	0.93	4.73	-3.80	0.68	106.46	-105.78	—	—	—

kg ha^{-1} yr^{-1}; and PO_4-P, from 0.06 to 0.09 kg ha^{-1} yr^{-1}. Net losses occur for Ca, Na, K, Mg, and SiO_2 in all watersheds, a reflection of the dominance of these net budgets by geochemical weathering processes (Henderson et al. 1978). The greatest variability between watersheds was observed for Ca (1.8–6.4 kg ha^{-1} yr^{-1}) and Na (3.9–9.9 kg ha^{-1} yr^{-1}), with larger net losses on higher elevation watersheds. Net losses of SiO_2 were also quite variable (58–106 kg ha^{-1} yr^{-1}), whereas net losses of K and Mg showed much less variability between areas. Chloride can provide a measure of reliability in the budget approach to characterizing the biogeochemistry of forest ecosystems, since biological accumulations of this ion are presumed to be small. Five of the seven controls at Coweeta appear to be in close balance (net loss or gain less than 1 kg ha^{-1} yr^{-1}), but WS 2 and WS 36 show larger net losses. These results could be due to several factors including the presence of chloride-bearing minerals in the bedrock underlying the two catchments, overestimates of unit area discharge because of subsurface water transfer from adjacent basins, or underestimates of watershed area precipitation. The budget for HCO_3 for the single watershed for which data are available shows a large net loss, a reflection of the biological production and dissolution of CO_2 within the soil.

The most striking set of budgets for all ions is for SO_4-S. All controls show large apparent accumulations of sulfate ranging from 6 to 11 kg ha^{-1} yr^{-1}. Because greatest stream SO_4-S concentrations occur on high-elevation watersheds (Tables 4.3 through 4.6), the lowest net SO_4-S gains occur at high elevations. As pointed out earlier, SO_4-S outputs calculated from weekly grab samples underestimate annual exports. Conversely, sampling methods used in these studies substantially underestimate total SO_4-S inputs; when taken together, the net gains for these watershed ecosystems are probably even larger than indicated. The primary mechanisms or processes of SO_4-S accumulation are soil SO_4-S adsorption on Fe and Al oxides (Johnson et al. 1980), and microbial incorporation of SO_4-S to organic S forms in litter and soil layers. Supporting data on both of these processes have been published for Coweeta soils and are reviewed and discussed by Fitzgerald et al. (Chapter 18).

Finally, although the focus of this chapter is on dissolved inorganic constituents, it is appropriate to place nutrient export via stream water particulates in perspective. Annual quantities of particulate cation export from control watersheds are insignificant as compared to dissolved export, while suspended and bedload sediments are the dominant forms of N loss (Table 4.12). It is also important to note that export of carbon

Table 4.12. Annual Dissolved and Sediment Exports of Select Nutrients From a 60 ha Mature Hardwood Covered Catchment at Coweeta Hydrologic Laboratory[a]

Export Form	Total	Organic Matter	N	Ca	K	Mg	Na
			(kg ha^{-1} yr^{-1})				
Sediment (bedload and suspended)[b]	258	6.16	0.52	0.16	0.006	0.01	0.003
Dissolved[c]	215	15.0 (DOC)	0.08	9.68	5.63	4.38	11.02

[a] After Swank and Swank (1984).
[b] Sediment cations extracted using double acid extraction procedure (0.025 N H_2SO_4 + 0.05 N HCl).
[c] The "Total" column for dissolved exports refers to total dissolved solids (TDS). Averages are based on an annual flow of 1185 mm and sediment sampling during 1974–1976; DOC for 1979–1980.

in the dissolved form exceeds sediment carbon export fivefold (assuming organic matter is ca. 50% C). Based on total sediment and dissolved outputs, it is apparent that solution export is an important form of fluvial denudation for relatively mature hardwood forests in the Southern Appalachians. Organic analyses of C, N, and S in bulk precipitation have also been made, but lack of information on aerosol impaction precludes making meaningful estimates of total inputs at this time.

Summary and Conclusions

Analysis of an eight-gage bulk precipitation network distributed over the Coweeta Basin shows that weekly arithmetic averages of ion concentrations across all gages provide a reasonable means of estimating bulk solute inputs to individual watersheds. Weekly grab samples of stream water appear to be sufficiently frequent to provide adequate estimates of solute export of all ions except SO_4, which is underestimated by grab sample methods.

The chemical composition of bulk precipitation is dominated by H and SO_4 ions and is characterized as a dilute solution of sulfuric and nitric acids buffered by base cations to produce a mean annual pH of 4.6. Composition of stream water representative of low-elevation watersheds shows that Na and HCO_3 are the dominant ions and that stream water is characterized as a cation-bicarbonate solution with a mean pH of 6.7. In streams draining high-elevation watersheds, SO_4 has replaced HCO_3 as the dominant anion, a result which indicates major differences in processes regulating the chemistry of low- and high-elevation watershed ecosystems. Differences among control watersheds in NO_3 concentrations reflect the occurrence of defoliation outbreaks, whereas variability in cations and SiO_2 is related to differences in bedrock mineralogy and weathering.

Dryfall can be a major source of chemical loading at Coweeta for some ions and exhibits substantial year-to-year variability. Annual sulfate dryfall inputs have decreased about 50% over a 9-year period; a slight tendency for a decline in Ca inputs over time is also apparent.

Sea salt aerosols are the major sources of Cl and Na and about 40% of Mg in bulk precipitation. Remaining ions are almost entirely of terrestrial origin. Based on stoichiometric calculations, SO_4 contributes about 74% to precipitation acidity and NO_3, 23%. The mean pH of 4.67 predicted by these stoichiometric calculations compares favorably with the measured mean pH of 4.58.

Analysis of long-term trends of mean annual precipitation and stream chemistry showed no significant trends of increasing or decreasing acidity, although annual stream water H ion concentrations were partially related to precipitation H ion concentrations. When all ions in both precipitation and stream water were examined, the most definitive annual trend was an increase in stream water SO_4 concentrations. This trend was present for all control watersheds, with an average increase of about 0.7 μeq L^{-1} yr^{-1}. Other trends in stream chemistry included an increase in anion deficit, and decreases in concentrations of HCO_3, Ca, and the sum of the four cations. The only significant trend in precipitation chemistry was a decline in concentrations of Ca and the four cations summed. Thus, Coweeta control watersheds may be in the initial phases of response to atmospheric inputs of air pollutants, even though total ion inputs to the Coweeta Basin have not increased over the period of record.

Monthly analyses of solutes in precipitation and stream water showed distinct seasonal trends for some ions. Concentrations of H and SO_4 ions in bulk precipitation peak in May through August, with lowest values occurring in late winter. Chloride concentrations in precipitation are highest throughout the winter in response to frontal storms and deposition of aerosols of marine origin. Calcium and K concentrations are minimum in winter and maximum in early spring with a second peak concentration in October. These patterns are associated with the timing of local agricultural and burning activities. Several ions (H, SO_4, K) tended to exhibit a negative relation between concentration and precipitation amount. Other ions showed no relation (NO_3, Ca) or a weak positive relation (Cl) between monthly concentration and precipitation amount. For these latter three ions a large fraction of the variability in bulk precipitation inputs was accounted for by precipitation amount, whereas this was not the case for the former three ions.

Streams draining undisturbed watersheds showed distinct seasonal trends in both concentrations and export of ions that are mainly regulated by watershed discharge. Most ions showed strong negative correlation of monthly concentrations with monthly flow. Furthermore, monthly flow accounted for at least 96% of the variation in monthly export of solutes. Nutrient export via sediments dominated the total export of nitrogen, but was relatively insignificant for other elements or ions.

Annual input-output budgets for control watersheds show net gains of NO_3-N, NH_4-N, and PO_4-P. Net losses occurred for Ca, Na, K, Mg, and SiO_2 in all watersheds. Chloride was in close balance for five of the seven control watersheds. Sulfate showed the most striking budgets with large apparent accumulations of SO_4-S ranging from 6 to 11 kg ha^{-1} yr^{-1}. Differences in net budgets among control watersheds reflect differences in bedrock geology within the Coweeta Basin and in the hydrologic responses and biological characteristics of high- and low-elevation watersheds.

As described in this paper, the long-term research on precipitation and stream chemistry for control watersheds at Coweeta provides a valuable and extensive data base for evaluating the integrated biogeochemical function of Southern Appalachian forest ecosystems, as well as for evaluating system-level responses to natural episodic events, deposition of anthropogenic materials, and forest management practices. Such system-level measures must be integrated with detailed process-level studies to understand factors regulating ecosystem-level behaviors and responses to disturbance.

5. Bedrock Geology and Regional Geologic Setting of Coweeta Hydrologic Laboratory in the Eastern Blue Ridge

R.D. Hatcher, Jr.

The Appalachian Highlands represent the last remnant of a more extensive mountain chain, which at one time was as lofty as many of the modern less eroded chains of the world such as the Alps, Andes, and North American Cordillera. The purpose of this chapter is to discuss the bedrock geology of Coweeta Hydrologic Laboratory, which is situated in the eastern Blue Ridge, and place it into the context of Blue Ridge geology (Figure 5.1A). In addition, some discussion of control of topography by structure and the nature of the Quaternary deposits will be made.

Blue Ridge geology has been of interest for many years. Geologic studies of the North Carolina-Tennessee-Georgia Blue Ridge began in the late 19th century with the U.S. Geological Survey folio mappers (e.g., Keith 1907a, 1907b, 1952). Subsequently, there were relatively few regional geologic field studies in the southern Blue Ridge of North Carolina and northern Georgia until the 1950s and 1960s. Several of these studies were related to distribution of economic mineral deposits, despite the fact that they produced geologic maps containing considerable detail (e.g., Van Horn 1948; Cameron 1951; Hurst 1955). The largest mapping project in this region was the study of the Great Smoky Mountains National Park and vicinity by the U.S. Geological Survey (Hamilton 1961; Hadley and Goldsmith 1963; King 1964; Neuman and Nelson 1965), which produced a large amount of useful data including the stratigraphic framework of the western Blue Ridge. During the 1960s and 1970s a large amount of detailed geologic mapping was completed in the eastern Blue Ridge and western Piedmont. Much of this information has been published as papers which discuss the stratigraphy and structure of the eastern Blue Ridge (Mohr 1975; Hatcher 1971, 1974, 1977, 1978).

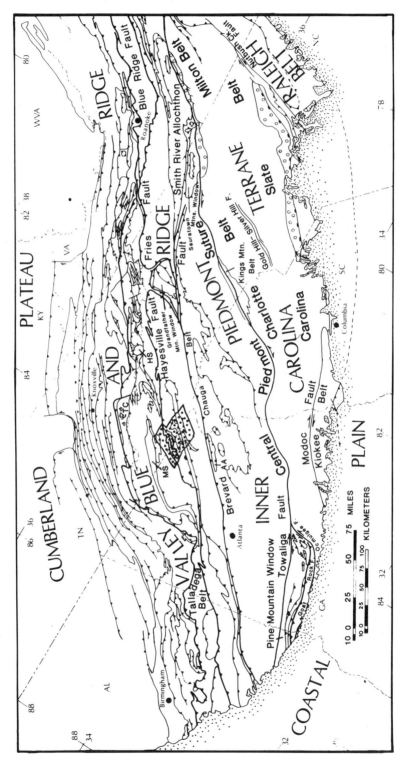

Figure 5.1. *A.* Index map of lithotectonic units in the southern Appalachians. Heavy stippled area is location of Figure 1*B.* MS, Murphy syncline; TC, Tuckaleechee Cove; AA, Alto allochthon; HS, Hot Springs Window.

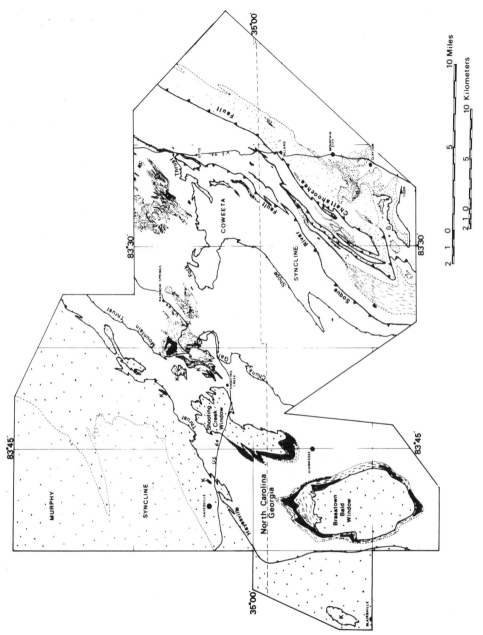

Figure 5.1. *B*. Major structures in the east-central Blue Ridge in southern North Carolina and northeastern Georgia, including the Coweeta Laboratory area.

Detailed geologic maps of the 155 km² Prentiss quadrangle and a more detailed map (1/14,400) of Coweeta Hydrologic Laboratory have been published (Hatcher 1980) along with compilations of detailed work on maps at smaller scale which are published in several regional guidebooks (Hatcher 1974, 1976; Hatcher and Butler 1979).

Hadley and Nelson (1971) published a map of the Knoxville 2-degree sheet at a scale of 1/250,000. This map provided new insight into the geology of the central and eastern Blue Ridge. Some of this map was compiled from published detailed maps of the Great Smoky Mountains National Park. Additional studies completed in the eastern Blue Ridge east of the Coweeta laboratory area from theses by Livingston (1966) and McKniff (1967) and Horton (1982) produced useful geologic maps and provided important information on the bedrock geology in the Highlands, Cashiers, and Rosman areas.

Several studies of the economic mineral deposits have been made in the southern Blue Ridge in the vicinity of Coweeta Laboratory in southern North Carolina and northeast Georgia. These provided additional information on the regional geology of the area. The study by Conrad and others (1963) of anthophyllite and related deposits of North Carolina identified soapstone and other altered ultramafic bodies in the Blue Ridge and Piedmont. Hunter (1941) described the olivine deposits of North Carolina and Georgia. The study of the Buck Creek dunite by Hadley (1949) provided much detailed information about this ultramafic body in the central Blue Ridge northwest of Coweeta Laboratory. Feldspar and mica deposits have been relatively important in this area. Lesure (1968) provided a useful summary of the geology of the feldspar and mica deposits of the Franklin area. Several abandoned mica deposits occur within the Laboratory area or immediately outside of it. The locations of these deposits are indicated in the Prentiss quadrangle (Hatcher 1980), while these and many other deposits in this area outside of the Prentiss quadrangle were located by Lesure (1968).

Geologic Setting of Coweeta Laboratory

Regional Setting

Coweeta is located in the eastern part of the southern Appalachian Blue Ridge. The Blue Ridge may be subdivided into two major subdivisions (Figure 5.1A) based upon the kinds of rocks and the rock associations present in each part. The western Blue Ridge consists of a sequence of pure to impure sandstones and shales with minor carbonate in the Ocoee Series. These sedimentary rocks were deposited on the 1.0 to 1.2 billion year old Grenville basement of eastern North America. They are therefore thought to be late Precambrian, since in the westernmost Blue Ridge they are overlain by fossiliferous Cambrian Chilhowee Group rocks. These rocks comprise the rifted continental margin sequence which formed when the proto-Atlantic (Iapetos) Ocean opened during the late Precambrian. The Ocoee Series was deposited in a series of rift basins located along the eastern margin of North America. The basins were gradually filled with sediment from the eroding Grenville Mountains and were then succeeded by cleaner Chilhowee Group sediments of Cambrian age. Late Precambrian rift basin sequences of continental margin sediments were deposited upon Grenville basement rocks from Georgia to Newfoundland within the Appalachians. In the western Blue

Ridge there is an obvious relationship between basement rocks and the sediments that rest upon them. Also, the sediments which occur here contain minerals which are related to the North American basement source from which they were derived.

The eastern Blue Ridge contains an assemblage of rocks which is in many ways different from that in the western Blue Ridge. Grenville basement rocks occur only in two small isolated masses in the eastern Blue Ridge: in the Toxaway dome along the South Carolina-North Carolina line (Hatcher 1977; Fullagar and others 1979) and in the Tallulah Falls dome in northeast Georgia (Odom et al. 1973; Hatcher 1976). Elsewhere in the eastern Blue Ridge, it is very difficult to relate the metasedimentary rocks to continental basement. In contrast to the western Blue Ridge, mafic and ultramafic rocks are very common here and all of the major granitic rocks of the Blue Ridge occur in this eastern zone. The ultramafic rocks of the eastern Blue Ridge may serve as a basement upon which the sediments of this area were deposited.

Separating the eastern from the western Blue Ridge is a major fault, the Hayesville thrust, which is a pre- to synmetamorphic thrust fault (Figure 5.1A). This fault was implaced prior to the thermal peak which affected the rocks of the eastern and western Blue Ridge, because the metamorphic isograds cross the Hayesville thrust without being offset or truncated along it.

Local Stratigraphy Near Coweeta Laboratory

The sedimentary rocks of the eastern Blue Ridge in southern North Carolina and northeastern Georgia consist of a sequence of metasandstones which are feldspar and biotite rich, and are interlayered with mafic volcanic rocks and aluminous schists. This is the Tallulah Falls Formation, which occupies a significant portion of the eastern Blue Ridge of northeastern Georgia, South Carolina, and southwestern North Carolina.

Overlying the Tallulah Falls Formation, or a possibly equivalent unit described below, is the Coweeta Group, named for exposure in the Coweeta Hydrologic Laboratory (Hatcher 1980) (Figure 5.2). This unit consists of a basal coarse-grained quartz diorite gneiss called the Persimmon Creek Gneiss, which is probably an orthogneiss. The Persimmon Creek Gneiss occurs extensively throughout the Coweeta area and in the adjacent parts of northeastern Georgia. Overlying the Persimmon Creek gneiss is a metasandstone and pelitic schist unit called the Coleman River Formation. This unit consists of medium to coarse grained feldspathic to quartzose metasandstone and interlayered muscovite-biotite schist. It also contains abundant calc-silicate quartzite, pin-striped metasandstone and some cleaner quartzite. The Coleman River Formation is overlain by the Ridgepole Mountain Formation, which consists of a sequence of cleaner quartzose metasandstone, along with garnetiferous muscovite and biotite schists, some conglomerate, and lesser amounts of calc-silicate quartzite than occur in the Coleman River Formation. The Ridgepole Mountain Formation is the highest unit in the Coweeta Group and is preserved in synclines which are present throughout the area. The Ridgepole Mountain Formation was named for exposures on Ridgepole Mountain occurring just to the south of Coweeta and west of Dillard, Georgia. The entire Coweeta Group stratigraphic section is exposed on the east slopes of this mountain. A

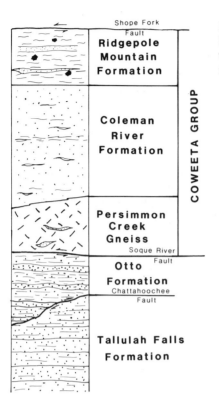

Figure 5.2. Stratigraphic section for the Coweeta Laboratory area in the eastern Blue Ridge. Note that each stratigraphic package is bound by major faults. Thicknesses are approximate.

more complete description of the stratigraphy and regional relationships of the Coweeta Group may be found in Hatcher (1979).

Another unit which is more closely related to the Coweeta Group sedimentary rocks in the immediate area of the Laboratory is a sequence of metasandstones and schist first described in detail by Hatcher (1980) in the Little Tennessee River valley just east of Coweeta. This unit consists of a sequence of relatively quartz-rich two-mica feldspathic metasandstones interlayered with aluminous schists. The name Otto Formation is suggested here for this unit because of its extensive exposure in the vicinity of Otto, North Carolina. The Little Tennessee River and Tessentee Creek area, along with hillsides close to Otto, provide excellent exposures to serve as the type area.

The Coweeta Group appears to be in fault contact with the Otto Formation. The Tallulah Falls Formation is likewise in fault contact with the Otto. On the east side of the Coweeta syncline, the Otto Formation underlies the Coweeta Group but is separated from it by a major thrust. To the west of the Coweeta syncline (Rainbow Springs area), there appears to be a relationship between a lower metasandstone unit, which is tentatively correlated with the Tallulah Falls Formation, and a higher, slightly cleaner, and possibly feldspathic metasandstone unit which is tentatively correlated with the Coleman River Formation. The latter is in turn succeeded by a garnetiferous muscovite-biotite schist unit, which is tentatively correlated with the Ridgepole

Mountain Formation. In order for these correlations to be correct, the Persimmon Creek Gneiss must pinch out toward the west and Ridgepole Mountain Formation must become more muscovite-rich and uniformly garnetiferous toward the west. This particular unit of schist and garnetiferous schist occurs in the Shooting Creek area and has been referred to as the Shooting Creek schist.

Metamorphism

Metamorphism in the western Blue Ridge is very low grade (anchizone) to nonexistent, but increases rapidly through the greenschist facies into the amphibolite and granulite facies in a classical Barrovian sequence (high pressure to moderate to high temperature) of metamorphic zones.

Metamorphic grade rises to the upper amphibolite and granulite facies to the west of Franklin. This relationship to very high metamorphic grade was first recognized by Force (1976). The metamorphic grade of rocks just north of the Coweeta Hydrologic Laboratory is in the upper amphibolite or possibly in the granulite facies. However, within the laboratory area metamorphic grade drops considerably to lower to middle amphibolite facies conditions. Index minerals present here include kyanite and staurolite. These are particularly abundant in the Otto Formation just to the east, but may also be observed in thin sections of Coleman River schists. As a consequence, it is clear that the rocks of this area reside in the amphibolite facies of regional metamorphism. Towards the east the metamorphic grade continues to remain in the kyanite or staurolite zone, but increases towards the south in northeast Georgia to the sillimanite zone once again, only to drop southeastward once more to the middle amphibolite facies kyanite zone near the Brevard fault.

Structure

The structure of the Blue Ridge consists of a series of large thrust sheets which transported the previously deformed and metamorphosed rocks of the central and eastern Blue Ridge towards the west over the rocks of the Valley and Ridge foreland of the Appalachians (Figure 5.1B). The Hayesville fault was mentioned above as a major boundary which separates eastern Blue Ridge sequences from those of the western Blue Ridge. This fault and a number of others formed in the early Paleozoic and were deformed again by later folds. Many of these folds, however, are very tight to isoclinal, which results in deformation of these early fault surfaces producing very sinuous outcrop patterns.

Another large early to premetamorphic thrust is the Shope Fork fault which occurs in the Coweeta Hydrologic Laboratory area and to the west. This fault apparently dips toward the west today and appears to overlie the Coweeta syncline. The Shope Fork fault is a premetamorphic fault, since it does not appear to be a brittle or extensively ductile boundary and does not juxtapose metamorphic isograds. The existence of this fault is based primarily upon stratigraphic criteria and truncations which occur along it. There is no evidence along the few exposures of the actual contact that extensive movement has taken place since metamorphism. Thus, the fault surface appears to be annealed, having formed and then overprinted by metamorphism. Its sinuous outcrop pattern in the Coweeta Hydrologic Laboratory area (Figure 5.3) attests to its early history. The Shope Fork joins another major fault, the Soque River fault, just east of the Coweeta Laboratory area. This fault separates rocks of the Coweeta Group from rocks of the Otto Formation.

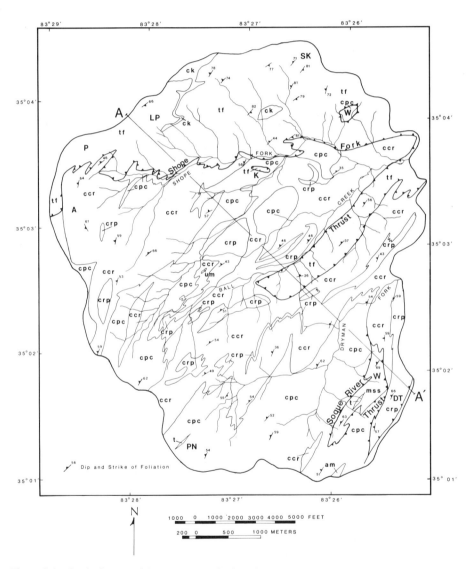

Figure 5.3. Geologic map of Coweeta Hydrologic Laboratory area. tf, Tallulah Falls Formation; ck, Carroll Knob mafic-ultramafic complex rocks; mss, Otto Formation; cpc, Persimmon Creek Gneiss; am, amphibolite; um, ultramafic rocks; t, trondhjemite; W, window; K, klippe; LP, Little Pinnacle Mountain; PN, Pickens Nose; DT, Doubletop Mountain; A, A'-section in Figure 5.4.

Rocks of the Coweeta Hydrologic Laboratory generally strike to the northeast and dip towards the northwest (Figure 5.4). This is an exception in most of the southern Appalachians, particularly in the Blue Ridge and western Piedmont. Most of the rocks of the Piedmont, Blue Ridge, and Valley and Ridge of the Appalachians dip to the southeast, while all of the rocks strike to the northeast. However, this panel of rocks in the eastern Blue Ridge is an extensive zone of northwest dip resulting either from later folding or from rotation of the early-formed foliations during later faulting.

The early faults of this area were refolded by more northeasterly trending folds (Figures 5.3 and 5.4). Earlier folds, which are present only as relict interference patterns, occur in this area and are represented on the geologic map as elongate domes and basins in the Coweeta Laboratory area (Figure 5.3). However, there does not appear to be a great deal of early recumbent folding, because most of the rocks in the Coweeta syncline appear to represent a reasonably upright sequence, with the Ridgepole Mountain Formation remaining at the top and not being exposed in antiforms throughout this area (as it might be if the rocks were extensively overturned by earlier recumbent folding). As a consequence, rocks of the Persimmon Creek Gneiss and the underlying Otto Formation appear to be exposed principally in antiforms and the Ridgepole Mountain Formation rocks appear to be preserved in synforms. These structures, which are very tight, do not appear to be greatly modified by later structures, although there are areas in the geologic map where later structures trending more or less north south may have affected the outcrop patterns. The earliest structures present here probably trended east west and were developed as very tight isoclinal structures, which were then refolded along northeast trending axes.

It may have been during the formation of these very early folds that the Hayesville, Soque River, and Shope Fork faults would have formed. These structures have trends which are not easily resolved by the utilization of the standard techniques of structural geology. This is particularly acute since the plunges in many of these folds appear to be fairly steep, on the order of 60° or more, and as a result produce interference patterns which do not express themselves well in the geologic map. Modified interference patterns appear particularly in the central part of the Coweeta Laboratory area, suggesting that these are remnants of earlier fold structures.

In addition to the folding which has affected these rocks, there appears to be very little in the way of later brittle faulting within this part of the eastern Blue Ridge. However, younger fractures (joints) are present in these rocks. These may be as young as Mesozoic, but may also have formed during the late Paleozoic.

Relationships of Structure and Lithology to Topography

Numerous examples of structural and lithologic controls of the present topography and drainage exist in the Coweeta Hydrologic Laboratory area. Most obvious and prevalent are the controls of drainage by fractures (joints). In addition, many of the northwest-facing slopes on northeast-trending ridges in the Coweeta Laboratory area south of Shope Fork Creek are dipslopes on their northwest flanks (since the rocks dip toward the northwest here) and scarp slopes on their southeast flanks.

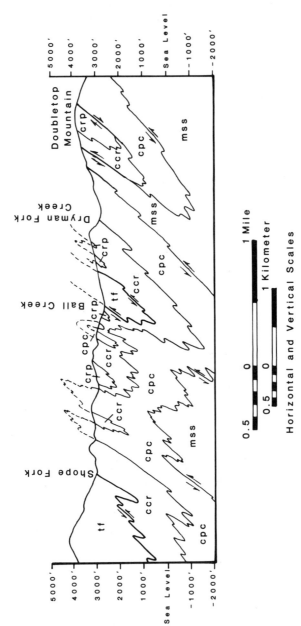

Figure 5.4. Cross-section A-A' (location Figure 5.3) showing the configuration of rocks and structures through Coweeta Hydrologic Laboratory. Abbreviations of rock units the same as in Figure 5.3.

A more subtle relationship exists between bedrock structure and the headwaters regions of several valleys. The headwaters areas of Falls Branch, and of Barkers and Commissioners Creeks (which drain south from the Coldspring Gap-Rockhouse Knob-Commissioner Gap area) involve either a change of strike of foliation paralleling the head of each valley, or the head of the valley is coincident with the strike with no change in orientation of foliation (Hatcher 1980). All of these valleys are cirquelike, but the steep scarp slope that developed here is related both to structure and mass wasting (slope) processes.

Ridges that are capped by Ridgepole Mountain Formation frequently develop a very steep upper scarp slope below which is a bench developed on Coleman River Formation rocks. Persimmon Creek Gneiss is a common cliff-forming unit owing to its massive homogeneous structure/texture and lack of fractures over most of its outcrop area.

Quaternary Deposits

Surficial deposits in the Coweeta Laboratory and vicinity that formed during the Quaternary include colluvium, alluvium, and stream terraces. Their areal distributions are shown in the maps by Hatcher (1980).

Colluvium is unsorted unstratified valley fill that serves as a good conduit for percolation of near-surface ground water in an area in which the ground water is confined to the weathered bedrock and saprolite, except for the few open fractures that may penetrate deeply into the otherwise impermeable bedrock. Its thickness ranges up to 5 to 7 m. This material may avalanche down slope if it becomes oversaturated and/or the toe of the deposit is removed. Slower rates of movement may also occur if the toe of the deposit is removed either naturally or artificially.

Alluvium and terraces occur along the major stream courses and range from a thin veneer up to 10 m in thickness. This material was deposited subaqueously and may be moderately to well stratified and sorted. Older terrace deposits (late Tertiary?) may be distinguished by the degree of weathering of pebbles and cobbles and by the compositions of clasts that survive the longest. Quartz clasts are the last to decompose and thus indicate the oldest alluvial deposits.

Conclusions

1. The Coweeta syncline occurs in the eastern Blue Ridge of southwestern North Carolina. Rocks of the eastern Blue Ridge are distinctly different from those of the western Blue Ridge because of their different composition and associations with particular rock types.
2. The name Otto Formation is suggested for a distinctive sequence of metasandstone and aluminous schist that tectonically underlies the Coweeta Group in the Little Tennessee Valley near Otto, North Carolina.
3. The stratigraphic sequences in the Coweeta area consist of the Otto Formation, which is tectonically overlain by the Coweeta Group. Tallulah Falls Formation rocks

also occur here, and it is possible that Coweeta Group rocks stratigraphically overlie this unit.

4. Rocks of the eastern Blue Ridge were progressively metamorphosed during the middle Paleozoic to middle and upper amphibolite facies assemblages, and locally to granulite facies, in a Barrovian sequence. Rocks of Coweeta Hydrologic Laboratory area were metamorphosed to the staurolite-kyanite subfacies (lower-middle amphibolite facies).

5. The structure of the Coweeta area is dominated by two early major thrusts, the Shope Fork and Soque River thrusts, which appear to overlie and underlie (respectively) rocks of the Coweeta syncline and were refolded by northeast-trending tight to isoclinal folds and whose axial surfaces dip towards the northwest.

6. Subtle controls of drainage and topography by joints, dip and strike of foliation occur here.

7. Extensive quaternary deposits of colluvium and alluvium serve as surficial aquifers, since they are composed of unconsolidated unstratified or stratified material. Colluvium also moves down slope rapidly in the form of debris avalanches, and slowly in the form of creeping intact masses.

6. Weathering and Soil-Forming Processes

M.A. Velbel

Weathering of rocks and minerals is one of the most important processes operating at the surface of the earth. On a microscopic scale, mineral weathering reactions occur at the interface between solids and solutions. On a grander scale, weathering is a process which operates at the interface between the earth's endogenic and exogenic cycles. The earth's internal forces determine the distribution of rocks, minerals, chemical elements, and relief in time and space. The exogenic system then transforms rock-forming minerals and their constituent elements into residual and secondary solids and dissolved products, and redistributes them physically, chemically, and biologically. These interfacial interactions between the lithosphere, hydrosphere, atmosphere, and biosphere profoundly alter both the surface of the solid earth and the chemistry of its fluid envelopes.

Weathering also influences a variety of human activities. Of particular interest here are mineral transformations during weathering which form soil minerals, releasing both beneficial (nutrient; e.g., K, Mg, and Ca) and detrimental (e.g., Al) elements into terrestrial ecosystems and subsurface and surface waters. In addition, primary and secondary minerals exhibit a range of abilities to ameliorate adverse anthropogenic impacts on our environment. Primary rock materials, for instance, have widely differing abilities to buffer atmospherically deposited acidity and thereby determine the sensitivity of a landscape to acid precipitation. Weathered rocks and soils also have different capacities to store water and therefore buffer the sensitivity of regions to flash floods. Finally, weathering profiles have a wide range of susceptibilities to erosion, thereby linking chemical processes of weathering and soil formation with the long-term geomorphic evolution of regional landscapes.

The purpose of this chapter is to summarize recent research on the chemical weathering of rocks and minerals in selected watersheds at Coweeta. Because of the importance of mineral weathering in supplying inorganic nutrients to the dissolved load of streams, and the possible importance of soil minerals in supplying inorganic nutrients to the forest biota, this chapter emphasizes the chemistry of mineral weathering processes, and the rates at which these processes of weathering and soil formation proceed. The bedrock geology and tectonic setting of the Coweeta Basin and their relationships to geomorphic evolution are summarized elsewhere (Chapter 5), as are some of the major physical processes of landscape modification (Chapter 7).

The Weathered Regolith of Coweeta

The weathering profiles of the Coweeta Basin consist of two major subdivisions, soil and saprolite. The uppermost 30 cm of the weathering profiles are true, biologically active soils, mostly Ultisols and Inceptisols. Most soils are residual and occur over large areas of moderate slopes. Inceptisols occur in restricted areas of steep slopes, on colluvial parent materials occurring as valley fill, and in fluvial valley bottom sediments. The main difference is that the Inceptisols are developed on parent material which has been transported (by mass movement on steep slopes, or by fluvial processes in valley bottoms), thereby resetting the soil in terms of horizon development. The Ultisols form on a physically untransported, and therefore geomorphically older, substrate. This results in preservation of better developed horizon differentiation. The chemical and mineral properties of the Ultisols and Inceptisols at Coweeta are very similar; the major distinction between them is in soil morphology.

Beneath the soil at Coweeta is a considerable thickness of saprolite. Saprolite was defined by Becker (1895) as "thoroughly decomposed, earthy, but untransported rock." Over the years the term has come to mean a residual subsoil (regolith) developed on crystalline rocks in which some or all of the primary rock-forming minerals have been transformed in situ to weathering products. In saprolite, the spatial distribution of weathering products often mimics the distribution of parent minerals (that is, parent minerals are replaced by *pseudomorphs* consisting of weathering products), resulting in the preservation of parent rock textures, fabrics, and structures in the saprolite. Furthermore, because of the pseudomorphous nature of the mineral weathering reactions, a unit volume of crystalline parent rock weathers to a unit volume of saprolite (iso-volumetric weathering; Millot and Bonifas 1955). In general, mass is lost in the form of dissolved element removal by subsurface solutions, while volume is preserved, in part by hydration of parent minerals and formation of secondary minerals with more poorly crystallized and/or more open structures. Therefore, the bulk density of saprolite is considerably lower than that of fresh parent rock. At Coweeta, for instance, a typical value for the bulk density of fresh bedrock is around 2.8 g/cm³, compared with 1.6 g/cm³ for a typical saprolite (Berry, unpublished data). Thus, around 40% of the mass of a unit volume of rock is removed without changing the bulk volume, so the saprolite may have a void ratio (a rough estimate of porosity) of 40%, compared with negligible porosity in the fresh bedrock. This profound change in physical properties during isovolumetric weathering of rock to saprolite gives rise to saprolite's unique hydrologic and geomorphic behavior. The blanket of saprolite on the surface of the landscape, with its enormous volume of pore space, is generally believed to be a primary source of base flow and, to a considerable degree storm flow as well, at Coweeta (Hewlett 1961; Hewlett and Hibbert 1963, 1966) as in the rest of the southern Blue Ridge (Winner 1977). From a geomorphic standpoint, the distinction between rock and saprolite is equally important; porous, friable, unconsolidated saprolite is much more vulnerable to erosion by fluvial processes and landslides (Grant 1983 and Chapter 7) than is fresh bedrock.

Volumetrically, the saprolite portion of the weathering profile dominates over the soil. Although bedrock exposures as exfoliation surfaces, cliffs, and stream exposures are not uncommon at Coweeta, most of the landscape is covered by at least some soil and saprolite. Saprolite is thickest at drainage divides between watersheds, where 6 to 23 m of weathering profile was encountered by drilling before reaching fresh bedrock (Berry unpublished). Of this weathering profile, the true soil (A and B horizons) usually comprises no more than 70 cm; the remainder is "C" horizon material (saprolite). On the slopes between drainage divides, erosion and mass wasting keep saprolite thinner than on divides (Berry unpublished). Swank and Douglass (1975) report an overall average weathering profile thickness (depth to bedrock) of about 6 m. Of this total thickness, perhaps 5% is true (A and B horizon) soil—the remaining 95% is saprolite. This average thickness is used in many of the calculations discussed below.

The Stoichiometry of Major Mineral-Weathering Reactions

Two major lithostratigraphic units occur in the Coweeta Basin, the Tallulah Falls Formation and the Coweeta Group. Descriptions and major features of their origin and distribution are discussed by Hatcher (Chapter 5). From a chemical weathering perspective, the most important feature of the bedrock is its mineralogical composition. Five rock-forming minerals make up the bulk of bedrock in the Coweeta Basin. These are: quartz, biotite and muscovite micas, plagioclase feldspar, and almandine garnet. Of the five major minerals, quartz and muscovite mica are so stable in the weathering environment that they are not likely to contribute significantly to dissolved and botanical mineral nutrient budgets (e.g., Goldich 1938). The three remaining rock-forming minerals are both weatherable, and present in sufficient abundance to significantly influence dissolved mineral nutrient budgets. The remainder of this discussion is concerned with the stoichiometries (that is, the amount of each element released per unit amount of reaction) for the weathering reactions of these three minerals.

Biotite Mica

Biotite mica is an aluminosilicate mineral containing potassium, magnesium, and iron in a sheet structure. General aspects of biotite mica structure and weathering are summarized by Velbel (1984a). Weathering of biotite mica at Coweeta has been studied by

Velbel (1984b, 1985c), who found that the major weathering product of biotite in the saprolite of Coweeta is hydrobiotite, which consists of a regular alternation of 1.0 nm mica layers with 1.4 nm vermiculite or "pedogenic chlorite" layers. The more extensive weathering of biotite in the soil continues to transform biotite layers into vermiculite and pedogenic chlorite. Electron probe microanalysis (EPMA) of fresh biotites and their weathering products (Velbel 1984b, 1985c) have suggested that biotite weathering at Coweeta, as in most natural weathering environments, occurs by a "simple transformation" (Fanning and Keramidas 1977). In this transformation, the primary sheet silicate lattice is conserved; compositional change during weathering is due largely to removal and partial replacement of interlayer potassium by small amounts of calcium and sodium, accompanied by minor loss of magnesium from the octahedral layer. The combined EPMA and mineralogical data suggest the following weathering reaction for the formation of the 1.4 nm layers of hydrobiotite and pedogenic chlorite:

$$K_{.85}Na_{.02}(Mg_{1.2}Fe^{II}_{1.3}Al_{.45})(Al_{1.2}Si_{2.8})O_{10}(OH)_2 + 0.19O_2 + 0.078H^+ + 0.31H_2O +$$

$$0.016Ca^{2+} + 0.04Na^+ + 0.35Al(OH)^+_{2(aq)} \; \frac{1}{3} \; 0.3Fe(OH)^+_{2(aq)} \rightarrow$$

$$K_{.25}Na_{.06}Ca_{.016}(Mg_{1.1}Fe^{II}_{.5}Fe^{III}_{1.1})(Al_{1.2}Si_{2.8})O_{10}(OH)_2 \cdot 133Al_6(OH)_{15} + 0.6K^+ + 0.1Mg^{2+}$$

The weathering of biotite to hydrobiotite and pedogenic chlorite releases a significant proportion of the potassium and some of the magnesium present in the parent mineral, and results in the removal of a small amount of calcium and sodium from the ambient soil- or groundwater.

Almandine Garnet

The garnets are a compositionally complex group of silicate minerals in which silica tetrahedra are linked with octahedrally coordinated aluminum, chromium, or iron (rather than with other silica tetrahedra), with calcium, magnesium, iron, and manganese located in the interstices within the linked tetrahedral-octahedral structure. The almandine variety consists mainly of iron and aluminum silicate, although minor amounts of the other elements are usually present. Little is known about the weathering of the garnet group minerals, largely because of their extreme compositional complexity combined with the fact that garnet does not occur abundantly in most shallow crustal rocks.

Garnet weathering at Coweeta and the related pertinent literature have been studied by Velbel (1984a,b,c). Almandine garnet at Coweeta weathers by congruent breakdown of the parent crystal lattice, localized reprecipitation of most of the iron (as goethite) and some of the aluminum (as gibbsite), and removal of the remaining constituents in solution. The stoichiometry of the almandine garnet weathering reaction at Coweeta based on EPMA and mineralogical data is:

$$Ca_{.2}Mg_{.5}Mn^{II}_{.2}Fe^{II}_{2.1}Al_2Si_3O_{12} + 0.625O_2 + 2.5H^+ + 8.35H_2O \rightarrow$$

$$2FeOOH_{(goethite)} + 0.1Fe(OH)^+_{2(aq)} + Al(OH)_{3(gibbsite)} +$$

$$Al(OH)^+_{2(aq)} + 3H_4SiO_{4(aq)} + 0.5Mg^{2+} + 0.2Ca^{2+} + 0.2MnO_{2(s)}$$

The weathering of almandine garnet contributes significant quantities of silica, calcium, and magnesium to the soil and streams.

Plagioclase Feldspar

The feldspar minerals consist of aluminosilicates of potassium, calcium, and sodium, in which tetrahedra of silica and alumina are linked with one another in three dimensions, giving rise to a framework structure with the alkali and alkaline earth elements in the interstices. The two major compositional groups are the potassium feldspars, and a continuous solid-solution series between endmember calcium and sodium feldspars known as the plagioclase feldspars. The feldspars are the most abundant single mineral group in the earth's crust, and have therefore received considerable attention from geochemists interested in mineral weathering. Much of the recent literature on feldspar weathering is reviewed by Velbel (1984a, 1986b).

The transformation of the predominantly sodic plagioclase feldspars of Coweeta to their clay-mineral weathering products has been studied by Velbel (1982, 1983, 1984b, 1985c). The weathering of plagioclase feldspar at Coweeta is a two-stage process. The first stage is complete congruent dissolution of the parent feldspar, during which the constituents of the feldspar are released to the solution in the same ratios in which they occur in the fresh parent mineral. The general stoichiometry of this stage of the reaction can be written:

$$Ca_xNa_{(1-x)}Al_{(1+x)}Si_{(3-x)}O_8 + (6-2x)H_2O + (2+2x)H^+ \rightarrow$$

$$xCa^{2+} + (1-x)Na^+ + (3-x)H_4SiO_{4(aq)} + (1+x)Al(OH)^+_{2(aq)}$$

where $x = 0.25$ for rocks of the Coweeta Group and $x = 0.32$ for rocks of the Tallulah Falls Formation (determined by EPMA).

The second stage of the plagioclase weathering reaction is the precipitation of some of the dissolved constituents (released by the dissolution of the feldspar in the first stage of the reaction) as clay-minerals, which may occur either very close to the original site of dissolution (forming clay-mineral pseudomorphs after the feldspar) or after transport (as clay-mineral fracture-linings) some microns to millimeters or more from the original site of dissolution (Velbel 1982, 1983, 1984b, 1985c). The clay-mineral products of plagioclase feldspar weathering in the soils and saprolites of Coweeta are gibbsite and kaolinite, which precipitate by reactions:

$$Al(OH)^+_{2(aq)} + H_2O \rightarrow Al(OH)_{3(gibbsite)} + H^+$$

.and

$$2Al(OH)^+_{2(aq)} + 2H_4SiO_{4(aq)} \rightarrow Al_2Si_2O_5(OH)_4 + 2H^+ + 3H_2O$$

The presence of these cation-depleted clay minerals in the saprolites of Coweeta could indicate either prolonged weathering (that is, that the weathering profile of Coweeta is extremely old), or very intense weathering over a shorter time period, with the relative depletion of alkali and alkaline earth cations due to extreme rapid flushing of water through the porous and permeable weathering profile (see the discussion above on the hydrologic properties of saprolite). Given the extremely high rainfall and the occurrence of primary rock-forming minerals in the soils and saprolite of Coweeta, it appears that the kaolinite–gibbsite mineralogy of the soils and saprolites is due more to the intensity of weathering at Coweeta than to the antiquity of the weathering profile. As is discussed below, present day rates of mineral and chemical weathering of Coweeta rocks could produce the observed thickness of saprolite in several hundred thousand years, suggesting that the Coweeta landscape is much younger than many kaolinitic–gibbsitic landscapes beneath 10 to 60-million-year-old peneplains. Other clay minerals form in incipiently weathered bedrock below the saprolite and in outcrops, but as is shown elsewhere in this chapter, these weathering microenvironments involve minimal quantities of the subsurface water, and therefore do not contribute measurably to the dissolved load of surface or subsurface waters.

Geochemical Mass Balance of Small Watersheds

Most streamflow at Coweeta is fed by subsurface water percolating through the soil and saprolite (e.g., Hewlett and Hibbert 1966); as a consequence, only the weathering reactions which take place within the soil and saprolite influence the composition of streamwaters leaving Coweeta watersheds. This can be seen in the close correspondence between the geochemical character of the streamwaters and the clay mineralogy of the weathering profiles (Velbel 1982, 1984b, 1985a,c). The amount of water percolating into and through fractures in the bedrock is extremely small relative to the amount flowing through the saprolite, and mineral–water interactions taking place within these small bedrock fractures leave no detectable imprint on the geochemistry of Coweeta waters. Because only mineral transformations taking place within the soil and saprolite influence the measured output fluxes via Coweeta streams, only these reactions need to be considered in geochemical modelling of input-output budgets in small watersheds.

Geochemical mass balance studies (also known as input-output budgets) invoke a simple conservation-of-mass principle, the principle that "some of it plus the rest of it equals all of it." Put another way, if the flux of any element leaving a watershed (e.g., via streams), and the flux of that element into the watershed (e.g., via atmospheric precipitation) are known, the difference between the two can be calculated. This difference must be due to the sum of all reactions and transformations involving that element which took place within the watershed. Pioneering mass balance studies on weathering profiles and/or small watersheds include those of Garrels (1967; Garrels and Mackenzie 1967), and Cleaves and Bricker and their coworkers (Cleaves et al. 1970, 1974). Geochemical mass balance studies are widely recognized as the most reliable means of estimating mineral weathering rates in nature (Clayton 1979). The mathematical basis for watershed mass balance studies is developed and discussed by Velbel (1986a).

Early geochemical mass-balance studies placed primary emphasis on inorganic weathering. Cleaves et al. (1970), however, recognized that weathering reactions in a forested watershed produce larger quantities of mineral nutrients (especially calcium and potassium) than are observed in streams and that these excess quantities are probably taken up by the forest biota. Likens et al. (1977) also recognized that weathering products derived from the breakdown of soil minerals leave the weathering profile both via the dissolved load of streams and via biomass uptake, and that estimates of weathering rates (i.e., mineral breakdown rates) based only on stream fluxes might be as much as a factor of two lower than the actual value which includes biotic uptake.

Velbel (1984b, 1985b) developed a geochemical mass balance model for forested watersheds at Coweeta which extends previous models in two ways. First, mineral compositions in this study were determined empirically, and compositional data were combined with micromorphological data to arrive at the mineral weathering reactions described above. This is in contrast to many of the earlier studies, in which mineral compositions were assumed to be nearly ideal (essentially, estimated from geological occurrence and standard mineralogy texts). Secondly, botanical uptake of three major mineral nutrients (potassium, calcium, and magnesium) is explicitly included in the mass balance expressions, which permits direct estimates of biomass uptake.

Mineralogical and petrographic (micromorphological) data constrain the mineral weathering reaction stoichiometries described above. Combining these with previously published hydrogeochemical data (Hewlett 1961; Hewlett and Hibbert 1963, 1966; Swank and Douglass 1977; Velbel 1982, 1984b, 1985a), reveals that four transformations influence elemental budgets in the watersheds; the weathering of the three major weatherable rock-forming minerals (biotite mica, almandine garnet, and plagioclase feldspar), and the uptake of mineral nutrients by the forest biota. The stoichiometries of these reactions (that is, the amount of each element released per unit amount of reaction) are known for minerals (from the data and references discussed above) and can be approximated for biomass (using the data of Day, as reported in Boring et al. 1981), but the rates of the four reactions are not known. However, the input and output fluxes (mass of each major element per unit time) are known, as is the difference, which is the rate of total change of element abundance in the watershed per unit time. The difference between output and input fluxes for each element is due to the sum of the four transformations taking place within the watershed; we can therefore write a mass balance expression for each element, which has the following form:

$$\sum_{j=1}^{\varphi} \alpha_j \beta_{c,j} = \Delta m_c$$

where $\beta_{c,j}$ is the stoichiometric coefficient of element c in reaction j (known from mineralogical studies or published botanical compositions), Δm_c is the net flux of element c from the watershed (stream output minus precipitation input), and α_j is the number of moles of weathering reaction j occurring per unit area of watershed per unit time (the rate of the transformation, which is unknown). φ is the number of phases involved (i.e., transformations taking place); in this case, $\varphi = 4$. Each mass balance equation is an equation in φ unknowns; therefore $n = \varphi$ such mass balance equations are required to solve for the desired number of unknowns (φ). There are four unknowns in Coweeta watersheds, and therefore mass balance equations for four elements are needed. Fortunately, stoichiometric coefficients and net fluxes are known for sodium, magnesium, potassium, and calcium, so a system of four equations in four unknowns can be constructed, constrained, and solved for rates of mineral weathering and botanical uptake. Rates of clay-mineral formation can be incorporated by adding additional terms to the equations if we add an equal number of mass balance equations for addi-

tional elements, such as silica. Rates of mineral weathering and mineral nutrient supply
to the terrestrial biota have been calculated by this method (Velbel 1984b,c, 1985b) and
rates of weathering profile development and geochemical denudation of the landscape
have been calculated from these results (Velbel 1984b, 1985b).

Rates of Mineral Weathering

In previous studies weathering rates were calculated for individual minerals in seven
control (unmanipulated) watersheds of the Coweeta Basin, including WS 27 (Velbel
1984b, 1985b). Using volumetric estimates of mineral abundance (based on petrographic data), estimates of mineral grain size and geometry, and the average thickness
of the weathering profile, the results of the WS 27 mass balance calculations (in moles
of reaction/ha of watershed/yr) were transformed into moles/m² of mineral surface/sec,
the units in which laboratory experimental rate data are often reported. The results for
rates of garnet and plagioclase weathering are no more than approximately one order
of magnitude slower than rates determined in laboratory experiments under similar
hydrogeochemical conditions. Rates of biotite mica weathering are considerably different, due in part to the fact that a reasonably similar set of laboratory experimental data
does not yet exist.

The similarity of field and laboratory rates of plagioclase and almandine weathering
is quite good in geochemical terms; the two major sources of the remaining discrepancy
are (a) the character of artificially treated mineral surfaces in laboratory experiments,
which renders them more reactive than their natural counterparts (e.g., Holdren and
Berner 1979) and/or (b) difficulties in estimating the reactive mineral surface area in
natural systems. The second is probably a major problem (Velbel 1986b) although the
method of estimating mineral surface area in this study gives results which compare
favorably with laboratory data and other recent mass balance studies (Pačes 1983;
Siegel 1984). Several orders of magnitude of error are associated with the surface area
estimate, so the favorable correspondence between field and laboratory rates in this
and other mass balance studies may be somewhat fortuitous. Nevertheless, the relatively good accord between field and laboratory kinetic results is encouraging, and
suggests that (1) the two aforementioned sources of error *may* not be large, and (2)
refinement of laboratory kinetic data and field mineral surface area determinations
should rank high among future research priorities.

The same mass balance calculations which provide the basis for the above discussion
also provide quantitative estimates of mineral nutrient uptake by the forest biota.
However, the integration of these results with ongoing ecological studies is only beginning, and no conclusions can be safely drawn at present regarding the possible ecological significance of the model results. Assessment of this aspect of the weathering
processes and their rates is currently underway.

Chemical Weathering Rates, Rates of Saprolite Formation, and the Geomorphic Evolution of the Southern Blue Ridge Landscape

Geochemical export via the dissolved load of streams is used in various ways to estimate rates of weathering profile development and chemical preparation of the landscape for erosion. In saprolitic landscapes, chemical weathering is isovolumetric, and

involves no change in the volume of landforms or landscapes (Cleaves and Costa 1979; Costa and Cleaves 1984); it is therefore inappropriate to speak of "chemical denudation" of saprolitic landscapes. The rate of saprolite formation does, however, determine the rate at which rock is made susceptible to physical erosion, and is therefore a significant factor in the geomorphic development of saprolitic landscapes.

Berry (1977) calculated geochemical weathering rates using total dissolved solids (TDS) in streams as the measure of chemical mass export from the watersheds of Coweeta, and determined that a rate of landscape lowering of between 0.4 and 1.0 cm/1000 yrs would be in balance with the observed rate of chemical export of TDS. Two difficulties exist with the use of TDS as the measure of chemical weathering; first, the possible transfer of elements out of the minerals into compartments other than stream export (e.g., uptake and accumulation by biomass; Likens et al. 1977) is not allowed for, and, second, TDS includes mass from sources other than weathering in the landscape (e.g., atmospheric input). The geochemical mass balance approach allows for the incorporation of both nonweathering inputs and nonstream outputs from the mineral compartment; it is (partly) for this reason that mass balance models are considered the most reliable method for the determination of natural weathering rates (Clayton 1979).

The geochemical mass balance results of Velbel (1984b, 1985b) permit estimation of the volume of rock transformed into saprolite per unit area of watershed per unit time. The rate of saprolite formation from fresh bedrock in WS 27 is around 3.8 cm/1000 yrs. This value is very close to the long-term average denudation rate of 4 cm/1000 yrs estimated by Hack (1980) for the southern Appalachians based on a variety of independent means. These rates suggest that, in geologic time, the landscape of the southern Blue Ridge is in dynamic equilibrium, in which all components of the landscape and the rates of all geomorphic processes are essentially in steady state. In the (geologically) short term, however, erosion at the top of the weathering profile is much more sporadic and discontinuous than chemical weathering at the base of the weathering profile. In order for dynamic equilibrium to exist in the short term, sediment export rates from watersheds should equal 600 kg/ha/yr (assuming a soil bulk density of 1.5 g/cm^3). Normal erosion rates at Coweeta (as measured at weirs) range from about 30 kg/ha/yr for undisturbed forested watersheds to around 255 kg/ha/yr for a watershed clearcut and replanted with grasses (Monk 1975). This "normal" erosion load is therefore only about 5 to 45% of that required to maintain dynamic equilibrium (steady state with respect to chemical weathering). If long-term dynamic equilibrium is to be maintained, some 50 to 95% of the short-term physical erosion must occur as high-magnitude, low-frequency events (e.g., severe storms, landslides, debris avalanches), giving rise to brief episodes of abnormally high physical erosion not represented by "normal" sediment export loads. These findings are in excellent accord with the findings of Grant (1983, and Chapter 7) regarding the importance of debris avalanches in the geomorphic evolution of the Coweeta landscape.

Summary

1. Most weathering profiles of the Coweeta Hydrologic Laboratory comprise thin (to ~30 cm) A and B horizons atop substantial thicknesses (ave. ~6 m) of saprolite.

2. Extensive isovolumetric weathering of rock to saprolite results in the landscape being covered by a porous, permeable blanket of weathered material which supplies streamflow as water drains slowly from it, and which is vulnerable to erosion by fluvial processes and landslides.

3. Rock weathering to saprolite occurs primarily via weathering of three major rock-forming minerals: biotite mica, almandine garnet, and plagioclase feldspar. Furthermore, each of these mineral weathering reactions contributes to mineral nutrient budgets of the terrestrial ecosystems and watersheds of Coweeta. Biotite mica weathering releases significant quantities of potassium; magnesium comes from both biotite mica and almandine garnet; and calcium is derived from both plagioclase feldspar and almandine.

4. Geochemical mass-balance calculations permit quantification of mineral weathering rates and rates of geomorphically significant saprolitization. Mineral weathering rates determined for Coweeta WS 27 compare favorably with mineral weathering rates determined in laboratory experiments under similar hydrogeochemical conditions. One major source of uncertainty in this comparison of field with laboratory rates is the difficulty in estimating reactive surface area of the minerals in natural systems. Provided that future research resolves this difficulty, it should become practical to use carefully constrained laboratory experiments to estimate rates of elemental transfer in natural weathering systems.

5. Transformation of bedrock to saprolite (saprolitization) presently prepares the landscape for erosion at a rate equal to the long-term average denudation rate for the southern Appalachians (4 cm/1000 yrs), suggesting that dynamic equilibrium of the landscape prevails in the long term. In the (geologically) short term, erosion at the top of the profile is much more sporadic. Present rates of erosion by normal fluvial processes suggest that high-magnitude, low-frequency geomorphic events accomplish most of the short-term erosion required to maintain long-term geomorphic equilibrium of the southern Blue Ridge landscape.

7. Debris Avalanches and the Origin of First-Order Streams

W.H. Grant

Landslides in general and debris avalanches in particular are of concern to engineers who encounter them in the course of highway construction (Whitney et al. 1971). They are also of interest to botanists, because they offer an opportunity for the study of plant succession on naturally denuded slopes (Flaccus 1959). Catastrophic natural processes are becoming of greater concern as population pressures force humans to live on less and less desirable, steeply sloping land. Geologists have been concerned with avalanches at least since 1826, when an avalanche was reported in the White Mountains of New Hampshire. Lyell visited the site in 1845 and commented that the moving rock mass had removed the glacial striae from the bare granite surface testifying to the erosive power of the moving avalanche (Lyell 1875).

The literature on mass wasting is abundant in many parts of the world, especially in those regions of recent orogenic activity. In older mountain chains, with warm humid climates underlain by metamorphic rocks such as the southeastern Blue Ridge Mountains, data are especially sparse. Sharpe (1938) described avalanches as leaving long narrow tracks on steep mountain slopes. He also emphasized humid climates as a factor in their formation. Avalanches have regional characteristics which include climate as well as geology and topography (Krynene and Judd 1957). According to this concept slides within a geomorphic region possess similar characteristics. Data from the numerous (186) debris avalanches produced in Virginia by Hurricane Camille showed that the deluge, 28 inches of rain in 8 hr, was a factor (Williams and Guy 1973). They also concluded that debris avalanches follow preexisting depressions in hillsides generally steeper than 35°. Rock type does not appear to be a definitive factor in avalanche formation, although shales are probably more prone to slide than other rocks (Voight 1978). Chemical weathering and structural features are significant in some slides, but slides are common in areas of rapid mountain growth where chemical weathering is minimal (Voight 1978). In the southeast, chemical weathering is very prominent and is a factor in slide development. Slopes most likely to produce rock avalanches during earthquakes are: steeper than 25°, higher than 150 m, undercut by active streams, composed of intensely fractured rock, and at least one other factor such as: faults, bedding and foliation dipping down the slope, and weathering (Keefer 1984). Geologic structures are frequently invoked as factors in avalanche formation and control. These factors include jointing, bedding, faulting, and cleavage. They are usually part of the descriptive data (Inners and Wilshusen 1983). Factors in the development of joints in granite include topography, residual tectonic strain, and dilation. The unambiguous assignment of any joint set to a single cause is not possible (Chapman 1958). A strong correlation between jointing and folding is evident in the Inner Piedmont (Grant 1958).

In the Grandfather Mountain area of western North Carolina, circular patterns of large size occur. They are related to joints which are classified as either tectonic or exfoliation (dilation). Most tectonic joints match structural patterns inherited from a period of deformation (Hack 1966). The release of high confining pressure by removal of overlying rock is the most important cause of sheet jointing (dilation jointing) in granite (Hopson 1958). These views are adopted in this study, of which the major considerations are the relation between jointing and first order streams and between dilation joints, tectonic joints, chemical weathering, and debris avalanches.

The Study Area

The work was done in the area around the Coweeta Hydrologic Laboratory (Figure 7.1). The geologic mapping was done for both the laboratory area and the Prentiss Quadrangle (Hatcher 1980). The geomorphology is typical of the Blue Ridge Mountains. Locally, it is characterized by narrow sharp ridge tops flanked on either side by steep slopes (30° to 45°). These are interspersed by somewhat rounded ridges and gentler slopes at lower elevations. The valleys of the upper slopes are steeper and vee-shaped. They may contain large amounts of coarse rock debris, but little alluvium. Broad valleys at lower elevations are flanked by moderately steep slopes and have almost flat gently sloping bottoms filled with alluvium. For the laboratory area the maximum elevation is on Albert Mountain (1592 m). The minimum is near the confluence of Ball Creek and Shope Fork (675 m). The total relief is 914 m.

The soil cover on the mountains varies greatly in thickness and development. The steeper valley walls are bare rock or mantled with a thin creeping soil. The soil appears to sustain its continuity by being laced together by interlocking plant roots. This condition creates a ruglike integrity for the soil mass. These soils are not necessarily residual. They may overlie rock or saprolite from which they are not derived. In some areas, especially steep ones, bare rock with limited minor plant growth is exposed. On the more rounded lower ridges, the more gently (<30°) sloping valley walls may be covered with residual soils. That is, the rock, saprock, saprolite soil sequence is complete and apparently undisturbed. These observations are supported by the occurrence

Figure 7.1. The major drainage system, the trends of two selected geologic formations (Hatcher 1980), and the directions of the tectonic joint maxima. The map center is approximately 35° 03′N by 83° 27′W.

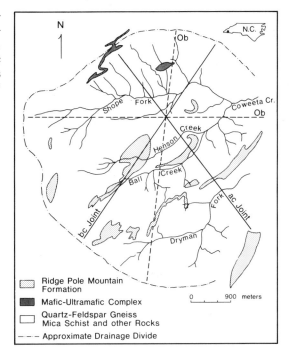

of both residual (underlain by saprolite) and transported (creep) soils in Rabun County, Georgia, which borders Macon County on the south (Carson and Green 1977).

Alluvium in the lower valley bottoms is composed of very coarse sand and granules but particle sizes may range upward to boulders which are commonly 30 cm to 1 m in maximum dimension and much larger ones occur sparsely. Small quantities of clay minerals occur in all stream sediments.

Avalanche material includes all particle sizes from clay to boulders, but is strongly skewed towards the larger particles. It is thought that the particle sizes as well as the water content of a moving avalanche is different from the final deposit. Mudflows occur occasionally in the high rainfall areas of the southern Blue Ridge Mountains (Hursh 1941). These mudflows carry large rock fragments and are the same as debris avalanches.

The underlying parent rock strata are late Precambrian to lower Paleozoic metasediments which have been intruded in places by late Paleozoic plutonic igneous rocks, mainly granites. The most common rock type is a micaceous quartzofeldspathic gneiss. The most common minerals are quartz, oligoclase, biotite, muscovite, and potash feldspar. Sillimanite, kyanite, staurolite, and garnet are less common but important indicators of metamorphic grade. These rocks have been recrystallized by regional metamorphism (Hatcher 1980). Deformation at elevated temperatures amd pressures tends to homogenize the fabric and to a limited extent decrease the geomorphic importance of old planar structures such as bedding. Several episodes of folding have occurred, but only the last appears to be of geomorphic importance. These folds trend roughly N38E–S38W (Figures 7.1 and 7.4) and are probably Alleghenian. Further geologic information is given by Hatcher in this volume (Chapter 5).

The Blue Ridge Mountains in the Coweeta area are steep. It is doubtful that such declivities could have survived 300 million years of mass wasting without some vertical movement. Data from Meade (1971) indicates on his map for crustal movement in the southeastern United States, vertical movement of between 3 mm and 4 mm per year in the Coweeta area. Schaeffer et al. (1979) show active stresses are still present in the northwestern South Carolina Piedmont. Bandy and Marincovich (1973) point out that uplift near Los Angeles has been irregular (178 m in the past 36,000 years), but that there has been little or no uplift in the last 50 years. Bollinger (1973) indicates that minor seismic activity has been observed since 1794 in western North Carolina. Thus, it is logical that modern landforms are the product of Holocene climate interacting with slow vertical uplift, but any indication of rate based on short term data is of doubtful value.

The annual precipitation and average temperature at Coweeta are adequate to produce a well developed weathering profile, soil and saprolite (Chapter 3). However, it is not as well developed as the warmer but somewhat drier Piedmont.

Methods

General rock, soil, and saprolite observations were made at stream and road cuts. The major source of data are surface exposures and appropriate core samples. Joints, hinge lines, and foliations were measured with a Brunton compass. Thalweg vectors, plunge, and bearing were measured, from the 1:7200 scale map of the Coweeta Hydrologic Laboratory (1972), along portions of first order streams which are straight for distances greater than 60 m. The restriction of the vectors to straight portions of streams

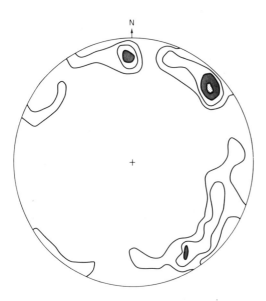

Figure 7.2. The bearing and plunge of 88 thalweg vectors measured from straight segments of first order streams. Contours 2, 4, 6, 8% per 1%.

Figure 7.3. Equal area distribution of 198 tectonic joint poles. Contours 1, 2, 3, 4, 6, 8, 10% per 1% net area.

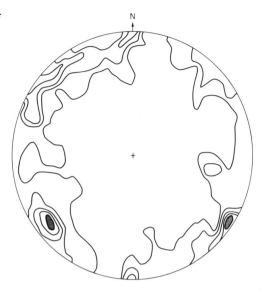

eliminates irregularities introduced after the formation of a structurally controlled channel. The thalweg vectors are shown on an equal area projection in Figure 7.2. Joint distributions are shown on an equal area projection in Figure 7.3. The hinge lines are shown in equal area projection in Figure 7.4. The data contain 16 lines from the present work and 25 from Hatcher (1980). All structural data are taken as available and are somewhat irregular but widespread in their areal distribution.

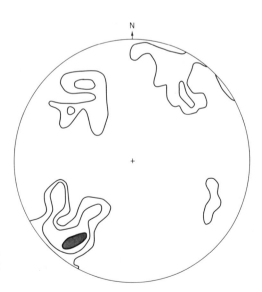

Figure 7.4. Bearing and plunge of 41 hinge lines. Contours 2.5, 5.0, 7.5% per 1% of net area.

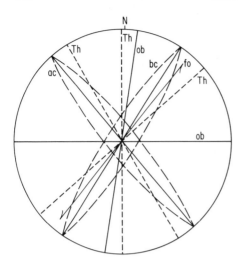

Figure 7.5. Generalized spatial relationship between hinge-lines (*Fa*), tectonic joints (*ac*, *bc*, and *ob*) and thalweg vectors (*Th*). Equal area projection.

Results

There are two, possibly three, fold trends shown in Figure 7.4. The strongest is approximately N38E–S38W. A similar trend is evident in the geologic units (Figure 7.1). The NE trend is taken as the major fold direction and the jointing is described on this basis. The joints are resolved into four major sets: an *ac* set approximately normal to the fold axis, a *bc* set approximately parallel to the fold axis, and two oblique sets approximately North–South and East–West (Figure 7.3). It is the correlation of jointing and folding which makes it probable that the joints are tectonic. These joint attitudes are similar to the Schaeffer et al. (1979) account of jointing in the South Carolina Blue Ridge.

A comparison of the straight streams containing the thalweg vectors with the tectonic joints (Figures 7.2 and 7.3) shows a close similarity in orientation. They are subparallel to three of the four joint sets. The N90E oblique set is an impossible direction for streams since it cuts across the hillsides. The relation between tectonic joints, thalweg vectors, and fold trends is summarized in Figure 7.5.

Dilation joints are subparallel to topography. Their dips may vary up to 90°. Most are in the 30° to 50° range. They are probably formed as the result of elimination of confining pressure by erosion of large amounts of overlying rock. They are much younger than the tectonic joints. Because of this dilation pressure, opening may take advantage of tectonic joints or foliation surfaces when they are appropriately oriented with respect to topography, thus substituting for dilation joints. When occurring independently, dilation joints are gently curved, producing lenticular cross sections. The major surficial function of these two kinds of joints is to admit, store and provide for circulation of ground water. Ground water is responsible for chemical weathering, feeding springs, and is a factor in avalanche formation (Figure 7.6).

An example of a recent avalanche is shown north of Ball Creek about 0.2 miles southwest of the Laboratory Administration Building (Figure 7.6). A water bearing joint was identified by drilling successive bore holes across the head of the chute. All holes

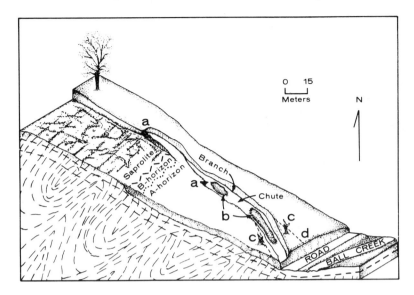

Figure 7.6. Coweeta debris avalanche map and section. The chute encloses remnants of the avalanche (b). Rock, thin saprolite and soil are shown. A small branch runs down the east side of the chute. Foliation, compositional layering, and four steep dipping tectonic joint sets and topographically subparallel dilation joints are indicated. Stippling along joints indicates the presence of water and weathering. The scarred trees (c) show that the avalanche upper surface was raised above general surface level during the movement period. The approximate width of the moving body is shown by dashed lines (d). The breakaway point is indicated at (a). The curvature in the lower part of the chute is probably caused by gravity.

stopped in rock at approximately 25 cm depth, except an aligned section which is water bearing. Here bore holes went down 1 m before striking rock. These deep holes have a bearing of N48W which is within the ac joint set. Along the whole length of the chute, except where avalanche debris remains, either bare rock is exposed or a thin, approximately 25 cm thick soil layer occurs. On either side of the chute drilling shows soil depths ranging from 1 to 1.5 m. Joint directions on rock outcrops are consistent with the data shown on the joint diagram (Figure 7.3). The exposed dilation joint surface is independent of the foliation which dips into the hillside. Evidence that the avalanche was a water inflated mass of rock, mud, and plant debris includes the scars on the upslope sides of surviving trees near the edges of the chute. The height of the scars, if projected across the chute, indicate that the avalanche was roughly lenticular in cross section, about 3 m thick and about 15 m wide during the time it was in motion. A small amount of the avalanche material still remains in the chute (Figure 7.6).

The contour plot (Figure 7.7) shows that most thalwegs, all of which are water bearing, plunge between 12° and 15°. This may have an effect on the permanence of avalanche generated streams, since at the time of formation they have chutes plunging about 30°. This situation may limit the groundwater reserve for sustained flow.

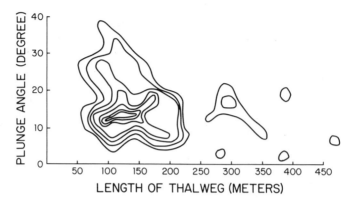

Figure 7.7. Contour diagram showing lengths and plunges of thalweg vectors. Contour interval 1%/cm² maximum 8%.

Conclusions

Debris avalanches and underlying joint systems are responsible for some first order streams. Other mechanisms exist for both phenomena. The sequence for development of an avalanche associated with first order streams is as follows. A small depression is initiated by the subsoil intersection of two lenticular dilation joints with a downslope striking tectonic joint. Such a configuration could easily occur beneath a small valley. The depression accumulates ground water. Chemical weathering proceeds faster in the depression enhancing its water capacity. This process can continue indefinitely or be interrupted by a violent storm. The latter situation is discussed by Eschner and Patric (1982) who report a storm in May 1976 in the Coweeta area. Their data show 25 hrs of gradually increasing rainfall which peaked sharply at about 18 to 20 cm per hour. This peak coincided with a debris avalanche. The avalanche is initiated by hydraulic pressure fed through the water saturated subsoil joint sysem. At the storm peak, pressure is strong enough to break the adhesion between the rock and the water saturated soil and saprolite. The water inflated mass of rock and soil slides quickly down hill, leaving a chute as evidence of its passage. The chute is the locus of a new first order stream which needs only a sustained supply of water to develop into a steep-walled mountain valley.

   In summary, the major subsurface water controls are soil-rock, saprolite-rock interfaces, dilation, and subvertical tectonic joints. Metamorphism tends to promote the development of dilation jointing independent of foliation by reducing the anisotropy of the rock fabric. Thalweg vector plots support the relation between first order streams and joints. The optimum place for a new first order stream to form is on slopes of about 30° which are covered with thin soil and underlain by well jointed sap-rock.

8. Streamflow Generation by Variable Source Area

A.R. Hibbert and C.A. Troendle

For many years, hydrologists and engineers alike have attempted to describe and model the processes that generate streamflow from rainfall. Modern understanding of runoff processes expanded rapidly in the twentieth century. Horton (1933) proposed that the soil surface partitions rainfall into overland flow and ground water. Simply stated, when rainfall rate exceeds the infiltration capacity of the soil, the excess water flows over the surface to become the primary source of storm flow. Base flow, that which maintains streams between rains, issues from ground water aquifers sloping gently to the channels. During the 1930s and 1940s, Horton and his coworkers firmly established the idea that infiltration was dominant in the runoff process. The central criticism of the Hortonian concept is that too much emphasis was placed on rain that fails to infiltrate the soil, and not enough on that which does infiltrate.

The Hortonian concept of storm flow generation prevailed for many years, and is still valid when applied to land surfaces that do not accept water readily. Much of the western rangelands fall in this category. By the mid 1930s, however, foresters in the eastern United States were becoming increasingly aware that overland flow was not being generated on forest slopes in sufficient quantity to account for storm flows from first- and second-order streams (Lowdermilk 1933; Hursh 1936; Hursh and Brater 1941). Horton (1943) recognized that some soils were capable of absorbing all the water that fell on them. With reference to the Little Tallahatchie drainage basin in Mississippi, he observed that "owing to somewhat unusual conditions, surface runoff rarely occurs from soil well protected by forest cover."

Development of the Variable Source Area Concept

Much of the early work that shaped current concepts of streamflow generation on forest lands was done at the Coweeta Hydrologic Laboratory. In 1941, Hursh and Brater systematically analyzed the sources of storm flow on Coweeta Watershed 13 (WS 13), a 16-ha catchment covered by mixed hardwoods. Streamflow was perennial, with annual yields similar to larger basins in the region. They found that characteristic flood hydrographs were produced by heavy rains, even though no overland flow was observed. Two types of storm flow hydrographs were analyzed. The first was produced by a short, intense rainfall of 18 mm in 25 min. Storm flow response was 1.1% of rainfall; all of it passed the gaging station within 2 hr of rainfall. They attributed this response primarily to channel precipitation, rain falling directly on the channel surface, which measured 1.2% of the total catchment area.

The second type of hydrograph was from a much larger storm (164 mm in 24 hr). Total storm runoff within 24 hr of rainfall cessation was 14%, even though antecedent flow rate was below the watershed average, a condition that would suggest a lower-than-average response. Again, 1% or more of rainfall was intercepted by the channel, but in this case they attributed the bulk of the runoff to subsurface storm flow, since no overland flow was observed. Conclusions by Hursh and Brater are summarized below:

1. Channel precipitation is the first contributor to storm flow (for short, intense storms it is the only or primary contributor).
2. Second to contribute are areas of shallow water table close to the stream (spring heads, seepages, and swamps that are quickly saturated by rainfall). Where such conditions occur along the stream, it is expected that there will be an *actual increase in the effective width of the channel* (emphasis ours) and subsequent increase in the amount of channel area available for precipitation interception.
3. Storm water moving through layers of porous soil material, colluvial fill, and talus slopes along streams and drainage ways could reach the stream in time to contribute to the storm hydrograph.

The concept of a subsurface source for storm flows requires a dynamic process. Lowdermilk (1933) conceived of dynamic subsurface storm flow in certain soil profiles, generating rapid hydrologic response that was not the result of either true overland flow or true ground water discharge. Hoover and Hursh (1943) and Hursh (1944) clearly described a "dynamic form of subsurface flow" that contributed significantly to storm flow. Roessel (1950) concurred that soil water discharge may form the major part of flood flows. He experimented with small sand troughs to determine how aquifer length affects outflow, and then used the results to explain apparent anomalies in storm flows on the Emmental watersheds in Switzerland. Through the last 30 years, this thinking has been expanded into a more dynamic and integrated concept of streamflow generation on forest lands, of which the Hortonian overland flow concept is but a part.

Slope Processes

If in defining the concepts of flow generation we can begin with the simplest case, a hillside, then Hewlett (1961a) probably described it best in his perception of the

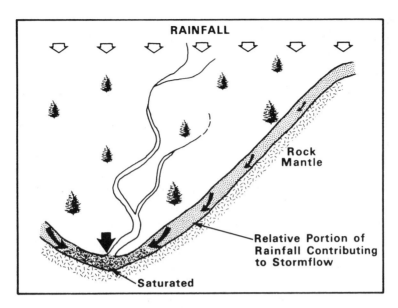

Figure 8.1. Schematic cross-section of a mountain watershed, showing how the relative contributions of rainfall to storm flow vary with position on slope. (Hewlett 1961A.)

processes operating on steep headwater catchments at Coweeta (Figure 8.1 accompanied the original report):

When rain falls on porous forest soil, it enters the ground and either begins to migrate to the nearest stream or is held as 'retained' water by the soil particles . . . Whether it migrates or is held in place depends chiefly on the character and wetness of the soil, which in turn is usually related to its depth and position on slope. . . . Where it sinks in near a stream and consequently can contribute more to immediate rises in streamflow, it generally will move faster than if it enters the drier slopes and ridges above . . . Rainfall influence in producing immediate runoff obviously diminishes with distance from the stream channel. . . . Under prolonged and heavy rainfall, the stormflow-contributing area contiguous to stream channels may grow wider and wider, depending on the nature and depth of the earth mantle. But what happens to the remainder of the water-portion not reaching stream channels a day or two after rainfall? Of course a great deal of this evaporates or is transpired by plants and hence is lost to streamflow. But while this is happening, a substantial portion continues to migrate downward, eventually appearing as clear springs or streamflow. Thus the soil mantle is able to moderate erratic rainfall into continuous outflow between storms. The deeper the soil, the better the moderation. . . .

In lowlands or wide valley areas, base flow is partly fed by the slow depletion of free-water, underground aquifers, i.e., the saturated material comprising or lying below a gently sloping water table. But in mountain country such as Coweeta, the soil mantle is sloped too steeply to retain large bodies of ground water in water tables as commonly pictured. . . . Coweeta catchments are underlain by massive, water-tight material; and it seems unlikely that deep fissures in underlying rock although possibly holding some water, are a major source of base flow . . . Accordingly, it was conjectured that unsaturated soils and moisture in the field capacity range must be supplying most of the dry-weather base flows. The deeply-weathered Coweeta soils are of variable depth averaging about 6 feet in most catchments; and after a heavy rain they can hold temporarily up to 30 area-inches of water (42% by volume). Perhaps drainage at almost imperceptible rates from this huge soil mass operating for long periods after recharge might produce enough water to sustain base flows.

This concept of saturated/unsaturated drainage from soil profiles as processes generating storm flow and base flow in mountain streams is supported by the literature. Nielsen et al. (1959) and Remson et al. (1960) noted that continuous drainage occurs from what Meinzer (1942) called "no-man's land of hydrology," the unconsolidated regolith above bedrock or a water table. Rather than considering this portion of the soil mantle as "dead storage," a more realistic concept is "dynamic storage," since the water within is in a constant state of flux. Edlefsen and Bodman (1941), Ogata and Richards (1957), and Robbins et al. (1954) also noted from field experiments that unsaturated soils can supply streamflow.

Hewlett (1961b) and Hewlett and Hibbert (1963) verified these concepts experimentally in artificially packed soil slabs. The second of three experiments is shown in Figure 8.2. A 91 cm × 91 cm × 14 m inclining concrete trough was constructed on a 40% slope and packed with locally excavated sandy-loam soil to a bulk density of 1.3. After thorough soaking, the surface was covered with plastic to prevent evaporation. Outflow was measured at the base (Figure 8.3), and soil water content and tension within the soil were monitored for 145 days.

Piezometers in the floor of the trough indicated, through a pressure change from positive to negative, that the large soil pores were substantially emptied in 1.5 days, during which outflow rate (L day^{-1}) decreased according to the expression,

$$\text{Outflow} = a_1\, T^{-b_1}$$

Figure 8.2. Coweeta soil model used to verify slow rates of drainage from moisture in the field capacity range. Photo taken of a later run, when in grass cover. (Hewlett and Hibbert 1963.)

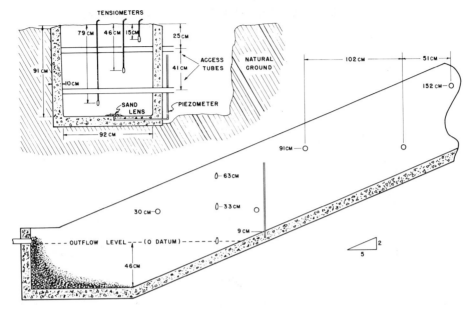

Figure 8.3. Cross and longitudinal sections of the lower portion of the inclined soil column. Three additional sets of tensiometers and piezometers are located at 3 to 4 m intervals upslope. Except where denoted by arrows, numbers indicate vertical distance above the outflow level. (Hewlett and Hibbert 1963.)

where T is days since artificial rainfall application was stopped and a_1 and b_1 are coefficients of regression. After a 5-day transition period (Figure 8.4), the logarithm of outflow for the next 80 days was again linearly related to the logarithm of time, but with new coefficients, a_2 and b_2, expressing the contribution of the unsaturated soil mass. Depletion rates of soil moisture content and tension substantiated the theory that the entire unsubstantiated soil mass was contributing through the lower saturated portion of the slab to outflow throughout the experiment.

The authors concluded that unsaturated outflow from soil is both an important and virtually immediate source of sustained base flow in mountain streams. Even after 2 months of drainage, the unsaturated soil volume (10.85 m³) was yielding 1.42 L of water per day, with no change in the water table. In terms of yield per unit area, this rate is similar to minimum flows from Coweeta watersheds after more than a month without significant recharge.

Hillslope Studies

Since the early 1960s there have been numerous hillslope studies attempting to define the processes described earlier. Dunne (1978) and Troendle (1985) have reviewed these efforts. Several of the more pertinent hillslope studies are reviewed here. The reason for beginning with a hillslope in describing streamflow response is that the hillslope, if based at the perennial stream, approximates the smallest unit on which the entire range of flow-generating processes occur. Attempts have been made to correlate watershed

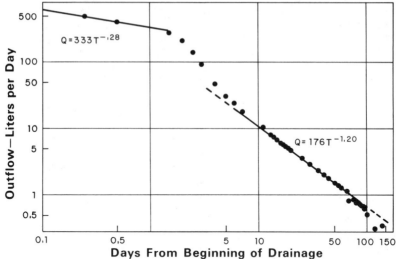

Figure 8.4. Outflow since beginning of drainage, showing log to log relation to time. (Hewlett and Hibbert 1963.)

area with hydrologic response (Langbein 1947; Dunne 1978), but the correlation is usually poor in basins less than 260 km² (Woodruff and Hewlett 1970). However, since flow velocity increases an order of magnitude or more once water enters the channel, the keys to hydrologic response are (1) how far water must travel (slope length) to get to or influence the channel flow, and (2) the mechanics by which it is translated. Storm flow dynamics are a function of slope length and pathway and are, therefore, better equated with drainage density than basin area. During a storm event, drainage density and slope length do in fact become quite dynamic.

Weyman (1970) studied the contribution to streamflow from a 12° hillslope along a 270 m reach of stream near Somerset, England. Streamgages were located in the channel above and below the slope segment so that its contribution could be determined by differences. In addition, surface and subsurface flow contributions from differing soil horizons were monitored on small study plots. Weyman observed no overland flow during the 16 storms of the study period, one of which was 7 cm with maximum rainfall intensity of 0.3 cm hr⁻¹. The A, B, and B/C horizons represent layers 0 to 10, 10 to 45, and 45 to 75 cm below the surface. The primary contributor to storm flow was what he termed "throughflow" in the B horizon.

Figure 8.5 demonstrates the strong correlation between Weyman's total basin discharge, control section discharge, and the soil plot data. The sharper "spikes" apparent in the total basin hydrograph probably reflect direct channel response to rainfall, although Weyman attributed the bulk of the flood peak to the headwater zone. Weyman concluded that:

Water moves vertically and laterally in unsaturated soil. The lateral component of flow may be increased by a drop in vertical permeability. This may occur at the soil/bedrock interface or at

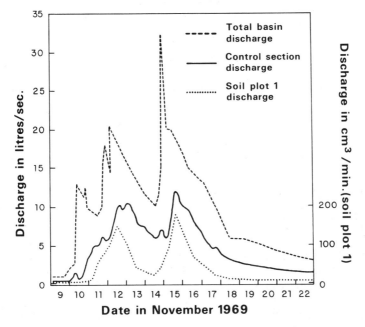

Figure 8.5. Total basin discharge, control section discharge, and throughflow discharge for soil plot 1 (different scale) for 9–22 November, 1969. (From Weyman 1970.)

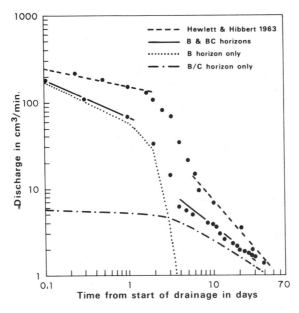

Figure 8.6. Throughflow drainage over time. Results of Hewlett and Hibbert (1963) compared with total discharge from soil plot 1 and separate horizon drainage. (Start of drainage is peak for each horizon.) (From Weyman 1970.)

some boundary within the soil. Soil at the base of the slope must be saturated for discharge from the slope to occur. This saturated zone is supplied by unsaturated flow from upslope. Instantaneous discharge from the profile is directly related to the hydraulic head in the saturated zone, which in turn is a function of the upslope extent of saturated conditions. Assuming that saturated permeability of the soil is constant upslope, the extent of the saturated zone is dependent only upon unsaturated supply rate. When supply exceeds lateral permeability, the saturated zone grows upslope and discharge from the slope base increases.

Weyman's observations were similar in nature and response to those observed by Hewlett and Hibbert (1963) (Figure 8.6).

Mosley (1982) studied lateral water movement on small plots in South Island, New Zealand. He noted that (1) a restricting layer was needed to shunt infiltrating water laterally downslope, (2) the portion of the catchment contributing to storm flow depends on antecedent precipitation and event size, (3) up to 100% of the basin could contribute if the event is large enough, (4) contribution of subsurface storm flow is less under lower rainfall or deeper soils, and (5) macropore movement requires saturation.

Beasley (1976) monitored surface and subsurface flow from two small plots (.054 and .089 ha) on the upper slopes of two forested watersheds in Mississippi. Streamflow was also monitored from the 1.86 and 1.62 ha watersheds on which the plots were located. Overland flow and flow through the shallow subsurface layer (above B horizon) were negligible during the study period. Beasley felt macropore conductance (throughflow) was responsible for streamflow generation rather than displacement or translatory flow.

Harr (1977) studied soil water flux on a steep forested slope in the Western Cascade range in Oregon. He observed no overland flow from storms that ranged from 4 cm to 18 cm. Of the 38% of the precipitation that on average became storm flow, 97% was generated by subsurface flow from the macropores in the deeper horizons. The remaining 3% was channel precipitation. He could not specify the nature of the flow mechanisms, but concluded that more than 38% of the watershed had to be contributing to the response. He noted continuous saturation only in the lower 10 to 15 m of slope, so he felt the lower slope had to be supplied by macropore drainage and unsaturated flow from upslope. His findings support some form of translatory flow.

A number of others, including Bren and Turner (1979), Whipkey (1965), Dunne and Black (1970a, 1970b), Pilgrim et al. (1978), Bonnel and Gilmour (1978), and Stephenson and Freeze (1974) have also studied pathways of water movement on hillside plots. Observations have been basically consistent with Hewlett's (1961a) description of the runoff process.

The central precept of the variable source area concept as applied to forested land is that water generally infiltrates undisturbed forest soils, migrates downslope, and maintains saturation or near-saturation at lower slope positions (Figure 8.1). These lower slope positions readily contribute subsurface flow to storm flow as the zone of saturated soil surface expands laterally and longitudinally. The degree to which saturation and subsequent expansion would occur for a given slope varies as a function of antecedent soil moisture conditions, precipitation volume, and duration of input.

There is a significant distinction between the variable source area concept and the true "Hortonian" overland flow concept. The latter represents a scenario that lies at one extreme, while total "throughflow" domination (Kirkby and Chorley 1967; Harr 1977;

Beasley 1976) lies at the other. However, throughflow describes only one pathway of movement rather than the process. The variable source area concept attempts to incorporate the entire range of hillslope processes, variable in time and space, integrating all pathways, and ranging between the extremes.

Studies of Watershed Processes

The wealth of information concerning water movement of hillslopes in forested environments generally supports the basic tenet of streamflow generation as proposed by Hursh and other early foresters. Describing hydrologic response on a hillside plot does not necessarily define the dynamics of watershed response, however. From the three basic building blocks — the convex, the concave, and the uniform slope segments that express either convergent or divergent flow — we can attempt to dissect and analyze the whole basin response.

The dynamics of the variable watershed response were described by Tichendorf (1969) for a small forested catchment in the Georgia Piedmont. After studying 55 individual rainstorms, he found that about 65% of them generated a storm flow volume that was equivalent to expanded channel interception. During larger storms, or storms on wet soil, lateral subsurface flow caused a rise in the perched water table in the vicinity of the perennial springheads and a subsequent streamflow response. He noted that the deep groundwater table was unaffected by these rainfall events. Most of the

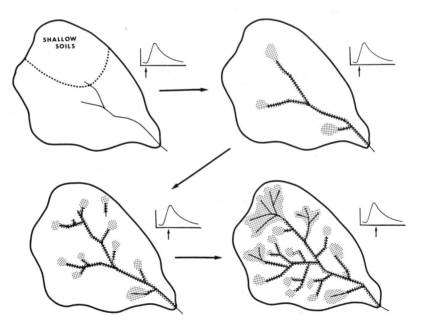

Figure 8.7. The small arrows in the hydrographs show how streamflow increases as the variable source extends into swamps, shallow soils, and ephemeral channels. The process reverses as streamflow declines. (Adapted from Hewlett and Nutter 1970.)

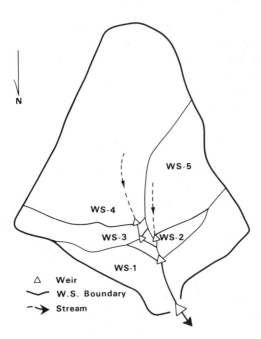

Figure 8.8. Schematic of a subdrainage, Fernow Experimental Forest, West Virginia. Shows location of weirs dissecting the 4.8 ha subdrainage.

response to rainfall was attributed to the upper 0.9 m of soil, where most of the change in soil moisture occurred. Soil water recharge, or change in soil moisture content caused by rainfall, increased with slope distance from the channel. He thereby concluded the near-channel areas were contributing most directly to storm flows in excess of channel interception because they were initially wetter, and that these "source areas" were fed between storms by drainage from upper slope positions. His somewhat qualitative description of a real basin reflects the translatory processes that Hewlett and others described earlier. Figure 8.7 illustrates the variable source area expansion for a simple case of a permeable basin with dendritic drainage. The expansion processes reverse as flow recedes.

Betson and Marius (1969) used subplots, observation wells, and piezometers to define the source areas of storm runoff from a small basin. They also found that the source area usually represented a small portion of the catchment, and the location and extent varied with rainfall intensity, antecedent moisture, and depth of the A horizon. The shallower the depth to an impeding layer, the quicker the response. Storm flow was conceived to be generated by saturated "interflow" exfiltrating in the source areas and augmented by direct precipitation.

Troendle and Homeyer (1971) used five stream gauges to dissect a 4.8 ha basin on the Fernow Experimental Forest near Parsons, West Virginia, into several smaller drainages to examine the source areas of storm flow. Unit 1 in Figure 8.8 is a drainage with perennial channel of known width and length and fed by two concave, convergent slopes; Units 4 and 5 are headwater draws or coves with ephemeral channels and convergent concave/convex slopes that concentrate surface/subsurface flow in and below the ephemeral channels; and Unit 2 is a minimally contributing divergent inter-basin area.

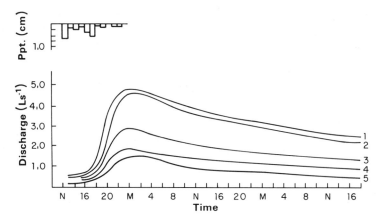

Figure 8.9. Hydrographs of flow for one storm on the nested watersheds depicted in Figure 8.8.

Hydrographs for five of the six weirs for an average 2.5 cm storm on wet soils are shown in Figure 8.9. Hydrographs 4 and 5 are independent of each other and all other hydrographs. Hydrograph 3 includes the flow from Units 4, 6, and the intervening area. Hydrograph 2 includes the flows from all except Area 1. And, finally, Hydrograph 1 is the total for the basin. Figure 8.10 contains hydrographs from a moderate storm on wet soils. Most of the total flow came from headwater coves 4 and 5. The delay in response, relative to precipitation, indicates the predominance of subsurface delivery to an expanding (lengthening) ephemeral channel. The spike in hydrograph 1-2 (Figure 8.10)

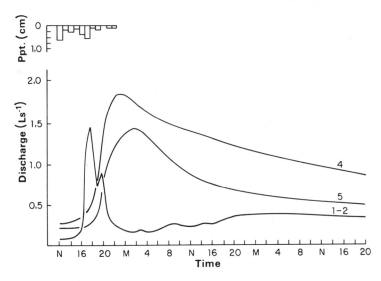

Figure 8.10. Relative contributions to flow from different watershed segments shown in Figure 8.8 for a moderate storm on wet soils.

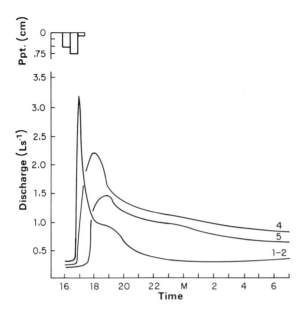

Figure 8.11. Relative contributions to flow from different watershed segments shown in Figure 8.8 for a moderate storm on dry soils.

is a reflection of channel and near-channel interception between the first two weirs (Figure 8.8); it represents only a small portion of the total flow from the 4.8 ha basin. An interesting observation is the secondary peak in hydrograph 1-2 (Figure 8.10), which is attributed to delayed subsurface flow from the long concave hillslopes feeding the lower reach of perennial stream. Again, Figure 8.11 represents the contribution from the same components for a smaller storm on dry soils. The relative contribution of the channel interception component is much greater, and the secondary peak is less well-defined.

Table 8.1 (Troendle and Homeyer 1971) summarizes storm flow responses to a number of precipitation events of approximately 2.5 cm magnitude for antecedent soil moisture conditions from very dry to very wet. Wetter soils generate larger storm flows. The observations in Table 8.2 allow comparison of the relative contribution to storm flow from a long divergent slope adjacent to a perennial channel (Unit 1, Figure

Table 8.1. Subbasin Response to Uniform Storms (2.5 cm) Under a Variety of Antecedent Conditions

Antecedent Soil Moisture	Percent Precipitation Watersheds (Nested)				
	1	2	3	4	5
Very dry	1.3	0.8	0.6	0.3	0
Dry	10	6	4	3	5
Moderately dry	13	14	7	6	22
Moderately wet	16	16	13	8	30
Wet	39	47	37	25	46
Very wet	57	64	61	47	68

Table 8.2. Comparative Storm Flow Response to Precipitation Under Varying Antecedent Conditions on a Watershed Segment Dissected with a Perennial Channel (Unit 1, Figure 8.8) Versus a Headwater Cove Without a Perennial Stream (Unit 5, Figure 8.8)

Antecedent Soil Moisture	Percent Precipitation	
	With Perennial Channel	Without Perennial Channel
Very dry	3	0
Dry	21	5
Moderately dry	8	22
Moderately wet	17	30
Wet	22	46
Very wet	34	68

8.8) and a convergent slope feeding an ephemeral channel (Unit 5, Figure 8.8). It is apparent, that as the system becomes charged the convergent slope is better able to concentrate flow and yield a greater storm flow, an observation also made by Anderson and Burt (1978). The uniform slope has a relatively fixed contributing area, and unless the event overloads the system (which a 2.5 cm event would not do) the slope tends to dampen response in a manner described by Hewlett (1961a). The interfluvial areas, such as Unit 2 in Figure 8.8, contributed primarily to base flow throughout the entire study period. Hewlett et al. (1977) noted that about 20% of storm rainfall is returned as storm flow for precipitation events equal to or in excess of 2.5 cm in the central Appalachians. Fernow WS 4 averaged 35% response; under wet conditions it can exceed 60% in extreme events.

The Variable Source Area Model

Hewlett and Troendle (1975) described the conceptual development of the variable source area model for the first- or second-order watershed. The following paragraph from that paper summarizes what has been presented, and sets the stage for the model and simulators that were developed later:

Thus we would appear to have four theories to explain source area hydrographs in forest and wildland basins. First we had the Horton (1945) overland flow concept, which should be fairly well relegated to cities and intensively cultivated fields by now. Second, we had Hursh's 1944 concept of piping flow, which is often noted on road cuts and streambanks as a visible phenomenon but which no one has been able to relate to the hydrograph or prove as a general process accounting for storm flow. Third, the variable source area concept arose from work at Coweeta about 1960. Fourth, the partial-area concept was expounded by Betson in 1964, and has been the subject of a number of papers since. It is obvious that these are different ways of looking at the same complex process in the first-order basin, and it will come as no surprise that our thesis here is that each of the other three are merely special cases of the variable source area concept. No rigorous analysis is necessary to prove that the partial area must expand and shrink as rainstorms vary in size and intensity, nor that, where local piping systems occur, they too must operate in an expanding-shrinking mode. No rigorous demonstration is anymore needed to prove that infiltrated water is the source of most storm flows from well-vegetated forest and wildlands in humid regions. Surface water flow from impervious areas (roads, rock outcrops, cultivated or trampled fields, and saturated depressions) may or may not reach the stream as overland flow, but when it does, there seems little question that the overland source will also vary with increasing precipitation.

There have been numerous attempts to model variable source area dynamics. Onstad and Brakensiek (1968) developed a surface runoff simulator that divided the drainage into strips or segments perpendicular to the contours. These strips were further subdivided into elements by soil characteristics, land use, etc. The simulator described a watershed as a system of finite elements that are not subjective, are simple, and allow expression of flow in only one dimension, independent of other units. Yarimanoghu and Ayers (1979) defined a one-dimensional flow strip with elements defined in the quick response zone. The model included some subsurface flow and represented a simple approach to the variable source area model described by Hewlett and Troendle (1975). The basic model was an overland flow model with infiltration functions dominant. In contrast, Beven and Kirkby (1979) developed a forecasting model based on empirically derived response functions representing the flow-generating processes. Although the hydrographs simulated are basically overland storm flow or channel precipitation, such models have tried to use basin morphology to describe source areas, and have attempted to simulate soil water depletion. These are among the many efforts to model portions of the variable source area concept.

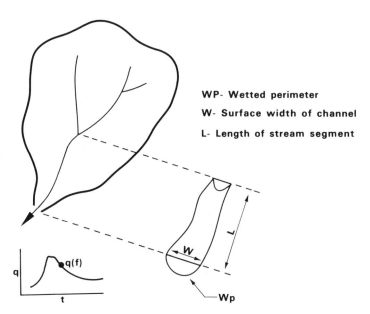

WP- Wetted perimeter

W- Surface width of channel

L- Length of stream segment

Figure 8.12. The conceptual basis for the Variable Source Area Model for surface/subsurface water routing. (Hewlett and Troendle 1975.)

The Variable Source Area Conceptual Model

Any point on a hydrograph of flow represents instantaneous streamflow [$q(t)$], (Figure 8.12). We can logically divide $q(t)$ into only two component rates: surface and subsurface storm flow. Surface storm flow is the rate of delivery of rain water or snowmelt that has not infiltrated the soil surface or channel bottom at any point along its path to the gaging station. Surface storm flow may be further divided into channel precipitation (essentially the rate of rain or snowmelt directly onto an expanding or shrinking channel network), and overland flow from virtually impervious areas such as cultivated fields, roads, campgrounds, rock outcrops, and compacted pastures which deliver *directly* to the expanding channel. Since forest and wildland drainage basins are rarely

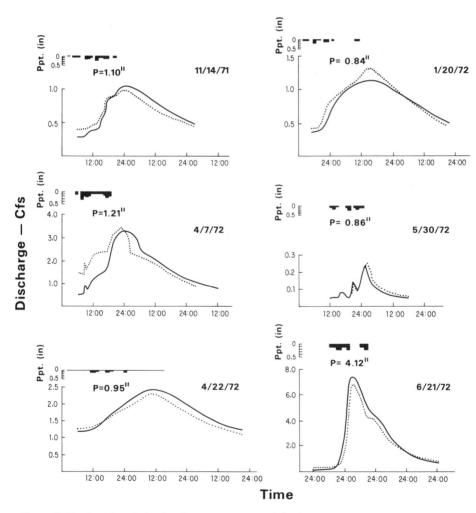

Figure 8.13. Actual and simulated responses to precipitation events on Fernow WS 5. (From Troendle 1979.)

dominated by impervious surfaces, it is justifiable to tentatively assign a constant value
to this area (A_3) in a wildland runoff simulator (Hewlett and Troendle 1975):

$$q(t) = [A_1(t) \cdot K^{dh}/dx] + [A_2(t) \cdot P(t)] + [A_3 \cdot P(t)] \tag{1}$$

The variables and notations in the above equation are defined with reference to the
channel components shown on Figure 8.12, where $A_1(t)$ is the saturated area of the
basin along the perennial, intermittent, and ephemeral channels, including radiating
seepage areas. Subsurface water exfiltrates through A_1 to the channel. The area A_1
varies with time, and can be estimated by multiplying the wetted perimeter by channel length.

$A_2(t)$ is the horizontally projected area of the saturated surface of the basin upon
which rain or snowmelt occurs. A_2 is time-dependent, and can be estimated by multiplying the channel width by channel length.

A_3 represents the virtually impervious areas (pavement, rock outcrops, bare soil
patches), but only those that contribute Hortonian overland flow directly to perennial
or expanding channels. Because forested areas seldom exhibit Hortonian overland flow,
these areas are treated as constant sources in this model. Surface routing models could
be appended if necessary.

$P(t)$ is the rainfall or snowmelt as a function of time, assumed to be weighted across
the basin's area.

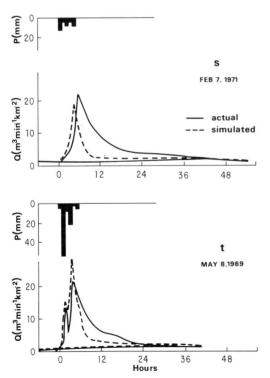

Figure 8.14. Actual and simulated responses, Whitehall Watershed, Georgia. (From Bernier 1982.)

K is the saturated hydraulic conductivity through the area A_1.

H is the hydraulic head, the algebraic sum of the pressure or matric head and gravity head (dH/dx is the difference in hydraulic head operating across a distance x).

Equation 1 is the simplest statement of the variable source area model. The first term comprises subsurface flow, assumed in the model to emerge from the soil over an expanding or shrinking area A_1. The estimation of K and dH at the soil–water interface is a key task in this or any other hydrograph model. The next two terms comprise surface storm flow. The term containing A_2 represents what is normally called channel precipitation. As the area A_2 expands or shrinks, the rain or snowmelt onto A_2 becomes a part of storm flow in a very short time. The product of A_2 and rainfall intensity across a small increment in time constitutes the rate of surface storm flow whenever A_3 is negligible, as it most often is in humid forest land.

The domain of the model is intended to be the first- to second-order forest or wildland catchment. It should include any vegetated catchment outside desert areas, with perennial flow. Within those size limits, it is assumed that channel routing procedures are reduced to a simple lag function based on records of average channel velocities in the region. Small-channel velocities are usually of the order of 0.3 to 0.6 ms^{-1}. Routing within small channels is not a problem until the basin is large enough to permit channel storage and lag to dominate the outflow hydrograph.

A first-generation computer program was developed by Troendle (1979) and tested on the Fernow Experimental Forest in West Virginia. Simulations of individual storm hydrographs from the 95-acre forested watershed were quite good (Figure 8.13). The moderately shallow soils and uniform slopes aided in the simulations. Lefkoff (1981) and Berneri (1982) had difficulty applying the simulator to the Whitehall watershed in Georgia. Bernier (1982) solved a number of computational problems in the earlier version, which improved the stability and conversion of the flow equations under the less homogeneous conditions present at Whitehall.

The second-generation model developed by Bernier appeared to work reasonably well, as evidenced by Figure 8.14. It should be noted that the storm hydrographs were not "fitted," or calibrated, in either the West Virginia or the Georgia application.

Both Bernier (1982) and Troendle (1985) noted that modeling the variable source area is far from complete. However, a simulator is available that addresses the flow dynamics described in this chapter.

9. Research on Interception Losses and Soil Moisture Relationships

J.D. Helvey and J.H. Patric

Few scientists understand better than H. L. Penman (1963) the complexity of finding out what happens to the rain:

"With few exceptions, they [i.e., field measurements] involve small differences between large quantities (of water) that cannot be measured with precision – rainfall, interception, surface runoff, streamflow, soil water content, stored water, deep percolation."

This list, of course, contains several components of the water balance equation. Earlier, Thornthwaite (1948) had hoped that methods to be perfected would help to quantify the then unmeasurable evaporative components of the water balance. When we – the authors – arrived at Coweeta almost a quarter of a century ago, we too became interested in learning more precisely what happens to the rain. Specifically, we wanted to find out how much rain reached the forest floor, and what became of the rain that infiltrated the forest floor and then soaked into the underlying mineral soil. Methods to quantify these components of the water balance were improving, but to this day are insufficiently perfected to permit close measurement.

Interception

By the mere fact of its physical presence, the forest will prevent a portion of the rain from replenishing soil moisture. Rain descending from clouds to forested land first encounters the tree canopy. Some of that rain wets the tree leaves and stems and is retained on them, but most of it drips from intercepting surfaces to reach the forest floor as throughfall. As rain continues, some of the water intercepted by stems will drain along them, ultimately running down the bole to reach the forest floor as stemflow. But the dead leaves, stems, and other litter on the forest floor must be wetted too, before throughfall and stemflow can soak into mineral soil. Rain intercepted by living leaves, stems, and the forest floor evaporates from those surfaces during and after the storm. These evaporative diversions are the forest's toll on descending rain, accounted for in the water balance as interception losses. Hence, soil moisture replenishment by precipitation necessarily is reduced by the total of the interception losses.

The preceding relations among components of the hydrologic balance were well known in 1960, even though the amounts of water involved could not be stated accurately. Coweeta's task of quantification had begun with studies of litter production (Kovner 1955) and of throughfall and stemflow (Black 1957). Then and now, methods to quantify those parameters of interception loss changed little since the beginnings of research in forest hydrology. Hoppe (1896) estimated canopy interception by comparing the catch in raingages beneath the tree canopy (throughfall) with the catch by other gages in forest openings (gross rainfall). This basic approach is subject to occasional variations in kind, number, and distribution of gages. Stemflow is estimated by diversion of water draining down sample trees into collectors. Volumes of measured stemflow are converted to depth units and compared to gross rainfall amounts.

Weights of forest floor samples, collected at field capacity and at selected times during subsequent drying, establish rates of water loss under varying climatic conditions. Likewise, samples collected as the litter is wetted by measured amounts of rainfall are used to define wetting curves. A bookkeeping procedure using rainfall records in conjunction with total litter weights, and wetting and drying curves, permits an estimate of water evaporation from the litter layer during a given time period.

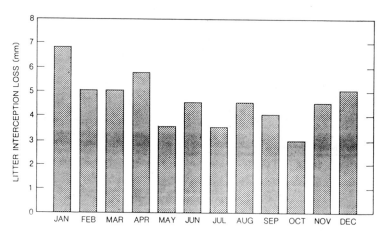

Figure 9.1. Litter interception losses by month. Note that monthly losses are greater during the winter months. (Adapted from Helvey 1964.)

Table 9.1. Equations[a] for Computing Seasonal Interception Losses in the Eastern Hardwood Region

Seasonal Value	Growing Season	Dormant Season
Throughfall	$\Sigma\,T = 0.901\,(\Sigma\,p) - 0.031(n)^{b}$	$\Sigma\,T = 0.914\,(\Sigma\,p) - 0.015(n)$
Stemflow	$\Sigma\,S = 0.041\,(\Sigma\,p) - 0.005(n)$	$\Sigma\,S = 0.062\,(\Sigma\,p) - 0.005(n)$
Litter interception loss	$\Sigma\,L = 0.025\,(\Sigma\,p)$	$\Sigma\,L = 0.035\,(\Sigma\,p)$
Net rainfall	$\Sigma\,R = 0.917\,(\Sigma\,p) - 0.036(n)$	$\Sigma\,R = 0.941\,(\Sigma\,p) - 0.020(n)$
Interception loss	$\Sigma\,I = 0.083\,(\Sigma\,p) + 0.036(n)$	$\Sigma\,I = 0.059\,(\Sigma\,p) + 0.020(n)$

Source: Helvey JD and Patric JH. Water Resour. Res. 1:193–206, 1965. Copyright by the American Geophysical Union.

[a] These are previously derived storm equations modified for computing total seasonal values.
[b] Number of storms per season.

Helvey (1964) used this method to estimate interception loss from the forest floor of a mixed hardwood stand. Annual losses amounted to 2 to 4% of annual precipitation. Monthly losses (Figure 9.1) were greater during the dormant season, when maximum litter accumulation was exposed to climatic conditions which favor rapid drying. This work also showed the importance of the dynamics of litter accumulation through the year. Just after leaf fall there were approximately 10 metric tons/ha of litter on the forest floor. By late summer, about half of this amount had decomposed. Since total litter amount is one variable in the computation of interception loss, accumulation and decomposition rates must be known.

Moore and Swank (1975) developed a mathematical model to predict water content and evaporative losses from the forest floor of a deciduous forest. Model variables were solar radiation, rainfall, mean daytime temperature, mean daytime relative humidity, and litter accumulation parameters. Simulated rates of evaporation for selected time periods agreed with Helvey's (1964) values within 13%.

An exhaustive review of the literature (Helvey and Patric 1965) established consistent results among 33 studies of throughfall, 11 of stemflow, and 10 of litter interception — all from the eastern hardwood region. Regression equations derived from all of these data describe relations among gross rainfall, throughfall, stemflow, and litter interception for eastern hardwood forests, for both growing and dormant seasons (Table 9.1). When these equations are solved for Coweeta's average annual precipitation and number of storms, an annual interception loss of 25 cm (13% of total precipitation) is indicated. Further analysis of data from that review (Helvey and Patric 1966) defined the variability of interception parameters and provided sampling designs for obtaining estimates to selected levels of probability for each parameter mean. Many of those instrumental and design techniques were applied in subsequent studies of interception by white pine (Helvey 1967) and loblolly pine (Swank et al. 1972), species not occurring naturally at Coweeta, but of considerable importance in surrounding areas. The white pine study was important because it provided insight as to what would happen to water yields as the white pine plantations on WS 1 and 17 developed. Our empirical studies of interception losses by hardwoods and white pine led us to predict

Table 9.2. Comparison of Computed Annual Interception Losses from Eastern White Pine and Mixed Hardwoods Under Average Climatic Conditions Prevailing at the Coweeta Hydrologic Laboratory

		Annual Interception Loss	
Species	Age	cm	Percent of Precipitation
Mixed hardwoods	Mature	25.4	12
White pine	10	30.5	15
White pine	35	38.1	19
White pine	60	53.3	26

Source: Helvey JD. Water Resour. Res. 3:723–729, 1967. Copyright by the American Geophysical Union.

that streamflow would decrease, perhaps as much as 28 cm/year, when the pines reached age 60 years (Table 9.2).

Soil Moisture

As early as 1960, it had become part of the Coweeta credo that overland flow was rare on well-forested land, and that rain almost always soaked into the forest floor as fast as it fell. "As anyone who has visited a well-vegetated watershed during a rainstorm knows, adequate infiltration and subsurface flows predominate during most storm events" (Hewlett and Hibbert 1966). This hydrologic behavior virtually eliminated overland flow as one possibility of what happens to the rain. Because unintercepted rain was expected to infiltrate completely, accurate accounting required following rain into and through the soil.

To that end, over 14,000 soil moisture samples had been taken at Coweeta between 1953 and 1959, a few based on neutron probe readings but most by gravimetric sampling. Plotting all of these results (Helvey and Hewlett 1962) showed that average monthly water content in the top 2.1 m of soil closely followed a sine function, with maxima and minima at the beginning of spring and autumn, respectively. The monthly range during the average year is from 48 cm in September to 66 cm in March. Also, this average annual cycle of soil moisture correlated almost perfectly with the long-term average monthly streamflow (Figure 9.2). The consistency of these relationships, however, was attainable only because variation among individual observations of soil moisture was minimized by averaging the enormous amounts of available data.

Hewlett and Douglass (1961) calculated that about 20 paired samples of bulk density and soil moisture were needed to quantify soil moisture at Coweeta within an accuracy of ± 3.0 mm of water per 30 cm of soil. This variability is inherent in the gravimetric method, wherein the investigator literally bores a hole into the soil profile. In the process, the network of roots and pores which permeates the forest soil is locally disrupted. Obviously, one cannot resample in the same hole and one should not attempt to sample adjacent to that hole, where atypical conditions of soil and roots are certain to prevail. Preferring typical conditions, the investigator moves to a nearby sampling

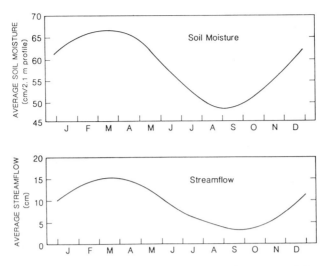

Figure 9.2. Soil moisture compared to total monthly streamflow averaged from the period of record 1953–1959. Such a close relationship between soil moisture and streamflow is rarely evident from a single year's record. (Adapted from Helvey JD and Hewlett JD. J. Forestry 60: 485–486, 1962. By permission.)

site, hopefully similar to the original, but where inescapable differences in soil and rooting inevitably produce the variation Hewlett and Douglass described.

The neutron meter provided a then-new method which overcame most such limitations of gravimetric sampling, allowing estimation of changing soil water content over space and time with great precision (Hewlett et al. 1964). Techniques were developed to determine the slope of neutron meter calibration curves (Douglass 1962) and to calibrate neutron probes (Douglass 1966). A method also was developed for installing neutron probe access tubes to the entire rooting depth of Coweeta soils, which sometimes exceeds 6 m. Use of this technique provided a means whereby soil moisture could be sampled nondestructively, permitting inexpensive, repeated, and, above all, accurate determination of changing soil moisture content at a given site, throughout the great rooting depth of forest trees.

The first large-scale application of the neutron technique was on an array of plastic-covered plots along a ridge above Barker's Cove (Patric et al. 1965). These plots were 15 m square; five were forested, one had no trees, and each contained eight access tubes (the deepest to 7 m), and eight tensiometers. Soil moisture was read weekly at 30 cm intervals, and tensiometers placed at 0.30, 0.76, 1.52, and 3.0 m deep were read weekly until they failed because soil moisture tension exceeded the limitations of the instrument. Record collection covered a 2-year period.

With soil moisture recharge prevented, water use by the trees started first in surface layers of the covered soils when trees leafed out in late April. Peak moisture withdrawal occurred during the longest summer days of June, but use from the deepest layers (>5m) did not occur until August (Figure 9.3). Total moisture extraction was 43.2 cm of water from the sampled profile, but matric potential never exceeded two bars, even at the 30-cm level.

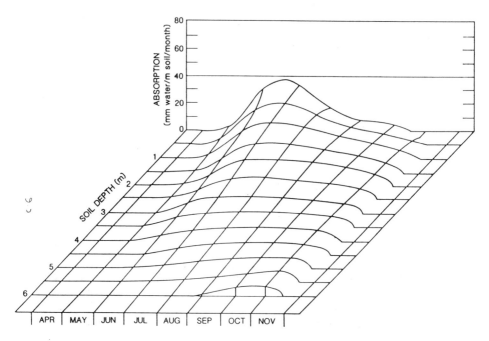

Figure 9.3. The patterns of soil water absorption from plastic-covered plots. (Adapted from Soil Science of American Proceedings, Volume 29, 1965, pages 303–308, by permission of Soil Science Society of America.)

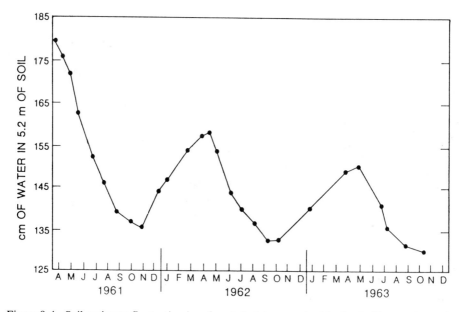

Figure 9.4. Soil moisture fluctuation in a forested plot covered with plastic film continuously for 3 years. (Adapted from Soil Science of America Proceedings, Volume 29, 1965, pages 303–308 by permission of Soil Science Society of America.)

Figure 9.4 illustrates seasonal moisture fluctuations within a plot covered continuously for 3 years. The water content decreased each summer and increased each winter, presumably by recharge from below. Even after 3 years, this ridgetop plot still contained 63.5 cm more moisture than the estimated value below which permanent wilting occurs. Midseason defoliation of trees on two of the plots immediately halted measurable absorption of water (Patric and Douglass 1965).

These experiments taught us a great deal about the source and rates of soil moisture use by forest trees. One inescapable conclusion of soil moisture studies up to that time was that hardwood forests, at least in the deep-soiled Coweeta Basin, may always have a supply of water adequate for potential transpiration rates.

Soil–plant–water relations under natural conditions (i.e., without plastic covering) were observed within the Coweeta Basin and at similarly forested sites within a 32-km radius of Coweeta (Helvey et al. 1972). Cove, midslope, and ridge positions were instrumented with access tubes permitting observation through 2.1 m of soil. There were 48 sampling sites, from which 60 sets of observations were obtained during all seasons between 1963 and 1967.

This study demonstrated the extreme variability in moisture content between sampling sites on a given date. We conventionally think of cove sites as wet and upper slopes as drier, but this generalization did not hold in the study area because there was no consistent relationship between soil moisture content and slope position. Average moisture contents for the period of record were 74.4, 64.8, and 75.7 cm of water per 2.1 m soil profile at the ridge, midslope, and cove sites, respectively. Thus, ridge sites were only slightly drier than coves. In fact, the wettest site in the study was a ridge top where the soil was high in clay content.

Models developed from this data set accounted for about 88% of the variations in soil moisture within the study area. Independent variables were weighted precipitation and easily measured topographic, seasonal, and soil physical factors. The models can be used to predict moisture content of watersheds. As reported by Helvey and Hewlett (1962), average soil moisture follows a sine function with a maximum in spring and a minimum in autumn. Amplitude of the sine function was greatest in the surface layers and least for the deepest layers. Also, it was smaller for cove sites than for corresponding soil depths on ridge or midslope positions (Figure 9.5).

Significance of These Studies

Interception

Occasionally it is prudent to pause in research, take stock of what has been learned, determine what is significant, and then to proceed with what is important and remains to be learned. In retrospect, it appears that such a time had come in the mid 1960s for studies of rainfall interception in eastern hardwoods. Our reviews provided equations as well as instrumental and design criteria that have seldom been overlooked in subsequent studies, in this country and abroad. All of these empirical studies, in conjunction with close examination of the physical processes involved, lead to the conclusion that we can indeed account for what happens to rain intercepted by forests. This knowledge

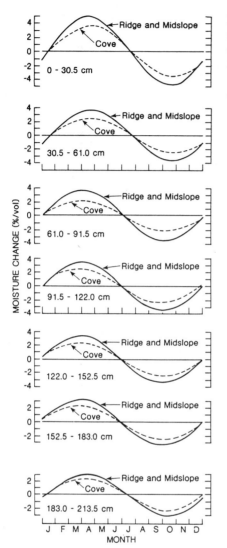

Figure 9.5. Computed soil moisture trends by slope position, assuming average rainfall distribution during the year. (Reproduced from Soil Science of America Proceedings, Volume 36, 1972, pages 954–959 by permission of Soil Science Society of America.)

may have led to quiet abandonment of the tired old canard that regional floods can be reduced by managing forests to maximize interception losses.

When our interception review was published, the primary interest in interception among hydrologists was to quantify the interception term of the water balance equation. More recently, those results have been useful in nutrient cycling studies where volumes of throughfall and stemflow must be known, in addition to nutrient concentrations, in order to define nutrient budgets on watersheds accurately. Thus, our results were quoted by several authors of the symposium on Mineral Cycling in Southeastern Ecosystems (Howell et al. 1975).

Soil Moisture

The unsaturated zone of a watershed has been called the no man's land of hydrology, and evapotranspiration a term of ignorance. Both concern a flux of soil water, the former its downward movement through and beneath the rooting depth of plants; the latter its uptake by plant roots and eventual loss by transpiration to the atmosphere. We have learned considerable information about the amount and timing of both fluxes, of soil moisture loss through trees, and of that same water escaping use by trees to ultimately become streamflow. In fact, fuller understanding of both fluxes were important in the formulation of the variable source area concept of water delivery to streams. Studies of soil moisture played a key role in the development of that concept, and suggested that the combination of deep soils and frequent rainfall at Coweeta may permit vegetation to transpire at near-potential rates at all times. This conclusion is important for the interpretation of water yield results and their applicability to other areas that receive less rainfall.

For example, Dunford and Fletcher (1947) tested the hypothesis that streamside trees and shrubs are unusually heavy water users because their roots penetrate the water table. However, they clearcut the riparian zone of WS 6 and found that although diurnal fluctuations in streamflow decreased, streamflow increase was no larger than might be expected from cutting an equal area elsewhere on the experimental watershed. Thus, the riparian effect, so important along channels in arid lands, appears to be lacking under Coweeta's perhumid climate. We believe that the riparian effect is of negligible consequence in areas where soil moisture remains readily available on all parts of the watershed throughout the year.

Other processes such as nutrient cycling, erosion rates, and rates of revegetation after disturbance must be considered in light of the soil moisture levels at the study site. Functional relationships existing at Coweeta may not apply to other areas where soil moisture relationships differ significantly.

3. Forest Dynamics and Nutrient Cycling

10.   Forest Communities and Patterns

F.P. Day, Jr., D.L. Phillips, and C.D. Monk

The vegetation of the Coweeta Basin is traditionally included in the oak–chestnut association (Braun 1972). However, since chestnut (*Castanea dentata*) has been lost as a dominant due to the chestnut blight, the area is probably more correctly classified as belonging to the oak–hickory association. The plant communities in the basin are typically diverse for the southern Appalachians and are distributed in a reasonably predictable mosaic over the highly varied topography in relation to moisture gradients (Day and Monk 1974). The composition and structure of many of these communities are apparently still changing and are dynamic. The predominant species composition is a mix of deciduous oaks with a commonly abundant evergreen undergrowth of *Rhododendron maximum* and mountain laurel (*Kalmia latifolia*). (Species authority throughout this paper follows Radford et al. (1964).)

Community composition has been affected by human disturbances (Chapter 2). The Coweeta Basin was occupied by the Cherokee Indians prior to 1842. Between 1842 and 1900 the main disturbances were light semiannual burning and grazing. Between 1900 and 1923 logging operations occurred over the entire basin, but cutting was heaviest on the lower slopes, valleys, and accessible coves. Since 1924, no major human disturbance such as burning, grazing, or logging has occurred within the basin except for restricted U.S. Forest Service experimental studies.

General Distribution Patterns of Plant Communities

Permanent quadrats were used to study the broad-scale vegetation patterns and changes over time in forest communities at Coweeta. During 1934 and 1935, 997 0.081 ha (0.2

acre) quadrats were established along 13 parallel transects spanning the Coweeta Basin.
Trees >1.3 cm (0.5 inches) DBH were tallied by species in 2.5 cm (1 inch) DBH
classes. During 1969 to 1973, the 403 permanent quadrats which occurred on control
watersheds, undisturbed since 1934, were reinventoried in the same way (W. T. Swank,
1969, unpublished). Percent slope, aspect, elevation, and slope position (ridge, upper
slope, middle slope, lower slope, and cove) were also recorded for each plot.

The permanent quadrats were plotted on axes of elevation and topographic position.
Using this direct gradient analysis approach, variations in vegetation composition
associated with changes in the environmental axes could be determined. Figure 10.1
outlines the position of four major community types on the landscape. The topographic
position axis runs from mesic coves to slopes to dry ridges. Within each of these, slope
aspects are arranged in order of most mesic (NE) to most xeric (SW), so the axis corresponds roughly to a topographic moisture gradient. The major community types
recognized are:

1. Northern Hardwoods: This forest type occurs at higher elevations, mostly above
   4000 ft on slopes and in coves. It is dominated by a variety of species including
   yellow birch (*Betula lutea*), basswood (*Tilia heterophylla*), buckeye (*Aesculus octan-*

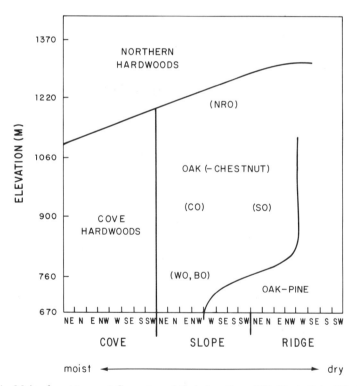

Figure 10.1. Major forest types at Coweeta and their locations. BO, Black Oak; CO, Chestnut
Oak; NRO, Northern Red Oak; SO, Scarlet Oak; WO, White Oak.

dra), northern red oak (*Quercus rubra*), yellow poplar (*Liriodendron tulipifera*), and black cherry (*Prunus serotina*).

2. Cove Hardwoods: Occurring in mesic coves and blending into the Northern Hardwoods at higher elevations is the heterogeneous Cove Hardwoods type. Yellow poplar and hemlock (*Tsuga canadensis*) are among the many dominants, which also include red maple (*Acer rubrum*), northern red oak, hickory (*Carya* spp.), black birch (*Betula lenta*), and formerly chestnut.

3. Oak (–Chestnut): Widely distributed over the slopes of the Coweeta Basin are forests dominated by oak, and formerly by chestnut before the chestnut blight. Chestnut oak (*Quercus prinus*) is the most widespread and important oak species in the Coweeta Basin. Its importance is perhaps greatest at middle elevations on slopes with mesic aspects. Scarlet oak (*Q. coccinea*) increases in importance on drier slopes and ridges. At higher elevations above 1070 m, northern red oak increases in importance, while at lower elevations below 820 m, white oak (*Q. alba*) and black oak (*Q. velutina*) are important species. Besides oaks, red maple and hickories are also significant components.

4. Oak–Pine: This community type is found on the ridges and drier slopes at low elevations. Pitch pine (*Pinus rigida*) and scarlet oak predominate here, along with chestnut oak.

Community Change

Prior to the introduction of the chestnut blight fungus [*Endothia parasitica* (Murr.) P.], the American chestnut was the dominant tree species over much of the Coweeta Basin. At the time of the first survey of the permanent quadrats in 1934 and 1935, chestnut was still present but dying from the blight. As shown in Table 10.1, the relative basal area of chestnut in 1934 to 1935 in the various forest types ranged from 46% on mesic slopes in the Oak (–Chestnut) type to 16% on the drier slopes and ridges of the Oak–Pine type.

In order to determine the replacement patterns for chestnut in these various forest types, basal areas and importance values for each species were compared at the time of the blight (1934 to 1935) and 35 years later (1969 to 1973). Table 10.1 shows the major changes in relative basal area over the period. Changes in importance value were similar and are not shown.

In most cases, the species which exhibited the greatest increases in relative basal area were those which were already codominants at the time of the blight, with at least 5% basal area. These included various species of oaks in the Oak (–Chestnut) type, scarlet oak in the Oak–Pine type, northern red oak in the Northern Hardwoods type, and yellow poplar and hemlock in the Cove Hardwoods (Table 10.1).

The death of chestnut by blight was a gradual process. First a few small limbs succumbed, then larger limbs, and eventually the whole tree in a process that took 2 to 10 years (Woods and Shanks 1959). Gaps in forest canopies may be filled in by (a) expansion of the crowns of adjacent codominant trees, (b) growth of advanced reproduction seedlings, and (c) new establishment of seedlings by germination. The first process may explain in part why the relative basal area of codominant species increased so much. Other studies have demonstrated increases in tree ring widths following the

Table 10.1. Changes in Relative Dominance in Response to Chestnut Blight from 1934–35 to 1969–73

Northern Hardwoods		Cove Hardwoods		Oak-Chestnut NE,N,E,NW,W		Oak-Chestnut SE,S,SW ridges		Oak-Chestnut <2700 ft		Oak-Pine	
Species	Change in % Basal Area	Species	Change in % Basal Area	Species	Change in % Basal Area	Species	Change in % Basal Area	Species	Change in % Basal Area	Species	Change in % Basal Area
CN	−36.7	CN	−37.6	CN	−46.1	CN	−33.9	CN	−34.5	CN	−15.7
YB	+9.9	(YP)	+9.8	(CO)	+13.4	(CO)	+9.2	(SO)	+8.2	(SO)	+5.7
(NRO)	+9.3	(HE)	+5.5	RM	+7.4	RM	+4.0	CO	+5.7	RM	+3.5
YP	+7.7	BB	+4.3	NRO	+5.3	YP	+3.9	(WO)	+4.9	SW	+3.2
CH	+5.7			SW	+3.0			(BO)	+4.8		
								RM	+3.3		
								SW	+3.3		

Only species with >3% increase in relative basal area are included. BB, black birch; BO, black oak; CO, chestnut oak; HE, hemlock; NRO, northern red oak; RM, red maple; SO, scarlet oak; SW, sourwood; WO, white oak; YB, yellow birch; YP, yellow poplar. Circled species were codominants in 1934–35, comprising >5% of the basal area.

death of an adjacent chestnut tree (Nelson 1955; Woods and Shanks 1959). However, almost every species which showed a significant increase in basal area also showed a significant increase in density. Therefore, seed germination and/or release of advance regeneration seedlings must also have contributed to this rise. This is not unexpected, because codominant species would have an advantage due to the already established advance regeneration seedlings and an available seed source.

In addition to the increased basal area and density of already codominant species, several other species increased significantly in importance. Yellow birch, yellow poplar, and red maple were among the most notable in the different forest types. All of these species are characterized by copious production of relatively small wind-borne seeds and rather rapid growth in openings. Thus, while the slow opening of gaps by dying chestnuts primarily allowed an increase in codominant species, more opportunistic species were also able to take advantage of the gaps.

Age structures for nine tree species (Spring 1973; Iglich 1975; Monk and Day 1984) collectively show periods of increased recruitment about 80 years ago and again 40 to 60 years ago. The first recruitment period coincides with early logging, while the second recruitment period begins in the logging period prior to chestnut blight introduction. Since the second period of recruitment includes the effects of both logging and chestnut blight damage, the relative importance of the two disturbances cannot be separated.

The data from 1934 to 1935 and 1969 to 1973 surveys of the permanent quadrats were also used to assess changes in tree regeneration over the time period and to evaluate the impact of *Rhododendron maximum* on these changes (Phillips and Murdy 1984). Plots in two subgroups of the Oak (–Chestnut) type on slopes were selected for the study: those between 884 to 975 m dominated by chestnut oak, and mixed oak stands between 701 to 792 m where white oak, black oak, and scarlet oak predominated. These in turn were subdivided into high density rhododendron (HR; at least 15% rhododendron basal area in 1969 to 1973) and low density rhododendron (LR; less than 2% rhododendron basal area in 1969 to 1973). Rhododendron was not tallied in the 1934 to 1935 survey. Density–diameter distributions for the five dominant tree species were determined from the 1934 to 1935 and 1969 to 1973 inventories of HR and LR plots.

Oak and red maple regeneration, which was abundant in 1934 to 1935 due to past disturbance, decreased by 1969 to 1973 as the canopy closed. Total tree reproduction was lower in HR plots than in LR plots and the magnitude of the difference increased with time (Table 10.2). There were no significant differences between HR and LR plots in species composition, canopy tree basal area, understory basal area, or basal area of chestnut killed by the blight. This suggests that the differential decrease in tree reproduction over time was due to an increase in the density and basal area of rhododendron. This in turn may have been fostered by a combination of logging, chestnut blight, and cessation of burning.

At the time of the later survey, chestnut oak and white oak reproduction was depressed in HR plots relative to LR plots, whereas red maple was only slightly affected. Scarlet oak and black oak regeneration was poor at all sites, regardless of rhododendron density.

Rhododendron and mountain laurel represent the two most important evergreen components in these "deciduous forests," though lesser amounts of hemlock, pitch pine,

Table 10.2. Total Density (stems/ha) of Tree Saplings in the 1.3–8.9 cm DBH Class

	LR	HR	Percent Difference	p
Mixed Oak type				
1934–1935	2324	2257	3	NS
1969–1972	1566	793	49	0.0024
% decrease	33	65		
p	0.0108	0.0009		
Chestnut Oak type				
1934–1935	2657	2343	12	NS
1969–1972	1198	660	45	0.0001
% decrease	55	72		
p	0.0005	0.0001		

Shrubs and small tree species with no individuals >11.4 cm DBH were omitted. LR, low rhododendron; HR, high rhododendron; NS, not significant ($p > 0.05$).

American holly (*Ilex opaca*) and dog-hobble (*Leucothoe axillaris*) may be present. These evergreen species (mostly rhododendron and mountain laurel) contribute between 20 to 35% of the total standing crop of leaf biomass (Monk and Day 1984). With one-fourth to one-third of the leaf biomass present as evergreen leaves, it becomes evident that some aspects of mineral cycling will be modified by their presence. Some of the rhododendron leaves are held for as long as 7 years. Thus, these two clonal evergreen shrub species may influence forest regeneration and the rate of mineral flow within the forest (Day and McGinty 1975).

Table 10.3. Composition of Woody Vegetation on WS 18

Species	Basal Area m²/ha	Relative Basal Area (%)	Density No. Stems/ha	Relative Density (%)
Quercus prinus	5.5	21.3	190.8	6.3
Acer rubrum	2.4	9.3	181.8	6.0
Quercus coccinea	2.0	7.9	44.5	1.5
Rhododendron maximum	1.9	7.4	887.0	29.2
Quercus rubra	1.7	6.8	21.4	0.7
Liriodendron tulipifera	1.6	6.4	53.7	1.8
Carya glabra	1.3	5.1	70.4	2.3
Kalmia latifolia	1.3	5.1	890.9	29.3
Quercus velutina	1.2	4.8	30.1	1.0
Oxydendrum arboreum	1.1	4.4	75.5	2.5
Nyssa sylvatica	1.0	3.7	70.0	2.3
Cornus florida	0.8	3.2	182.7	6.0
Betula lenta	0.7	2.7	62.1	2.0
Tsuga canadensis	0.4	1.4	41.0	1.3
Hamamelis virginiana	0.2	0.7	71.4	2.3
Others (27 spp.)	2.5	10.0	171.0	5.2
Totals	25.6	100.0	3044.3	100.0

Species arranged in order of contribution to percent basal area. Only species with ≥ 0.1 m²/ha basal area and ≥ 20 stems/ha are listed.

Species Distributions

An intensive study of the vegetation on an undisturbed watershed (WS 18) was conducted in 1970 by sampling 25 plots, each 25 × 50 m. The overstory on this watershed
was dominated by oaks (42.7% of total basal area) of which chestnut oak was the most
prominent, followed in importance measured by relative basal area by scarlet oak, red
oak, and black oak (Table 10.3). The hickories occurred as codominants in the overstory with pignut hickory (*Carya glabra*), mockernut hickory (*C. tomentosa*), and red
hickory (*C. ovalis*) comprising 5.1%, 2.2%, and 1.3% respectively of the total basal
area on the watershed. Red maple was also an important codominant in the overstory
(9.3% of total basal area). The oaks, red maple, and hickories constituted only 9.6%,
6%, and 3.2% respectively of the total number of stems >2.5 cm dbh. The most important understory species were mountain laurel and rhododendron (58.5% of the stems
>2.5 cm dbh), followed in relative basal area by dogwood (*Cornus florida*) and witch
hazel (*Hamamelis virginiana*). The herb layer was relatively sparse, with ferns compos-
ing much of the ground vegetation.

Distribution maps based on absolute basal areas of the major overstory species reveal
some of the general patterns on the watershed (Figure 10.2). The oaks and pignut hickory were distributed primarily high on the slope away from the stream. Red maple was
distributed over most of the watershed. In two areas, at the base and near the stream
halfway up the NW-facing slope, the composition of the overstory was atypical. Instead

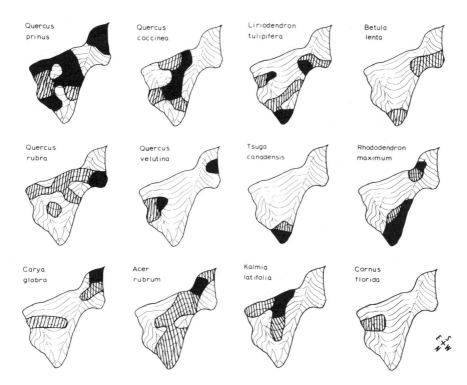

Figure 10.2. Species distribution patterns on WS 18 based on absolute basal area. Solid shading
= >5 m²/ha, slashed = 2–5 m²/ha, remaining area = <2 m²/ha.

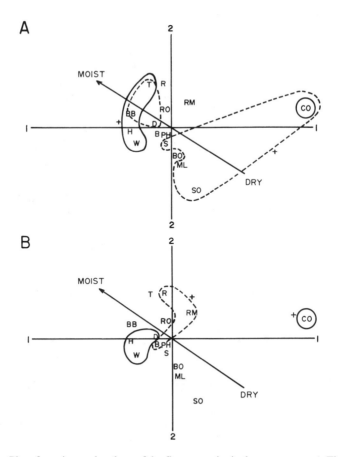

Figure 10.3. Plot of species on the plane of the first two principal components. *A*. The solid lines group species significantly correlated with distance from the water divide; the dashed lines group species correlated with distance from the stream. *B*. The solid lines group species significantly correlated with elevation; the dashed line groups species correlated with aspect. A + indicates that the group is positively correlated. The moisture gradient is arbitrarily sketched in. Species codes: CO, Chestnut Oak; RM, Red Maple; SO, Scarlet Oak; R, Rhododendron; RO, Red Oak; T, Tulip Poplar; PH, Pignut Hickory; ML, Mountain Laurel; BO, Black Oak; S, Sourwood; B, Blackgum; D, Dogwood; BB, Black Birch; H, Hemlock; W, Witch Hazel.

of being composed of oaks or hickories, much of the overstory consisted of tulip poplar, hemlock, and black birch. The patterns of understory species are striking (Figure 10.2) and were readily apparent in the field. Rhododendron was distributed primarily on the lower NE-facing slopes and areas near the stream, mountain laurel occurred near the ridges, and dogwood was concentrated mostly in an area near the stream. The understory species had very distinct distributions, and were virtually exclusive of each other in a given area.

In a principal components ordination of species based on basal area (Day and Monk 1974; Goldstein and Grigal 1972), the species significantly correlated with distance from the stream and distance from the water divide were grouped (Figure 10.3A).

Those species positively correlated with distance from the divide included tulip poplar, black birch, hemlock, and witch hazel. These species increased in basal area as the distance from the ridge increased, thus occupying moist sites. Overlapping this group was the group of species negatively correlated with distance from the stream; these included tulip poplar, black birch, and dogwood. Thus, those species occupying moist sites based on these two topographic gradients were grouped together in the ordination and were at one extreme of the ordination plot. At the other extreme were the species occupying the drier sites based on these two topographic gradients. The one species negatively correlated with distance from the divide was chestnut oak. Those species positively correlated with distance from the stream were chestnut oak, scarlet oak, mountain laurel, and sourwood. The two groups of species occupying the more moist sites were both located at one extreme of the arbitrary moisture gradient, while the two groups occupying drier sites were at the other extreme. Thus, the spatial arrangement of most species on the watershed as depicted by the ordination seems to align regularly with these two topographic gradients, which probably produce a moisture gradient.

The species grouped in Figure 10.3B were significantly correlated with aspect moisture value and elevation. There were no species negatively correlated with aspect moisture value. Those positively correlated with aspect moisture value, thus occupying more moist sites, were rhododendron, red maple, and blackgum. However, red maple had been observed widely distributed over the watershed, and blackgum was found on upper slopes and ridges. Also, densities of red maple and blackgum were positively correlated with elevation and distance from the stream channel, and blackgum density was negatively correlated with distance from the water divide. Therefore, we are less than satisfied with the manner in which aspect was quantified in the analysis. A relative moisture scale for aspect may be inappropriate for Coweeta; a measurement of solar radiation with respect to slope aspect would probably be a better quantification of this parameter. The only species positively correlated with elevation was chestnut oak, which was at one extreme of the gradient. Dogwood, hemlock, and witch hazel were negatively correlated with elevation, thus occupying the moist base of the watershed. Again, the groups associated with moist sites based on these two topographic gradients were located together at one extreme of the ordination plot and chestnut oak was at the other extreme.

Summary

The four major vegetation types at Coweeta (northern hardwoods, cove hardwoods, oak (–chestnut), and oak–pine) are apparently still undergoing change following major disturbances (logging, fire, chestnut blight). Over a 35 year period following loss of chestnut by blight, replacement was predominantly by species already codominant with chestnut. Chestnut oak should continue to be important, except perhaps in areas of high *Rhododendron* density, but there may be declines in white oak, scarlet oak, and black oak. Red maple will probably remain prominent as it does in many eastern forests following disturbance. The evergreen understory seems to be quite important in inhibition of regeneration of some canopy species. Individual species appear to be distributed along complex moisture gradients.

11. Biomass, Primary Production, and Selected Nutrient Budgets for an Undisturbed Watershed

C.D. Monk and F.P. Day, Jr.

The forests of lower elevations (700 to 1000 m) in the Coweeta Basin are similar to those studied elsewhere in the southern Appalachians (Whittaker 1966; Mowbray and Oosting 1968). Several species of oaks dominate these forests today; however, chestnut was the leading dominant prior to invasion by the chestnut blight. Spatial distribution of species on the slopes seems to follow topographic moisture gradients. A group of mesic species (tulip poplar, yellow birch, hemlock, witch hazel, dogwood, and rhododendron) are positively correlated with distance from the water divide and aspect, and are negatively correlated with distance from the stream channel and elevation. A group of xeric species (chestnut oak, scarlet oak, pignut hickory, mountain laurel, sourwood, red maple, and black gum) are positively correlated with distance from the stream channel and elevation, and are negatively correlated with distance from the water divide (Day and Monk 1974; Monk and Day 1984). A more detailed population and community description is summarized by Day et al. (Chapter 10).

The Coweeta Basin was inhabited by 1842 by white settlers, and from 1842 to 1900 the area experienced light semiannual burning and grazing. Logging operations were centered in the valleys, lower slopes and accessible coves from 1909 to 1923. Since 1924, no major disturbances have occurred except for restricted U. S. Forest Service experimental studies. By 1930, most of the chestnut trees in the basin were infected by chestnut blight. Resurvey of permanent plots established in the early 1930s by the U.S. Forest Service reveal that basal area today has recovered to post logging levels. It is clear from age structures for nine tree species studied by Spring (1973) and Iglich (1975) that there was a period of enhanced recruitment about 1905 and another recruitment period between 1925 to 1945. These peaks coincide with the initial selective logging period and the chestnut blight introduction. The later text supports that these forests have been altered by chestnut blight and past logging.

This paper will summarize data on standing crop biomass, net primary production, and selected nutrient budgets for a southern Appalachian hardwood forest. Table 11.1 gives the major sources for the data summarized in this chapter. Where possible, data given are for WS 18.

Biomass, NPP, and LAI

The standing crop of biomass in the oak forest of WS 18 is within the lower range reported for other temperate deciduous forests (Ovington 1965). While the total and above ground standing crops of biomass for the forest are toward the lower end of estimates for temperate deciduous forests, the estimate for belowground biomass ranges toward higher values. Within the southern Appalachians, aboveground biomass estimates in oak-dominated forests range from 126.0 M.T. ha^{-1} (Sollins 1972) to 420.0 M.T. ha^{-1} (Whittaker 1966). The 139.9 M.T. ha^{-1} (Figure 11.1) for WS 18 includes 5.6 M.T. ha^{-1} of leaf biomass. Root biomass is estimated to be 51.4 M.T. ha^{-1} with 27.2 M.T. ha^{-1} representing roots <2.5 cm diameter. McGinty (1976) reported root biomass estimates in the temperate deciduous forest of eastern North America to range from 28.3 M.T. ha^{-1} (Whittaker et al. 1974) to 51.4 M.T. ha^{-1} for the Coweeta forests.

Aboveground net primary production (NPP) estimates for the Coweeta forests are in the lower end of the range of values found for other temperate deciduous forests of North America. The range of aboveground NPP estimates are 5.4 M.T. ha^{-1} for an oak forest on the Georgia Piedmont (Monk et al. 1970), and 5.7, 14.6, and 24.1 M.T. ha^{-1} for oak, chestnut oak, and tulip poplar dominated forest, respectively, in the Smokies (Whittaker 1966). The 8.4 M.T. ha^{-1} aboveground NPP estimate for Coweeta falls in the lower portion of this range.

Table 11.1. Major Sources for Data Summarized in Chapter 11

Compartment	Data Source
Standing crop biomass	Day 1971, 1974; McGinty 1972, 1976; Day and Monk 1974, 1977a, 1977b
NPP	Day 1974; Day and Monk 1977a, 1977b; McGinty 1972, 1976
Leaf area index	Boring 1982
Litter production	Cromack 1973; Cromack and Monk 1975
Litter layer	Cromack 1973; Yount 1975; Cromack and Monk 1975
Decomposition	Yount 1975; Cromack 1973; Cromack and Monk 1975; McGinty 1976
Soil	Yount 1975; McGinty 1976; Best 1976
Throughfall	Best 1976; Mitchell et al. 1975; Henderson et al. 1978; Swank and Waide 1980
Atmospheric input	Chapter 4
Streamflow	Chapter 4
Canopy consumption	Schowalter et al. 1981c

Appropriate questions to ask are why the aboveground biomass and NPP estimates tend toward the lower portion of the range of values found for other temperate deciduous forests, and why the belowground biomass estimates tend toward the upper portion of the range. Assuming that the values are accurate and representative estimates, the answer probably involves the recent history of the forest. Logging in the early 1900s generally consisted of exploitive high-grading operations in which proper silvicultural regeneration was rarely practiced. Eroded slopes and degraded residue stands dominated by trees with crooked or otherwise poor forms often resulted from these practices. Chestnut blight further affected the southern Appalachian forests by the near elimination of the dominant tree species, although small chestnut sprouts are still common in chestnut oak and oak-pine forest communities due to the long persistence of their root systems. Presumably a forest with this type of disturbance history might maintain a relatively large residue root mass relative to the reduced aboveground biomass. Root grafts and live stumps are well documented (Page 1927; Bormann and Graham 1959), where water, minerals, and food are translocated from one root system to another. Bormann and Likens (1979) estimated that 25% of the stumps were alive 10 years after cutting on a white pine forest.

The forest on WS 18 is aggrading, since its total NPP averages 14.6 M.T. ha^{-1} (Day and Monk 1977a; McGinty 1976), of which 6.0 M.T. ha^{-1} are attributed to roots, 4.4 M.T. ha^{-1} are lost as aboveground litterfall, and 0.2 M.T. ha^{-1} are consumed by canopy arthropods, leaving 4.0 M.T. ha^{-1} as net biomass accumulation. Since these forests have been selectively logged prior to 1930 and subjected to chestnut blight damage, extensive tracts have been left that probably are not good representatives of the original forests (Boring 1982). Surely, the selective logging practices and the loss of chestnut left deformed and poorer quality trees, which in later years yield less growth. Forest growth simulations by Shugart and West (1977) suggest that forests that suffered chestnut blight damage would yield less biomass than ones that did not.

The broadleaved forests of the southern Appalachians are usually considered to be deciduous, when in fact they have an important evergreen component (Day and Monk 1974; Day and McGinty 1975). Rosebay rhododendron (*Rhododendron maximum*), mountain laurel (*Kalmia latifolia*), eastern hemlock (*Tsuga canadensis*), pitch pine (*Pinus rigida*), American holly (*Ilex opaca*), and dog hobble (*Leucothoe axillaris* var. *editorum*) collectively produce 20 to 35% of the total standing crop of leaves (Monk and Day 1984). An estimate of the evergreen component is given by the difference (1.4 M.T. ha^{-1}) in the standing crop of leaf biomass (5.6 M.T. ha^{-1}) and leaf NPP (4.2 M.T. ha^{-1}). The three most important evergreen species (rosebay rhododendron, mountain laurel, and eastern hemlock) contribute 1.9 M.T. ha^{-1} to leaf biomass in WS 18. Perhaps another relative estimate of evergreenness would be total leaf NPP less leaffall (4.2 M.T. ha^{-1} − 2.8 M.T. ha^{-1} = 1.4 M.T. ha^{-1}). When leaf dry weights (Day and Monk 1974) are coupled with leaf surface area/dry weight ratios (Boring 1982), the leaf area index (LAI) is estimated to be 6.2 m^{2}m^{-2}.

Leaf area index (LAI) for temperate forests tends to range between 5 and 8 m^{2}m^{-2}. The 6.2 m^{2}m^{-2} estimate for the Coweeta forest compares favorably with other estimates in the area (Whittaker 1966). The fact that the estimate is in the lower portion of the range for other warm deciduous forests is in part related to the sclerophyll-evergreen character of these forests. Rhododendron and mountain laurel collectively contribute 32.1% to leaf biomass, while their surface area/dry weight ratio (75 cm^2 g^{-1}) is only about half that of the deciduous species (Boring 1982).

Litter

Annual litter production on WS 18 is 4.4 M.T. ha^{-1}, with leaves (64%) being the major component. This is in the lower range of estimates for other warm temperate forests (Bray and Gorham 1964; Duvigneaud and Denaeyer-DeSmet 1970). Duvigneaud and Denaeyer-DeSmet (1970) estimate a Belgian oakwoods had 3.4 M.T. ha^{-1} of leaves, with a leaf litter production of 3.2 M.T. ha^{-1} yr^{-1}. This compares with 5.6 M.T. ha^{-1} of leaf biomass at Coweeta and a leaf litter production of 2.8 M.T. ha^{-1} yr^{-1}. The differences again illustrates the evergreen nature of the Coweeta forests. Nonleaf litter in the two forests is 1.6 M.T. ha^{-1} yr^{-1} at Coweeta versus 2.1 M.T. ha^{-1} yr^{-1} for the Belgian forest. The Belgian oakwoods are young and even-aged (70 to 75 yrs.), while the Coweeta forests are readjusting from selective logging prior to the 1930s and chestnut blight damage in the 1930s. Thus, low litter production may be related to a high percentage of evergreen leaves or an aggrading successional status.

The annual standing crop of litter is estimated to be 8.5 M.T. ha^{-1}. With an annual litter production of 4.4 M.T. ha^{-1} yr^{-1}, the turnover time of the litter layer is approximately 2 years.

The soil organic matter estimate of (145.6 M.T. ha^{-1}) is equivalent to 76% of the total standing crop of biomass. Another 8.5 M.T. ha^{-1} is present in the litter layer, giving a total organic matter estimate of 345.4 M.T. ha^{-1}. This figure is similar to the 287.8 M.T. ha^{-1} for Belgium mixed oak forest by Duvigneaud and Denaeyer-DeSmet (1970). Their estimate includes 121 M.T. ha^{-1} in aboveground biomass (3.5 M.T. ha^{-1} leaves), 35 M.T. ha^{-1} in root biomass, 125 M.T. ha^{-1} for soil organic matter, and 6.8 M.T. ha^{-1} of litter. These two forests have essentially identical percentage breakdowns for their organic matter compartments.

Nutrient Pools and Transfers

Budgets of nutrient atmospheric inputs and streamflow losses given by Swank and Waide (Chapter 4) show either net gains or relatively small losses of most nutrients.

Standing stocks for selected elements are given in Figures 11.1 through 11.3 for several ecosystem components. It is obvious that these elements are not distributed in the same proportions within the watershed. If the system is divided into vegetation (leaves, bark, stems, roots) and soil (including litter), it becomes clear that Ca and K are equally divided between the two. Also, Mg and N are more abundant in the soil, while P is more abundant in the vegetation. The soil data represent extractable, and not total, nutrient content (with the exception of N, whose values are estimates of the total soil N pool). Large pools of these nutrients (except N) will be available for forest growth over time, due to weathering processes within the regolith (See Velbel, Chapter 6).

The litter layer contains only 1.2 to 7.6% of the total standing stock of nutrients. The importance of the litter layer is not measured by the amount of nutrients that it contains,

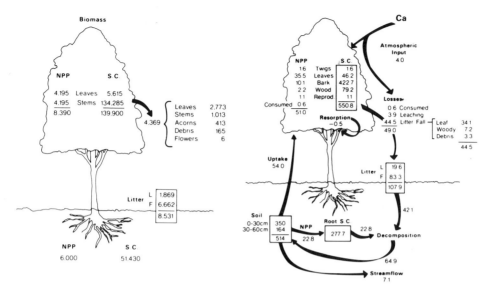

Figure 11.1. Biomass and calcium dynamics for a hardwood forest (WS 18). Data are given as kg ha^{-1}.

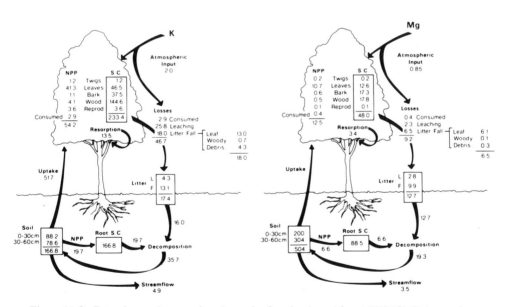

Figure 11.2. Potassium and magnesium dynamics for a hardwood forest (WS 18). Data are given as kg ha^{-1}.

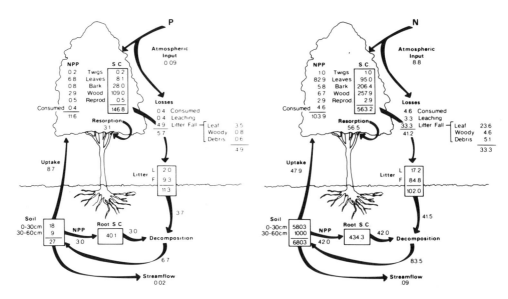

Figure 11.3. Phosphorus and nitrogen dynamics for a hardwood forest (WS 18). Data are given as kg ha⁻¹.

but rather by the nutrient flow through it. Annually, sufficient amounts are added to the litter layer through litterfall to equal annual decomposition. Assuming a stable litterfall rate, a standing crop of 8.5 M.T. ha⁻¹ of litter and an annual production of litter of 4.4 M.T. ha⁻¹ yr⁻¹ gives a turnover time of 1.9 years. The turnover time for the elements are Mg 1.4 yrs, K 1.5 yrs, N 2.2 yrs, Ca 2.6 yrs, and P 2.9 yrs.

The standing stocks of elements in the vegetation represent a significant proportion of the total nutrient pools. Although the distribution within the plant varies with the element, its proportional distribution is not related to the biomass distribution. The distribution of five elements within the vegetation compartment is as follows: Ca, bark > roots > wood > leaves; K, roots > wood > leaves > bark; Mg, roots > wood > bark > leaves; P, wood > roots > bark > leaves; N, roots > wood > bark > leaves. The turnover rates of these plant parts then become an important aspect in the cycling of any particular element. The nutrient concentrations, compartment sizes, and turnover rates are crucial.

Nutrient resorption from leaves prior to abscission has been studied in several forest types. Witherspoon et al. (1962), working with ¹³⁴Cs (an analog to K) in white oaks, estimated that 51% of the leaf Cs returned to woody tissue prior to leaffall while 36% was lost as leaf litter and 13% leached. Wells and Metz (1963) estimated that about one-half of the N, P and K was removed from loblolly pine needles prior to leaffall. Also, ¹³⁷Cs has been placed on the flowers of tulip poplar, and its movement followed into the seed and seedling measured (Witherspoon and Brown 1965). The pathway of ¹³⁷Cs movement seemed to parallel those of sugars and other translocated organic substances.

Thomas (1969) demonstrated that all of the [45]Ca in dogwood leaves was lost in leaffall, leaching, stemflow, or insect damage, and that none was returned to the woody tissue. Henderson and Harris (1975) estimated that about 22% of the N requirement in a southern Appalachian forest was internally cycled. Resorption in loblolly pine accounts for 60% (P), 39% (N), and 22% (K) of its annual requirements (Switzer and Nelson 1972). Ryan and Bormann (1982) reported that resorption could provide 34%, 30%, 5%, and 2.4% of the annual NPP for N, P, K, and Mg in a northern hardwood forest.

An estimate of resorption from leaves to woody tissue prior to leaffall is given by the difference in nutrient content of leaf net primary production in August (Day and Monk 1977b) and the nutrient content of leaf litterfall (Cromack 1973, Cromack and Monk 1975). These differences are 1.4 kg ha^{-1} yr^{-1} for Ca, 28.3 kg ha^{-1} yr^{-1} for K, 4.6 kg ha^{-1} yr^{-1} for Mg, 3.3 kg ha^{-1} for P and 59.3 kg ha^{-1} yr^{-1} for N. Adjusted for leaching losses (August to October), a more realistic estimate of resorption would be: Ca $-$ 0.5 kg ha^{-1} yr^{-1}, K 13.5 kg ha^{-1} yr^{-1}, Mg 3.4 kg ha^{-1} yr^{-1}, P 3.1 kg ha^{-1} yr^{-1}, and N 56.5 kg ha^{-1} yr^{-1} (Table 11.2). If these values are correct, then the forest internal nutrient pool is larger than the net annual growth, with the exception of Ca, which seems not to be internally recycled.

The standing crop of leaf biomass serves as a nutrient pool and resorption represents a pathway of renewing that pool. The uptake values in Figures 11.1 through 11.3 represent the quantity of nutrients annually removed from the soil nutrient pool. When these two nutrient sources are separated, one finds the following uptake contribution from the soil-litter pool: Ca 100%, K 79%, Mg 75%, P 74%, and N 48%.

The relative amounts of nutrients annually removed from the soil-litter pool as uptake (NPP plus herbivory and leaching) varies greatly between elements (P 30.0%, K 19.1%, Ca 8.8%, Mg 2.9%, N 1.7%). This suggests that between 2 to 30% of the soil-litter nutrient pool may be involved in cycling during any one year. On the other hand, annual returns (litter production, herbivory, leaching) account for a large proportion of annual uptake from the soil (Ca 89%, Mg 62%, K 58%, N 46%, P 42%). The difference in uptake and returns should include nutrients in the net annual growth of the forest plus

Table 11.2. Annual Fluxes

Fluxes	Ca	K	Mg	P	N
			kg ha^{-1} yr^{-1}		
Net annual accretion	4.5	5.0	1.0	3.0	6.7
Herbivory	0.6	2.9	0.4	0.4	4.6
Litterfall	44.5	18.0	6.5	4.9	33.3
Leaching	3.9	25.8	2.3	0.4	3.3
Resorption	−0.5	13.5	3.4	3.1	56.5
	(0)	(21)	(25)	(26)	(52)
Uptake	54.0	65.2	13.6	11.8	104.4

Net annual accretion = NPP minus NPP of leaves, minus nonleaflitter herbivory. Leaching during autumn = Ca 1.9 kg ha^{-1} yr^{-1}, K 14.8, Mg 1.2, P 0.2, N 2.8. Resorption = NPP net annual accretion + herbivory + litterfall + leaching + resorption. The numbers in parentheses are % resorption contribution to total nutrient uptake.

nutrients removed from leaves to woody tissue prior to leaffall. Nutrient content of leaves on the trees in August (Day and Monk 1977b) differ greatly from the nutrient content of leaf litter later in the year (Cromack 1972; Cromack and Monk 1975). Actually, a significant amount of the nutrients used for growth each year may come from this internal pool rather than from the soil-litter pool. At this time we must assume that annual resorption does not vary greatly from year to year. If that is the case, then the soil-litter nutrient pool annually loses an amount equal to net annual growth. Net annual growth (4.0 M.T. ha^{-1} yr^{-1} biomass) is measured as total NPP minus leaf primary production and nonleaf litter. The elemental content of net annual biomass accretion is Ca 4.5 kg ha^{-1} yr^{-1}, K 4.9 kg ha^{-1} yr^{-1}, Mg 0.9 kg ha^{-1} yr^{-1}, P 3.0 kg ha^{-1} yr^{-1} and N 6.7 kg ha^{-1} yr^{-1} (Table 11.2). These represent 8.8%, 9.0%, 7.2%, 25.9% and 6.4% of the total nutrient content of NPP, respectively. Assuming that an amount equivalent to net annual accretion comes from the soil-litter nutrient pool, there would be a static nutrient supply of 138 years Ca, 95 years K, 517 years Mg, 12.5 years P, and 1031 years N. The amounts returned each year through leaching and litter production represent an amount equivalent to >8 years of net annual accretion, with the exception of P, where annual returns equal about 1.6 years of net annual accretion. Further, the aboveground litter decomposition rate is >3 years of net annual nutrient accretion, again with the exception of P, where the decomposition rate is equal to 1.2 years of net annual accretion. If any of these nutrients are potentially limiting to forest productivity, one would suspect P.

The evergreen component of these forests may contribute up to 35% of the total standing crop of leaf biomass (Day and Monk 1974) and between 14 to 32% of the total leaf nutrient standing stocks (Day and Monk 1977b; Monk and Day 1984; Monk et al. 1984). The nutrient concentrations in the evergreen leaves are lower than those found in the deciduous ones; however, with 1.9 M.T. ha^{-1} of leaves, the evergreen component assumes an important role in mineral cycling. Leaf litter production in the two most important evergreen species, rhododendron and mountain laurel, is about 0.3 M.T. ha^{-1} (Monk et al. 1984), giving a winter standing crop of evergreen leaves of about 1.6 M.T. ha^{-1}. Only Ca (32%) approximates the 35% contribution that the evergreen leaves make to leaf biomass. Evergreens possess 21% of the P in the leaves, Mg 20%, 17% N and 14% K. Between 1 to 5% of the aboveground nutrient standing stocks are found in the evergreen leaves.

A good case can be made on a global scale for the regulation of leaf life span and leaf size by climate, through its effects on carbon gain, thermal regulation of leaf temperature, and water use efficiency (Parkhurst and Loucks 1972; Givnish and Vermeij 1976; Chabot and Hicks 1982). Large leaves are not as thermally regulated as small ones in hot or cold environments. Optimal leaf size regulates leaf temperature, keeping it near the optimum for photosynthesis while preventing thermal damage. Size may then be related to profit-cost ratio. The admixture of evergreen and deciduous species on a local level cannot be attributed strictly to climatic factors. In many locations, the interplay of the evergreen and deciduous nature of vegetation has been shown to be related to edaphic variables. For some time and for many parts of the world, researchers have reported differential response of deciduous and evergreen species to local nutrient gradients (Richards 1952; Beard 1955; Beadle 1966; Monk 1966; Chapman 1967; Webb 1968; Goldberg 1982; Chapin and Kedrowski 1983; Milewski 1983). The general response has been the association of deciduous species with nutrient-rich sites and evergreen species with nutrient-poor sites. This "nutrient theory" of evergreenness basically states that the evergreen leaf has selective advantages over the deciduous leaf on nutrient-poor sites through the creation of a more conservative nutrient cycle (Monk 1966, 1971).

In the southeast and particularly in north central Florida, Monk (1966) demonstrated a close relationship between community evergreenness and soil fertility. Harper (1914) reported differences in the distribution of evergreen species, and contended that soil K was the factor related to the distribution of evergreen species. Loveless (1961, 1962) showed a proportional increase in the degree of sclerophylly with leaf P below 0.3%. Evergreen leaves usually have lower or similar nutrient concentrations than deciduous ones. Rarely are nutrient concentrations higher in evergreens (Rodin and Bazilevich 1967; Chabot and Hicks 1982). They key elements probably include N, P, Ca, K, and Mg.

The data presented in this paper mostly reflect values for a single watershed at a low elevation rather than for specific sites. Soil P on WS 18 is quite low when compared with other elements and insufficiently available for uptake by many species. If P limits the productivity of any species then it should be more limiting to the deciduous component than to the evergreen component (McGinty 1972). This may in part explain the extensive distribution of evergreen species in the Coweeta Basin and their competitive success relative to deciduous species in responding to past forest disturbances.

12. Dynamics of Early Successional Forest Structure and Processes in the Coweeta Basin

L.R. Boring, W.T. Swank, and C.D. Monk

Clearcutting is a prevalent silvicultural practice used to regenerate mixed hardwood forests in the southern Appalachians and to improve stands degraded by prior poor selection-cutting practices. Its widespread application is due in part to the economics of harvesting, but primarily due to its potential for regenerating economically desirable species (Smith 1962; McGee and Hooper 1970).

Research on the effects of clearcutting in the southern Appalachians has addressed silvicultural aspects of early forest regeneration (McGee and Hooper 1970, 1975; Trimble 1973) and subsequent hydrologic responses (Swank and Helvey 1970; Swift and Swank 1981), but major gaps still exist in our knowledge about regenerating southern Appalachian forests. These gaps include understanding how initial patterns of species establishment affect stand composition in later years and how regeneration trends vary with different sites and forest communities.

Perhaps more importantly, there is a need to couple early successional changes in forest structure with associated changes in nutrient cycling processes. Biogeochemical studies subsequent to forest cutting have indicated basic changes in forest floor processes and nutrient uptake by vegetation (Marks and Bormann 1972; Swank and Douglass 1977; Bormann and Likens 1979). Johnson and Swank (1973) hypothesized that after clearcutting southern Appalachian forests, vegetation rapidly accumulates biomass, recovers nutrient uptake quickly, and consequently conserves nutrients that could otherwise be leached from the soil. Some of these mechanisms were shown to be important in reducing nutrient losses in early successional northern hardwood forests, although nutrient loss by soil leaching (especially NO_3-N) was observed immediately following forest removal (Marks and Bormann 1972; Marks 1974). However, at Coweeta nutrient losses of early successional forests are relatively conservative and it was hypothesized that the recovery of nutrient uptake rates are relatively rapid (Johnson and Swank 1973; Swank and Douglass 1977). Changes in decomposition rates and forest floor processes were also hypothesized to result in nutrient conservation (Seastedt and Crossley 1981; Abbott and Crossley 1982).

A hardwood forested watershed at Coweeta was clearcut as part of an interdisciplinary study of the physical, chemical, and biological effects on both terrestrial and aquatic components of the ecosystem. Our specific objectives addressed here were: (1) to examine differences in forest regeneration trends among former cove, chestnut oak, and xeric scarlet oak–pine sites on the clearcut watershed; (2) to compare species composition, leaf area index (LAI), biomass, net primary production (NPP), nutrient uptake and nutrient accretion over the first 3 years of regeneration with values for an adjacent, uneven-aged, mixed hardwood forest; and (3) to relate regeneration of forest structure to fundamental ecosystem processes of nutrient uptake, immobilization, and transfers.

Materials and Methods

Study Areas

Watershed 7 (WS 7) was the primary site for this study. On the basis of an earlier study (Williams 1954), the watershed was stratified into four plant communities, similar to those identified in recent studies of vegetation classification in the Coweeta Basin (Chapter 10). Based upon these previous studies and the delayed completion of site preparation on xeric sites, three sampling strata were identified: (1) a cove hardwood community found at lower elevations and along ravines at intermediate elevations; (2) a chestnut oak community on mesic southeast- and north-facing slopes at intermediate elevations; and (3) a scarlet oak–pine community on xeric southwest and south-facing slopes at intermediate to upper elevations and ridgetops, which combines the two xeric vegetation types from the previous studies.

Timber cutting and yarding with a mobile cable system began in January 1977 and was completed the following June. Tractor skidding was used on about 9 ha where slopes were less than 20%, and the remainder was yarded with a mobile cable system. Most of the ridgetops and xeric slopes were cut for the purpose of the experiment, but were not logged due to an insufficient volume of marketable timber. The site preparation treatment, completed in October 1977, consisted of clear-felling all remaining stems > 2.5 cm dbh. This treatment was completed on the xeric half of the watershed 6 months later than on the mesic half, thus introducing an age difference in hardwood regeneration between the two halves of the watershed.

Comparative data for net primary production (NPP) and element standing crops for an uneven-aged hardwood forest were taken from WS 18 (Day and Monk 1977a,b), because these measurements were not made on WS 7 prior to clearcutting. Detailed NPP information is available for WS 18, a 12.5 ha watershed which primarily differs from WS 7 in having a northern aspect. Although species dominance varies among the oaks on these two watersheds, there were no major differences in the overall tree species composition, basal area (25.6 vs. 25.3 m^2 ha^{-1}), density (3044 vs. 3058 stems ha^{-1} >2.5 cm dbh), and aboveground biomass (139 vs. 130 t ha^{-1}) (Boring 1979).

Forest Composition and Production

Standard dimension analysis techniques were used to estimate biomass, NPP, and LAI of woody species (Whittaker and Woodwell 1968; Whittaker et al. 1974; Phillips and Saucier 1979). Individual young hardwoods (mostly of sprout origin) were sampled each August from 1977 to 1979 from randomly chosen sample points within the study area to establish regression equations (Table 12.1). Sampled individuals were cut at the ground or at the point of sprout origin on the stump. Diameters were measured at 3 and 40 cm from the base. The 3 cm measurement gave the best fit for slow-growing species, and 40 cm was best for the fast-growing species. All leaves were removed, bagged, dried to a constant weight at 70°C, and weighed. Stems and all branches were similarly dried and weighed. Wood weights included all aboveground woody components.

An untransformed linear equation using diameter and basal area as independent variables and a logarithmic transformed equation using log_{10} diameter were compared for relative goodness of fit using r^2 and Furnival's index of fit. Log_{10} regression equations based upon pooled 1 to 3-year-old stem diameters provided the best fits and were selected to predict leaf and wood biomass for eleven species (Table 12.1). Bias from logarithmic transformation was corrected using a base 10 modified Baskerville technique (Baskerville 1972).

Prior to clearcutting, the vegetation was inventoried from 142 0.08 ha plots systematically located over WS 7. Following clearcutting, thirteen of the original plots were randomly selected among mesic sites and sampled for regrowth, including five plots in the cove and eight chestnut oak plots (Figure 12.1). In 1978, eleven additional plots were also randomly selected in the xeric scarlet oak–pine area. Two quadrats were located in opposite corners of each 0.08 ha plot. Hardwood sprouts and seedlings were sampled in each quadrat with subplots of 7 × 7 m and 3 × 3 m, respectively, and values were pooled for each pair (Shimwell 1971).

Herbaceous vegetation was destructively sampled in August of each year from one randomly placed 1 m^2 subplot within each quadrat, separated by species, or groups of species, and oven-dried to constant weight at 70°C. Total herb NPP was estimated by equating it with August standing crop biomass, since a study in the first year showed that most species attained peak biomass at that time (Boring et al. 1981).

LAI estimates were determined from leaf surface area to dry weight ratios measured on at least 20 leaves for each of 21 woody species (including *Rubus* spp.) in both 1977 and 1979. These ratios were established by subsampling leaves from several individuals of each species, measuring leaf area with a LI-COR portable leaf area meter (Lambda Instrument Company, Omaha, Nebraska), and then drying and weighing the leaves. For herbs, whole plant biomass to leaf surface area ratios were determined for three dominant composite species and a miscellaneous category.

At the end of the growing season, sprout and seedling densities were recorded separately by species and diameter class on each 7 m × 7 m sample plot. For the first year, diameter classes were designated by 0.5 increments up to a maximum of 3 cm. For

Table 12.1. Sample Sizes, Diameter Ranges, Regression Equations and Correction Factors (K) for Estimating Leaf and Wood Biomass of 1 to 3 Year-Old Hardwood Sprouts (Y = Dry Weight Biomass (g) and X = Diameter (mm))

Species	n	Range (mm)	Leaf Biomass	K^a	r^2	Wood Biomass	K^a	r^2
Group A (Diameter at 40 cm)								
Acer rubrum	26	3–57	$\log_{10}Y = -0.565 + 2.083 \log_{10}X$	1.036	0.98	$\log_{10}Y = -0.990 + 2.605 \log_{10}X$	1.031	0.99
Castanea dentata	16	3–27	$\log_{10}Y = -0.329 + 1.799 \log_{10}X$	1.095	0.91	$\log_{10}Y = -1.022 + 2.527 \log_{10}X$	1.090	0.96
Liriodendron tulipifera	25	5–77	$\log_{10}Y = -0.759 + 2.140 \log_{10}X$	1.070	0.96	$\log_{10}Y = -1.444 + 2.799 \log_{10}X$	1.082	0.97
Quercus prinus	20	5–42	$\log_{10}Y = -0.267 + 1.871 \log_{10}X$	1.076	0.93	$\log_{10}Y = -0.688 + 2.476 \log_{10}X$	1.091	0.97
Quercus rubra	14	5–33						
Robinia pseudo-acacia	19	3–76	$\log_{10}Y = -0.308 + 1.968 \log_{10}X$	1.069	0.97	$\log_{10}Y = -0.922 + 2.636 \log_{10}X$	1.032	0.99
Group B (Diameter at 3 cm)								
Carya ovalis, C. tomentosa	18	6–36	$\log_{10}Y = -0.844 + 2.212 \log_{10}X$	1.121	0.89	$\log_{10}Y = -2.479 + 3.487 \log_{10}X$	1.197	0.92
Cornus florida	26	3–36	$\log_{10}Y = -0.796 + 2.168 \log_{10}X$	1.052	0.97	$\log_{10}Y = -1.317 + 2.758 \log_{10}X$	1.046	0.98
Kalmia latifolia	24	2–12	$\log_{10}Y = -0.556 + 1.814 \log_{10}X$	1.080	0.89	$\log_{10}Y = -0.803 + 2.162 \log_{10}X$	1.048	0.95
Nyssa sylvatica	20	4–41	$\log_{10}Y = -0.741 + 2.087 \log_{10}X$	1.071	0.96	$\log_{10}Y = -1.666 + 2.972 \log_{10}X$	1.055	0.98
Oxydendrum arboreum	20	3–40	$\log_{10}Y = -0.732 + 2.054 \log_{10}X$	1.070	0.96	$\log_{10}Y = -1.285 + 2.694 \log_{10}X$	1.064	0.98
Rhododendron maximum	20	4–18	$\log_{10}Y = -1.176 + 2.555 \log_{10}X$	1.115	0.87	$\log_{10}Y = -2.027 + 3.218 \log_{10}X$	1.121	0.94

[a] Correction factor using the base 10 modified Baskerville correction (Baskerville 1972) where: $K = \text{antilog}\,[(S^2/2)\log_e 10]$, and Y (corrected) $= [\text{antilog}(a+b \log_{10}X)]K$.

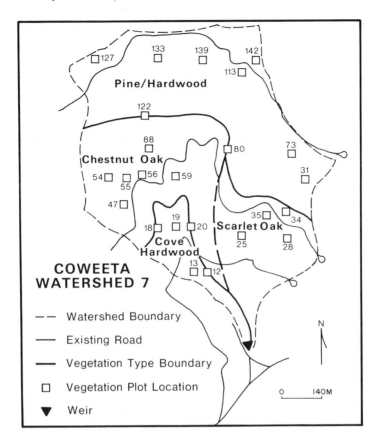

Figure 12.1. Map of Coweeta WS 7, vegetation types, and the vegetation plot locations.

the second and third years, classes were in 1 cm increments up to 8 cm. Aboveground leaf, wood and total biomass (kg ha^{-1}) were determined for each species by multiplying the stem densities for each size class by midpoint biomass values estimated by regression analysis. Summation of the wood and leaf biomass from each size class yielded total biomass per species for each plot. LAI was determined by multiplying the leaf area/dry weight ratios times leaf biomass for each species.

For watershed estimates of biomass and leaf area, values for each species were calculated by multiplying its mean biomass or leaf area in a forest type by the relative area of that vegetation type: (1) equating first-year biomass with NPP; and (2) for the second and third years, subtracting the previous year's woody biomass and evergreen leaf biomass from total biomass.

Element Concentrations and Standing Stocks

Element analyses for N, P, K, Ca, and Mg were conducted on wood and leaf tissue for all dominant hardwood sprout, seedling, herb, and vine species for 1977 to 1979. A total of 6 to 12 individuals per species were sampled, oven-dried and ground in a Wiley mill through a 425 μ mesh screen. Three to six replicate 0.5 g tissue samples from all species were ashed at 400°C for 4 hr, digested in 10 ml of 20% HNO_3, and analyzed for P, K, Ca, and Mg on a Jarrell-Ash plasma emission spectrograph (Jones 1977; Fassel 1978). Total N content was determined on a Coleman Micro-Dumas nitrogen analyzer (Bremner 1965). Orchard leaf standards (United States National Bureau of Standards) were utilized for quality control. All element data, expressed as mg kg^{-1}, were tested for significant differences in mean concentrations by SAS General Linear Models programs and ranked by Duncan's multiple range test (Sokal and Rohlf 1969; Helwig and Council 1979).

Results and Discussion

Species Regeneration Patterns

Prior to clearcutting, the highest biomass (166 t ha^{-1}) was in the cove hardwood community, followed by the scarlet oak–pine (133 g ha^{-1}) and chestnut oak (119 t ha^{-1}) communities (Table 12.2). These results are not surprising, since the site quality of the cove forest was highest and the scarlet oak–pine sites had the least logging activity in the past. However, these biomass figures represent stands which have a complex disturbance history, to include high-grade logging and high mortality of American chestnut (*Castanea dentata*) (Chapter 11). Old growth cove and chestnut oak forests on similar sites in the Great Smoky Mountains may exceed 400 t ha^{-1} (Whittaker 1966).

Although northern red oak (*Quercus rubra*) and hickories (*Carya* spp.) comprised 36% of the cove forest biomass before clearcutting, in the second year after cutting they decreased to 12% of the woody species biomass, while yellow-poplar (*Liriodendron tulipifera*) continued to comprise 18% (Table 12.2). Red maple (*Acer rubrum*), flowering dogwood (*Cornus florida*), and yellow-poplar dominated the regeneration, although vine and herbaceous biomass comprised 47% of the total community biomass. Most of the oaks and hickories in the cove likely originated from seedlings that were established prior to clearcutting (Sander 1972; Trimble 1973). The numerous yellow-poplars were predominantly seedlings (18,444 ha^{-1}) that germinated following clearcutting (Table 12.3). Although silver-leaf grape (*Vitis aestivalis* var. *argentifolia*) sprouts were faster growing than seedlings, the latter were more numerous, with 2244 ha^{-1} vs. 16,222 ha^{-1}.

Prior to clearcutting, oaks and hickories composed 60% of the biomass in the chestnut oak (*Q. prinus*) community, followed by red maple and yellow-poplar (22%) (Table 12.2). Following clearcutting, black locust (*Robinia pseudo-acacia*), flowering dogwood, yellow-poplar, and red maple comprised 57% of the woody regeneration, with oaks and hickories reduced to 15%. Although their biomass was relatively low, many mountain laurel (*Kalmia latifolia*) sprouts and grape seedlings were present (Table 12.3). Herbaceous vegetation and blackberries (*Rubus* spp.) comprised 29% of the total biomass.

Except for yellow-poplar and silver-leaf grape, the dominant woody species generated primarily from stump or root sprouts. The scarlet oak–pine (*Q. coccinea/Pinus rigida*) community was dominated by chestnut oak, scarlet oak, black oak (*Q. velutina*)

Table 12.2. Aboveground Biomass for Woody Species Before and 2 Years (2.5 Years for Scarlet Oak–Pine) After Clearcutting on Former Cove Hardwood, Chestnut Oak, and Scarlet Oak–Pine Sites on WS 7

Species	Cove Hardwood				Chestnut Oak				Scarlet Oak–Pine			
	Postcut		Precut		Postcut		Precut		Postcut		Precut	
	t ha⁻¹ Biomass	% Biomass	t ha⁻¹ Biomass	% Biomass	t ha⁻¹ Biomass	% Biomass	t ha⁻¹ Biomass	% Biomass	t ha⁻¹ Biomass	% Biomass	t ha⁻¹ Biomass	% Biomass
Acer rubrum	0.6	27	4.3	3	0.2	6	11.7	10	0.7	18	8.4	6
Betula lenta	<0.1	<1	14.2	9	<0.1	1	0.6	1	0	0	<0.1	<1
Carya spp.	<0.1	1	26.6	16	<0.1	1	18.9	16	<0.1	1	5.2	4
Castanea dentata	0	0	0	0	0.1	2	0.5	<1	0.2	5	1.2	1
Cornus florida	0.3	14	3.1	2	0.8	21	6.5	6	0.1	3	2.7	2
Kalmia latifolia	0	0	0.2	<1	0.3	8	2.5	2	0.8	20	15.2	11
Liriodendron tulipifera	0.4	18	29.4	18	0.3	7	13.6	12	<0.1	1	4.0	3
Nyssa sylvatica	<0.1	1	0.9	<1	0.2	5	2.6	2	0.2	6	4.9	4
Pinus rigida	0	0	0	0	0	0	0.4	<1	0	0	3.4	3
Quercus alba	0	0	10.0	6	0	0	0.3	<1	<0.1	1	6.3	5
Quercus coccinea	0	0	1.4	1	<0.1	1	10.5	9	0.5	13	20.2	15
Quercus prinus	0	0	3.9	2	0.3	8	26.6	22	0.4	10	35.1	26
Quercus rubra	0.2	11	34.0	20	0.2	5	5.1	4	<0.1	2	0.6	<1
Quercus velutina	<0.1	<1	0.1	<1	<0.1	<1	10.8	9	<0.1	<1	15.6	12

(Continued)

Table 12.2. (Continued)

| Species | Cove Hardwood | | | | Chestnut Oak | | | | Scarlet Oak–Pine | | | |
| | Postcut | | Precut | | Postcut | | Precut | | Postcut | | Precut | |
	t ha⁻¹ Biomass	% Biomass	t ha⁻¹ Biomass	% Biomass	t ha⁻¹ Biomass	% Biomass	t ha⁻¹ Biomass	% Biomass	t ha⁻¹ Biomass	% Biomass	t ha⁻¹ Biomass	% Biomass
Rhododendron maximum	0.1	4	5.3	3	<0.1	1	2.5	2	<0.1	1	0.4	<1
Robinia pseudoacacia	<0.1	1	0	0	0.9	23	0.8	1	0.4	10	2.9	2
Vitis aestivalis	0.2	8	–	–	0.2	6	–	–	<0.1	1	–	–
Others	0.3	15	32.7	20	0.2	5	5.2	4	0.3	8	7.6	6
Total for woody species	2.1	100	166.1	100	3.7	100	119.1	100	3.8	100	133.8	100

[a] Postclearcutting data include all stems, and precut data include all stems ≥2.5 cm dbh.

Table 12.3. Stem Density by Species Before and 2 Years After Clearcutting on Former Cove Hardwood, Chestnut Oak, and Scarlet Oak–Pine Sites on WS 7[a]

| | Cove Hardwood | | | | Chestnut Oak | | | | Scarlet Oak–Pine | | | |
| | Postcut | | | Precut | Postcut | | | Precut | Postcut | | | Precut |
Species	Sprouts/ha	Sdlgs/ha	% Density	% Density	Sprouts/ha	Sdlgs/ha	% Density	% Density	Sprouts/ha	Sdlgs/ha	% Density	% Density
Acer rubrum	2,876	6,222	12	6	3,172	4,861	7	7	6,956	5,389	13	6
Betula lenta	816	3,000	5	4	765	628	1	<1	0	0	0	<1
Carya spp.	306	1,222	2	4	1,010	278	1	7	1,285	356	2	3
Castanea dentata	0	0	0	0	1,112	0	1	3	1,938	0	2	4
Cornus florida	7,589	111	10	14	17,901	350	16	16	3,029	0	3	4
Kalmia latifolia	0	0	0	2	13,903	139	13	17	40,555	0	42	56
Liriodendron tulipifera	510	18,444	24	5	1,836	6,739	8	6	176	1,211	1	1
Nyssa sylvatica	408	222	<1	1	5,314	767	6	5	5,773	250	6	4
Pinus rigida	0	0	0	0	0	0	0	<1	0	0	0	1
Quercus alba	0	0	0	1	0	0	0	<1	571	406	1	1
Quercus coccinea	0	0	0	<1	316	1,600	2	2	2,009	3,839	6	2
Quercus prinus	0	0	0	2	3,121	417	3	7	3,121	606	4	5

(Continued)

Table 12.3. (Continued)

Species	Cove Hardwood				Chestnut Oak				Scarlet Oak–Pine			
	Sprouts/ ha	Postcut Sdlgs/ ha	% Density	Precut % Density	Sprouts/ ha	Postcut Sdlgs/ ha	% Density	Precut % Density	Sprouts/ ha	Postcut Sdlgs/ ha	% Density	Precut % Density
Quercus rubra	1,142	1,444	3	1	979	1,389	2	1	120	656	1	<1
Quercus velutina	0	889	1	<1	13	211	<1	2	65	556	<1	1
Rhododendron maximum	4,488	0	6	46	2,499	139	2	19	1,867	0	2	2
Robinia pseudoacacia	408	444	1	0	2,122	1,322	3	<1	1,275	200	2	1
Vitis aestivalis	2,244	16,222	23	—	2,387	28,000	27	—	1,234	1,417	3	—
Others	7,610	2,832	13	13	4,764	3,412	7	10	9,723	1,967	12	9
Totals	28,397	51,052	100	100	61,214	50,252	100	100	79,697	16,853	100	100

[a] Postcut data include all woody stems, and precut data include all stems ≥2.5 cm dbh. The % density of postcut stems includes both seedlings and sprouts.

and mountain laurel, which comprised 64% of the biomass prior to clearcutting (Table 12.2). Afterwards, mountain laurel and red maple were dominant (38%), followed by black locust (10%), American chestnut, chestnut oak, and scarlet oak (28%).

In the third year following clearcutting, there was little difference in herbaceous biomass between cove and chestnut oak sites, but both were higher than the scarlet oak–pine sites (Figure 12.2). Blackberries were dominant, ranging from 2026 kg ha^{-1} on cove sites to 83 kg ha^{-1} on scarlet oak sites. This range in herbaceous biomass could be attributed to the desiccating conditions in the upper forest floor along the southwest-facing slopes and ridges.

The density of seedlings decreased along the gradient from cove to chestnut oak to scarlet oak–pine communities (Table 12.3). This trend was likely due to lack of propagules, as well as to high mortality resulting from the xeric forest floor microclimate along the south-facing slopes and ridges. High numbers of silver-leaf grape and yellow-poplar seedlings germinated on the rich, moist cove sites (up to 16,222 and 18,444 seedlings ha^{-1}, respectively). The scarlet oak–pine sites were occupied only by 1211 and 1417 seedlings ha^{-1} of the two species.

Woody sprout densities increased from cove to chestnut oak to scarlet oak–pine communities, an inverse trend observed for seedling density (Table 12.3). Large red oaks

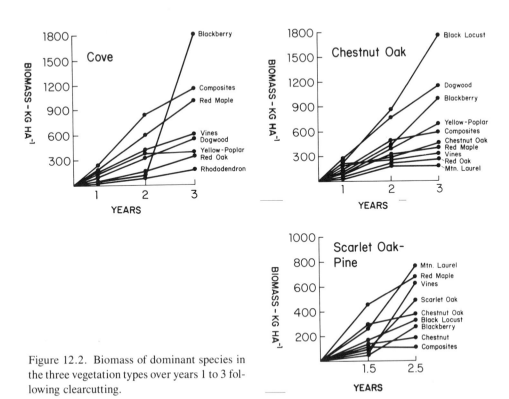

Figure 12.2. Biomass of dominant species in the three vegetation types over years 1 to 3 following clearcutting.

and hickories in the cove plots had poor sprouting capabilities (Sander 1972; Johnson 1975; McGee 1978), and sprout biomass was relatively low. The chestnut oak and scarlet oak–pine sites were occupied by species with superior sprouting abilities such as black locust, flowering dogwood, red maple, mountain laurel, and blackgum (*Nyssa sylvatica*). The scarlet oak–pine community lacks the rapid growth that characterizes cove and chestnut oak communities with tall, dense stands of black locust, yellow-poplar, chestnut oak, and dogwood. However, the scarlet oak–pine community had a higher density of small diameter individuals, including mountain laurel, red maple, oak, blackgum, and American chestnut. American chestnut has retained its ability to sprout vigorously, despite the introduction of chestnut blight in the mid 1920s.

This 3-year study may be placed in a longer-term perspective by comparing it to a nearby stand on WS 13, which was sampled at 9, 13, and 23 years following clearcutting (Swift and Swank 1981, Figure 12.3). Generally, the relative basal areas show early dominance of oak, red maple, black locust, and American chestnut. After 23 years, the oaks and red maple maintain dominance, but black locust and chestnut drop in significance. Yellow-poplar and flowering dogwood increase their dominance. There was a 73% recovery of basal area in 23 years following the first clearcut and an 80% recovery in 21 years following a second cutting (Leopold and Parker 1985). In a related study, Parker and Swank (1982) found that regeneration response varied by species and physiographic position. After two clearcutting treatments on WS 13 they found no overall change in the number of species present, but rather an increased density of a few species; notably yellow-poplar, red maple, and chestnut oak. The other remaining species, except various oaks, declined in number.

Following clearcutting, the successional sequences (Figures 12.2 and 12.3) show that black locust quickly sprouts, grows faster than other species, and attains early

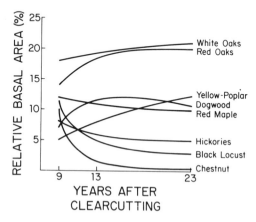

Figure 12.3. Relative basal area of dominant species on WS 13 from 9–23 years following clearcutting. (Data from Swift and Swank 1981.)

dominance. Similar establishment patterns have been documented throughout the southern Appalachian region (Beck and McGee 1974; McGee and Hooper 1975). However, the dense stands eventually decrease in vigor, suffer high mortality concomitant with locust stem borer attack (*Megacyllene robiniae*), and decline, leaving N-enriched soil and organic matter for exploitation by other species (Craighead 1937; Boring and Swank 1984a).

Early thinning changes the structure of the coppice forest as codominant and subdominant individuals fill gaps left by shorter-lived early dominants such as black locust and American chestnut. The relatively understory-tolerant oaks, hickories, and other species (*A. rubrum, Oxydendrum arboreum*) in the stand should eventually increase in relative density and biomass. It is well documented that yellow-poplar increases in density and basal area following clearcutting due to prolific seedling establishment and sprouting ability (Smith 1963; Trimble 1973; Beck and Della-Bianca 1981). This is clearly the pattern on WS 7 and WS 13 (Parker and Swank 1982). The species is long-lived and could conceivably occupy these sites for several hundred years (Buckner and McCracken 1978), or certainly until the end of the next harvest rotation. Yellow-poplar eventually attains a massive size (Lorimer 1980), and young stands in eastern Tennessee have been documented to have very high productivity, approximating 12.0 t ha^{-1} yr^{-1} of NPP (Sollins 1972).

Recovery of NPP and LAI

The recovery of aboveground NPP was rapid (Figure 12.4). The third year mesic site NPP was 5.4 t ha^{-1} yr^{-1}, or 62% of the NPP estimated for the uneven-aged mixed hardwood forest on WS 18 (Day and Monk 1977a). After clearcutting, NPP on mesic sites was greater than on xeric sites, due to rapid sprout growth and an abundance of herbaceous vegetation. The xeric site NPP was 3.0 t ha^{-1}, or 34% of the value for the older, uneven-aged forest on WS 18; although sprout density was high, the dominant species was mountain laurel, which is relatively slow growing. However, before logging it is likely that the NPP on these mesic sites was higher and xeric site NPP was lower than the weighted watershed mean for WS 18. Although the NPP appears to recover quickly, it may not represent an actual "recovery" of NPP. The root systems of the rapidly growing sprouts were intact to function in water and nutrient uptake, as well as to provide stored photosynthates and essential elements.

LAI comparisons with other sites in the Coweeta Basin show a rapid recovery on both the mesic and xeric sites (Figure 12.4). Including herbaceous leaf area, LAI in the third year was 4.2 m^2 m^{-2}, or 68% of the estimate for a mature mixed hardwood forest on WS 18 (LAI = 6.2 m^2 m^{-2}; Chapter 10). The LAI on xeric sites was 48% of the mature forest in 2.5 years. Although there was a rapid recovery of LAI, it does not necessarily follow that evapotranspiration will return rapidly to pretreatment levels. Other Coweeta research on WS 13 has shown increases in streamflow for 23 years following clearcutting (Swank and Helvey 1970). There are likely different rates of evapotranspiration for early successional woody species and herbs in comparison to the uneven-aged mixed hardwoods which dominate mature watersheds.

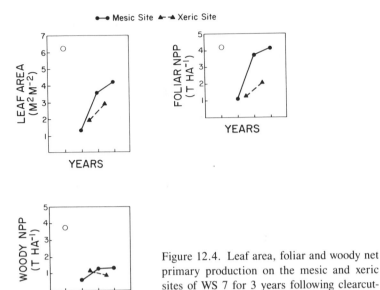

Figure 12.4. Leaf area, foliar and woody net primary production on the mesic and xeric sites of WS 7 for 3 years following clearcutting. Values for adjacent mature hardwood forest are also indicated (○).

The patterns of early regeneration in southern Appalachian hardwood forests differ somewhat from those observed following clearcutting of northern hardwood and Pacific northwest coniferous forests (Marks 1974; Bicknell 1979; Gholz et al. 1985). The prolific hardwood sprout, herb, and vine growth shown in this 3-year study results in more rapid biomass accretion (4.3 to 8.4 t ha^{-1}) than reported elsewhere. Pacific northwest coniferous forests at the H.J. Andrews Forest have profuse herbaceous regrowth, but Douglas fir seedling establishment and early growth are slow, resulting in aboveground biomass of only 2.6 t ha^{-1} in the third year (Gholz et al. 1985). Sprouting was less important and regeneration was slower for the first three years in the Hubbard Brook northern hardwood stands. A mixed northern hardwood stand 3 years old had a comparative biomass of 4.7 t ha^{-1} (Bicknell 1979). By 4 years, however, pin cherry (*Prunus pennsylvanica*) was well established from dormant seeds in the forest floor, resulting in increased accretion rates of biomass and nutrients, especially in dense pin cherry stands (Marks 1974). Although southern Appalachian mesic hardwood forests at Coweeta initiate regrowth and accumulate nutrients more rapidly, rapid growth rates in the intermediate-aged northern hardwood stands at Hubbard Brook may later minimize these early differences in vegetation processes (Marks and Bormann 1972).

Nutrient Accumulation and Cycling

In the clearcut watershed, significantly ($p < .05$) higher concentrations of N and P were found in leaves of early successional woody species in comparison to other woody species. These species included black locust (high N, P, and K), yellow-poplar (high N, P, Ca and Mg), and silver-leaf grape (high N and P). Comparisons among growth forms (herbs, vines, woody species) revealed several trends which were highly significant ($p < .05$): higher leaf K in herbs, and higher leaf Ca and P in sprouts. Most differences among species were significant in all 3 years, although the first year leaf and stem concentrations were greater than the second or third year for most species.

As a result of the high element concentrations in early successional woody species and herbs, as well as their rapid growth rates, there was a relatively large quantity of nutrients immobilized in the aboveground biomass by the third year in comparison to the control mixed hardwood forest (Table 12.4). Quantities of N, P, K, Ca, and Mg on the mesic sites were 27, 13, 36, 10, and 31% of the control forest, respectively. The estimates for the xeric sites after 2.5 years were 16, 10, 20, 5, and 17%. The relative amounts of nutrients contained within the foliar fraction is very high. Sprout leaves and herbs comprise 70 to 78% of the total standing stocks of all elements on mesic sites and 62 to 67% of all elements on xeric sites. Although large quantities of nutrients are immobilized each growing season, there is a proportionately large amount in leaf biomass, resulting in an autumnal flux back to the forest floor as litterfall and leaching (White 1986). These values, along with an estimate of retranslocation, are not currently available for comparison in these stands, but this flux likely comprises the majority of the nutrients in the foliar standing stocks. There is a need to understand more fully the transfer and turnover of nutrients in early successional stands.

Annual aboveground NPP on the 3-year-old mesic sites contained a large amount of nutrients relative to the mature, uneven-aged mixed hardwood forest (Figure 12.5). The N, P, K, Ca, and Mg were 93, 57, 141, 91, and 105% of the values for the mature forest. The xeric sites after 2.5 years were estimated to contain approximately half of the nutrients contained on the mesic sites. The species responsible for most of the NPP had high ratios of leaf/stem production and high tissue concentrations relative to the dominant later-successional species in the uneven-aged hardwood forest. This is especially true for herbaceous and vine regrowth, but is also representative of the fast-growing black locust, yellow-poplar, and other woody opportunistic species.

The early recovery of nutrients in NPP is likely a result of the ability of early successional species to quickly respond to forest floor nutrient availability following clearcutting. Sprouting hardwoods have extensive residual root systems to facilitate water and nutrient uptake. Also, studies of root uptake kinetics (Haines 1986) have shown that certain early successional southern Appalachian plant species have the physiological potential to extract nutrients more rapidly from solution than can later successional species.

Although decomposition was initially slowed by the hot and dry forest floor microclimate which followed clearcutting (Abbott and Crossley 1982; Seastedt and Crossley 1981), there was nutrient enrichment of throughfall as water passed through the woody logging residue (W.T. Swank, unpublished data). In the third year (Figure 12.5), the decrease in the rate of biomass nutrient accretion on mesic sites may have been attributable to depletion of readily leachable nutrients from this source, or to increased internal conservation of nutrients within individuals. A corollary hypothesis is that the aboveground standing stocks of nutrients may have originated from the intact

Table 12.4. Nutrient standing crops (kg ha^{-1}) in aboveground biomass from mesic (cove and chestnut oak) and xeric (scarlet oak and pine hardwood) sites on WS 7 following clearcutting.

Element	Growth Form	Mesic Sites Years			Xeric Sites Years		WS 18[d]
		1	2	3	1.5	2.5	Uneven-aged
N	Sprouts						
(kg ha^{-1})	Leaf	12	31	45	18	36	95
	Stem	5	9	17	8	10	310
	Seedlings	1	9	8	2	8	—
	Vines	3	8	3	2	3	—
	Herbs	8	31	35	5	6	—
	Total	29	88	108	35	63	405[a]
P	Sprouts						
(kg ha^{-1})	Leaf	0.5	2	3	2	3	8[c]
	Stem	0.5	1	1	1	1	52[b]
	Seedlings	<1	1	1	<1	1	—
	Vines	<1	0	<1	<1	<1	—
	Herbs	1	3	3	<1	<1	—
	Total	3	7	8	3	6	60
K	Sprouts						
(kg ha^{-1})	Leaf	5	18	26	9	22	47
	Stem	4	6	10	4	7	183
	Seedlings	1	6	5	1	5	—
	Vines	2	7	3	1	3	—
	Herbs	13	49	38	6	8	—
	Total	25	86	82	21	45	230[c]
Ca	Sprouts						
(kg ha^{-1})	Leaf	8	13	17	12	15	47
	Stem	4	5	9	5	6	504
	Seedlings	1	4	4	1	3	—
	Vines	3	7	3	1	2	—
	Herbs	4	16	21	2	3	—
	Total	20	45	54	21	29	551[c]
Mg	Sprouts						
(kg ha^{-1})	Leaf	2	3	5	3	4	13
	Stem	1	1	2	1	1	35
	Seedlings	<1	1	1	<1	1	—
	Vines	<1	2	1	<1	1	—
	Herbs	2	7	6	1	1	—
	Total	6	14	15	5	8	48

[a] From Mitchell, Waide and Todd 1975.
[b] From Duvigneaud and Denayer-DeSmet 1969.
[c] From Day and Monk 1977b.
[d] Comparative estimates from WS 18 average all sites on the control watershed.

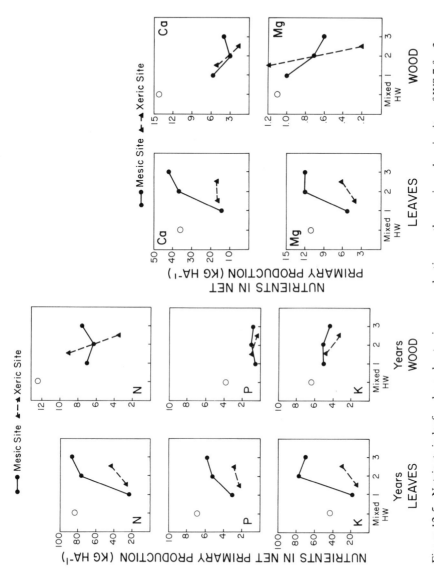

Figure 12.5. Nutrients in leaf and wood net primary production on the mesic and xeric sites of WS 7 for 3 years following clearcutting. Values for adjacent mature hardwood forest are also indicated (o).

roots of sprouting hardwoods in conjunction with stored photosynthate. The third year may represent the point at which hardwood sprout reserves were tapering off and uptake was the sole source of nutrients. For a few years, a large percentage of the elements concentrated by early successional woody species may be retranslocated and internally recycled to maintain growth beyond the initial period of high nutrient availability (Ryan and Bormann 1982). Other potential factors affecting the decreased accretion rates in the third year include annual variations in NPP due to climate, and early mortality resulting from competition.

Black locust symbiotically fixes at least 30 kg N ha^{-1} yr^{-1} (Boring and Swank 1984b), accumulates large quantities of N in its biomass (Boring and Swank 1984a), and maintains elevated concentrations of other nutrients in both leaves and wood (Boring et al. 1981). The biomass of this short-lived species may be regarded as a nutrient storage sink for an intermediate span of time, approximating the life of the stand (15 to 40 years). In plantation studies, elevated soil N from black locust has been shown to accelerate the growth of associated vegetation over the long-term (Chapman 1935). The high annual transfers of N in litterfall, throughfall, and root turnover are apparently the source of elevated NO_3-N in soil underneath black locust stands (Boring and Swank 1984a); intuitively, the transfer and availability of N should be maximal at the time of stand senescence. Of additional importance, the N ultimately transferred with stand senescence may potentially alter C/N ratios and affect decomposition and the availability of other macronutrients.

Vegetation Influences

Other investigators at Coweeta have found that clearcutting results in first-year forest floor temperatures up to 47°C, severe wetting and drying cycles in forest floor surface horizons, alteration of microarthropod activity in litter (Seastedt and Crossley 1981), and slow first-year decomposition of woody litter, especially on xeric sites (Abbott and Crossley 1982). The return of a high LAI by the third year results in shading, amelioration of the harsh forest floor microclimate, and dampening of environmental effects on forest floor biota and their processes. With early canopy closure, there should be a concomitant increase in the decomposition rates of woody and leaf litter. By the sixth year following clearcutting, there is a large cumulative amount of decomposition in large woody litter, with decreases in wood density of some species exceeding 50% (Mattson 1986). Woody debris can act as a nutrient sink at least during early phases of decomposition (Abbott and Crossley 1982), and the gradual reduction of this material (as well as dead roots) to humus should partially account for observed soil organic matter (0 to 15 cm depth) increases during the first few years following clearcutting (Chapter 16). The increase in soil organic matter increases storage pools of N and P in the soil, likely through increases in the exchange capacity and the amount of exchangeable nutrients that may be immobilized.

In summary, the effects of early successional forest vegetation upon ecosystem nutrient cycling processes in the southern Appalachians may be grouped into three major categories: (1) direct seasonal immobilization of nutrients in biomass, and hypothesized high annual throughfall and litterfall nutrient transfers to the forest floor as a result of accelerated nutrient uptake and concentration; (2) enrichment of N through symbiotic N fixation, immobilization, and eventual transfer of N to the forest floor with potential influence upon the rates of availability of other nutrients; and (3) early shading and the recovery of forest floor temperature and moisture conditions, resulting in the rapid recovery of biological activities which control decomposition, mineralization, and the potential mobility of nutrients.

13. Comparative Physiology of Successional Forest Trees

L.L. Wallace

Secondary forest succession occurs after some force causes an opening in the canopy. Large canopy openings, or gaps, may result from the death of one or more individual dominant trees, blow-downs, or selective cutting. These openings play a significant role in the Coweeta basin and in the eastern deciduous forest as a whole (Loucks 1970), as do large openings that result from clearcutting, forest fires, and widespread insect infestations or diseases. During succession, shade-intolerant or shade-intermediate trees will usually occupy a gap first. Later, an understory of shade-tolerant trees will form.

Tree response to canopy opening depends on the size of the opening and the resulting environment within that opening. Also important is the physiological age of the individuals that remain following opening, as represented by saplings, suppressed understory trees, or stump sprouts.

In this chapter, the abiotic changes that occur in gaps after openings of various sizes will be reviewed first. These parameters have been measured extensively at Coweeta and elsewhere. Next, physiological responses of several major forest species found at Coweeta will be examined in relation to adaptive characteristics. Although suites of physiological adaptation are associated with shade-tolerance or intolerance, the lack of one or more of the responses may limit tree distribution during succession.

Abiotic Changes Following Canopy Opening

The magnitude of the changes observed after canopy opening depends primarily on the size of the opening (Lee 1978). Larger openings produce greater incident light, and more variability in diurnal temperature fluctuations, relative humidity, and CO_2 levels.

Trees remaining after removal of a single tree will experience minimal changes in these physical factors. Soil moisture levels could also increase with the loss of a large canopy tree. The greatest change that these trees will experience is increased incident light intensity and a shift in the wavelength distribution of that light (Table 13.1).

Within a small canopy opening, a greater proportion of the total incident light will be direct beam radiation as compared with diffuse radiation (Hutchinson and Matt 1976, Figure 13.1). Diffuse radiation under a leafy forest canopy is relatively enriched in infrared wavelengths (Gates 1980). Therefore, as the size of the canopy opening increases, the relative enrichment by infrared will decrease. This enrichment may alter the response of systems regulated by phytochromes (Jacobs 1979). Within multi-tree gaps, microorographic effects could actually reduce rainfall within the opening relative to the amount falling in a clearcut (Lee 1978). However, the evaporative demand would also be minimal in this size of opening.

In a larger opening, a higher evaporative demand will exist. This will result from higher leaf and air temperatures, lower relative humidity, and higher wind speeds. Surface layers of the soil can dry out substantially (Swift et al. 1975). Simulations showed that the soil surface in clearcuts was at -15 to -20 bars for an average of 64 days per summer, compared with zero days for either oak-hickory or white pine watersheds. However, below the upper 30 cm of the soil, soil water content was most dependent upon the position of the site on the slope and on rainfall during previous weeks (Helvey et al. 1972). Many trees, including stump sprouts, can utilize water from these lower soil layers. In large clearcuts, the amount of rainfall reaching the soil could be greater than under closed canopies due to reduced interception (Swank and Miner 1968).

Physiological Responses

Light levels show the largest relative increase following canopy opening. For some time, forest trees have been classified with respect to shade tolerance. Shade-intolerant species such as *Liriodendron tulipifera*, *Prunus serotina*, and *Fraxinus americana* were classified as such since they are seldom found in the understory (Barrett 1980, Jackson 1967). Species of intermediate tolerance are found either as suppressed understory trees or as members of the overstory. These species, which include *Quercus alba*, *Q. nigra*, *Q. rubra*, *Fraxinus pennsylvanica lanceolata*, and *Acer rubrum*, may be found

Table 13.1. Relative changes in abiotic parameters after canopy openings of various sizes

Abiotic Parameter	Single-tree Gap	Multi-tree Gap	Large Gap
Incident PAR[a]	80%	90%	97%
Leaf Temperature at solar noon[a]	10%	10%	18%
Diurnal air temperature range[a]	28%	30%	35%
Relative humidity[a]	0%	−3%	−4%
Wind Speed[a]	0%	5%	8%
Soil Moisture 0–60 cm[b]	3%	−10%	−14%

Values of parameters are relative to those in an understory environment, which have arbitrarily been assigned values of zero. Increased levels are positive; decreased levels are negative.
[a] Wallace 1978
[b] Lee 1978

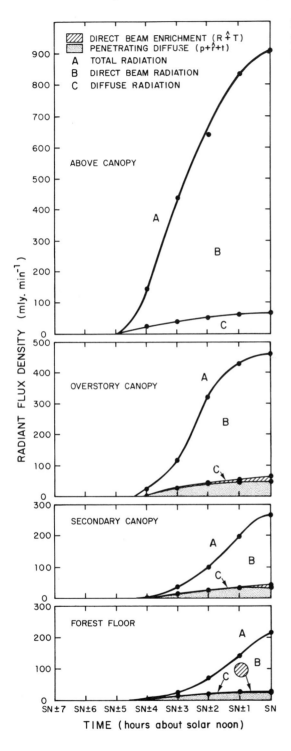

Figure 13.1. Solar radiation budgets in and above the winter leafless forest. (From Hutchinson and Matt 1976.)

in high abundance in forests whose major disturbance is individual or multi-tree gaps (Barden 1981). Shade-tolerant species, such as *Cornus florida*, *Diospyros virginiana*, and *Carpinus caroliniana* rarely occupy the overstory canopy and are primarily found as important understory components. The primary difference between these latter two classifications is whether or not a species can occupy the overstory.

Only recently have the physiological constraints operating within these species been examined. Three factors are of prime importance: the ability to harvest additional incident light, the ability to utilize this additional energy in CO_2 fixation, and the ability to orient the canopy and leaves to avoid high temperatures and/or photobleaching. Species which are of intermediate shade-tolerance have exhibited great plasticity in all three traits (Wallace and Dunn 1980).

Morphological Responses

Light-harvesting capability is regulated by the proportion of total biomass invested in leaves and by the size and number of photosynthetic units (PSU) within those leaves (Wallace and Dunn 1980). Generally, more shade-tolerant species have a greater proportion of biomass devoted to leaves. McGinty (1972) found that at Coweeta the more shade-tolerant *Rhododendron maximum* had 38.7% of its total biomass as leaves, whereas *Kalmia latifolia*, the shade-intermediate species, had 44.9% of its total biomass as leaves. However, leaf surface area averaged only 621 m² ha⁻¹ for *K. latifolia* and 1515.6 m² ha⁻¹ for *R. maximum*. Ericaceous species can have high fiber and lignin contents in their leaves (Mooney 1983); hence surface area is a better measure of light-harvesting ability, in these species. However, in the nonsclerophyllous species, percent investment in leaf biomass follows shade tolerance closely (Table 13.2).

Leaf orientation and canopy structure will modify the microenvironment of individual leaves and may ameliorate extremes in light intensity. Horn (1976) described the different canopy orientations assumed by understory and overstory trees. Shade-intolerant species typically exhibited a great deal of self-shading, whereas shade-tolerant species had monolayer canopies. Therefore, in a low-light environment intolerant species will have very little light reaching their leaves. Shade-intermediate species can, in many cases, switch from a monolayer to a self-shading canopy (Wallace 1978).

The angle at which the leaves are held can also significantly reduce incident light intensity (Gates 1980). Wallace and Dunn (1980) found that the shade-intolerant

Table 13.2. Proportion of Total Biomass Invested in Leaf Production in Species of Various Shade-Tolerance

Species	Shade-Tolerance	Percent Biomass in Leaves	Surface Area (m² ha⁻¹)	Source
Liriodendron tulipifera	Intolerant	39.8		Day and Monk 1977a
Kalmia latifolia	Intermediate	44.9	1515.6	McGinty 1972
Acer rubrum	Intermediate	46.5		Day and Monk 1977a
Rhododendron maximum	Tolerant	38.7	621.0	McGinty 1972
Cornus florida	Tolerant	65.0		Day and Monk 1977a

species, *Liriodendron tulipifera*, maintained a horizontal leaf display regardless of light environment. Both the shade-intermediate and shade-tolerant species had horizontal leaf displays at low light, but had leaf angles 30 to 35° from below horizontal under high light. This translates to a decrease in incident light intensity of 13 to 18%. Wallace and Dunn (1980) hypothesized that canopy shape and leaf display were under rigid genetic control in *L. tulipifera*. This, coupled with a low investment in leaf biomass, could restrict this species to high light environments.

Light-Harvesting Systems

Shade-adapted plants typically have larger photosynthetic units (PSU) to more effectively harvest the low quantum flux in the understory (Boardman 1977). However, with larger PSU, fewer PSU and electron transport chains will fit on each thylakoid. Because of lower ATP and NADPH potential, shade-adapted plants photosynthetically saturate at low light intensities (Figure 13.2). Wallace and Dunn (1980) found that the shade-tolerant species *Cornus florida* was incapable of increasing PSU number or decreasing PSU size in the field. When *C. florida* plants in the laboratory were abruptly transferred from a low light to a high light regime, photobleaching occurred.

By contrast, high light-adapted plants have more, but smaller, PSU (Boardman 1977). With a greater potential generation of ATP and NADPH, photosynthesis does not light saturate until relatively high light intensities are reached. *Liriodendron tulipifera*, a shade-intolerant species, and *Acer rubrum*, a shade-intermediate species,

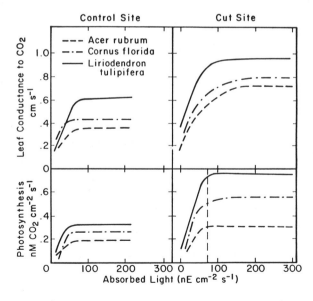

Figure 13.2. Light response curves of photosynthesis and leaf conductance for three tree species growing under a closed canopy (control site) and within a clearcut (cut site), from Wallace and Dunn (1980).

both showed an increase in PSU number and a decrease in PSU size within 1 month of being transferred to a high light regimen (Wallace and Dunn 1980). The results were quite variable, however, and may be the result of inactivation of PSU during extraction (Armond and Mooney 1978). It appears that both species may have fairly plastic responses with respect to PSU size and number. This may be a fairly general response among forest species. Given the shifting light regimes that occur seasonally, a plastic light harvesting system may be essential. Martinez (1975) determined the photosynthetic saturation points of both *Rhododendron maximum* and *Kalmia latifolia* in four different seasons. He found light saturation points increased as available light increased, indicating that both of these important understory species must have plastic light harvesting systems. Light saturation occurred at higher intensities for the shade-intermediate species *K. latifolia*, however. This indicates that although the response of a species may be plastic, the potential range of its response may limit its effective use of extremes in light intensity. *Cornus florida* is unusual in its inability to shift this parameter.

Photosynthetic Responses

The ability to utilize increased incident light levels in CO_2 fixation is also dependent upon the amount and activity of carboxylating enzyme in the leaf (Boardman 1977). This can be determined directly by enzyme kinetics, or indirectly by measuring leaf protein content. In C_3 plants, RuBP carboxylase constitutes the bulk of leaf protein (Larcher 1975) and is found in low quantity in shade-adapted plants (Boardman 1977). Wallace and Dunn (1980) found no significant differences in leaf protein content of the three tree species they examined. However, Day and Monk (1977b) examined leaf nitrogen content of 23 different species. If their data are separated into shade-intolerant vs. shade-tolerant and intermediate species (Barrett 1980), the intolerant species have substantially higher leaf nitrogen contents than the species in the other two classes ($2.38\% \pm 0.06$ and $1.93\% \pm 0.02$, respectively). McGinty (1972) determined the leaf nitrogen contents of *Kalmia latifolia* (1.35%) and *Rhododendron maximum* (1.60%). Both of these important understory evergreens have lower nitrogen contents than the average for overstory hardwoods (2.25%).

Plants that will be important in the early stages of gap closure and succession have a higher nitrogen demand than the later, more shade-tolerant species. This correlates well with the findings of Boring and Swank (1984b), which show nitrogen fixation to be greatest during the early stages of succession. However, the initial response of released trees does not involve increasing leaf protein content within 1 month of exposure to high light (Wallace and Dunn 1980). This lag in increased nitrogen demand may be important, since some time is needed to increase soil nitrogen levels after gap opening.

Two additional physiological responses are involved in a plant's ability to effectively utilize increased incident light to fix CO_2. The leaf conductance to diffusion into the carboxylation sites must be greater than under conditions of low light. This process involves both stomatal opening and an increase in transfer conductance (Nobel et al. 1975). These two phenomena are also involved in the plant's response to the increased evaporative demand within an opening. Stomatal opening can be measured directly, while transfer conductance can be examined indirectly, using greater specific leaf weights as an indication of greater internal surface area and higher transfer conductance (Nobel et al. 1975).

Within 1 month after exposure to high light, Wallace and Dunn (1980) noted that stomatal conductance of the shade-intolerant and -intermediate species was the same as for plants grown continuously in a high-light laboratory environment. However, the shade-tolerant species, *Cornus florida* showed no acclimation because stomatal conductance of the transferred trees was significantly lower than for the high light-grown individuals.

Roberts et al. (1980) examined water relations of the same three tree species grown in the field that Wallace and Dunn (1980) studied. Leaf water potential consists of three components: (1) turgor potential; (2) matric potential; and (3) osmotic potential. Initial changes in water potential are due primarily to changes in turgor potential. As turgor approaches zero, further changes are the result of variations in osmotic potential. When larger water deficits (WD) occur, the matric component becomes important. The increased evaporative demand within gaps will not cause WD to become so great as to involve adjustments of matric potential. However, rapid adjustments of turgor and osmotic potentials must occur. Roberts et al. (1980) found that both *L. tulipifera* and *A. rubrum* could rapidly adjust osmotic potentials. *C. florida*, the shade-tolerant understory tree, was unable to adjust osmotic potential rapidly and therefore had to undergo large changes in total potential at high WD values. Because of this, *Cornus florida* may be restricted to low-light habitats. Open-grown trees of this species appear to create this sort of microclimate via the canopy adjustments discussed earlier. Because the top leaves in the canopy typically face more xeric conditions than do lower leaves (McConathy 1983, McGinty 1972, Zimmerman and Brown 1974), vertical growth of *C. florida* may be sufficiently restricted so that it never enters the overstory.

On the other end of the spectrum, the shade-intolerant and shade-intermediate species showed rapid stomatal opening in the high light environment. Woods and Turner (1971) found that *L. tulipifera* stomata were unresponsive in low light intensities. Stomatal conductance in *Acer rubrum* was nearly 50% greater than for *L. tulipifera* at comparable low levels of light. No difference existed between species at high light levels. *Quercus rubra* behaved similarly to *A. rubrum*. This is further evidence of the physiological plasticity of shade-intermediate species.

Because leaf thickness and specific weight generally increases in xeric environments regardless of incident light intensity, it is difficult to separate high light effects on transfer conductance (which increases proportionally with leaf thickness) from the effects of high evaporative demands. Jackson (1967) examined leaf blade thickness and the ratio of palisade mesophyll to spongy mesophyll in 21 species. Using a previous classification of relative shade-tolerance, he found that the shade-intolerant species had the thickest leaves and the highest ratio of palisade to spongy mesophyll. Shade tolerant species had thinner leaves with a lower ratio.

Wallace and Dunn (1980) found that all three species they examined were capable of increasing specific leaf weight (SLW) within one month of exposure to high light. Consequently there were no significant differences in the SLW of species growing in a gap environment, regardless of shade-tolerance. The mean specific leaf weight of all three species was 0.035 g cm^{-2}. However, just as WD is only an approximate indicator of leaf water status (Fitter and Hay 1981), SLW is only an approximation of internal conduc-

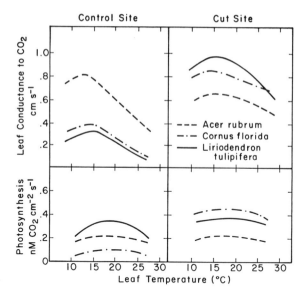

Figure 13.3. Temperature response curves of photosynthesis and leaf conductance for three tree species growing under a closed canopy (control site) and within a clearcut (cut site). (From Wallace and Dunn 1980.)

tance to CO_2. *L. tulipifera* had the highest intracellular conductance, followed by values for *A. rubrum* and *C. florida*, respectively. McGinty (1972) measured total leaf mass and leaf area of two important evergreen understory species, *Rhododendron maximum* and *Kalmia latifolia*. Using these data, the SLW for both species was calculated and according to expectations, the shade-tolerant species *R. maximum* had lower SLW than did *K. latifolia*, the shade-intermediate species (0.209 g cm⁻² vs. 0.173 g cm⁻²). When Martinez (1975) determined internal conductance empirically, *R. maximum* had a lower conductance (0.056 cm s⁻¹ vs. 0.063 cm s⁻¹). The values for SLW show a relative difference of 16.7%, whereas the values for conductance show a relative difference of 11.1%. However, both SLW and conductance values are consistent with shade tolerance theory (Boardman 1977).

In addition to changes in incident light and evaporative demand, leaf temperature will be slightly higher in open-grown trees. As opposed to the extreme photosynthetic temperature responses seen for species which experience large diurnal temperature fluctuations, Wallace and Dunn (1980) found that all the species they examined had broad, flat temperature response curves (Figure 13.3).

Of all the species theoretically able to fill gaps, only a few appear to be most successful in the Coweeta basin. The speed of response to gap opening is partially dependent upon the maximum attainable rate of photosynthesis (Table 13.3). The correlation between maximum photosynthetic rate and success rate in occupying the canopy gaps is far from perfect, however. For example, *Q. rubra* has the highest photosynthetic rate listed in Table 13.3. Barden (1981) found this species absent from single-tree gaps and in only 0.1% of the multiple-tree gaps in the Smokey Mountains of Tennessee. By contrast, he found *L. tulipifera* in 5% of the multiple-tree gaps and *A. rubrum* in 3.6% of the multiple-tree gaps. Both of these species had lower maximum photosynthetic rates.

Table 13.3. Maximum Measured Photosynthetic Rates (nM CO_2 cm^{-2} s^{-1}) of Trees of Different Shade Tolerance

Species	Shade-Tolerance	Maximum Photo-synthesis	Light Level	Source
Liriodendron tulipifera	Intolerant	0.672	Full sun	Wallace and Dunn 1980
Acer rubrum	Intermediate	0.382	Full sun	Wallace and Dunn 1980
Acer rubrum	Intermediate	0.5	1150 μE m^{-2} s^{-1}	Heichel and Turner 1983
Quercus rubra	Intermediate	0.8	1150 μE m^{-2} s^{-1}	Heichel and Turner 1983
Quercus alba	Intermediate	0.405	Full sun	Aubuchon et al. 1978
Kalmia latifolia	Intermediate	0.463	1500 μE m^{-2} s^{-1}	Martinez 1975
Rhododendron maximum	Tolerant	0.324	1500 μE m^{-2} s^{-1}	Martinez 1975
Cornus florida	Tolerant	0.556	Full sun	Wallace and Dunn 1980

The physiological plasticity of the latter species is further evidenced by its presence in 3% of the single-tree gaps as compared to *L. tulipifera* occurring in only 1% of these openings. Ehrenfeld (1980) noted that *C. florida* had essentially the same density in gaps and in undisturbed forest, but this species was found only in the understory. *Acer rubrum* dominates the understory of northeastern oak forests and invades the overstory whenever possible. This ability is not based upon maximum photosynthetic rates alone, but depends on other physiological and morphological features of the plant. This perplexing aspect of forest tree physiological ecology will be discussed further below.

Respiration Responses

Trees support a great deal of nonphotosynthetic respiring tissue, thus making respiration an important component of growth. Of course, to take advantage of a canopy opening an individual must outgrow its competitors. Rapid growth depends upon the plant maintaining a net carbon balance and upon ability to use stored carbon rapidly.

Within the leaves of shade-tolerant plants, respiration must be relatively low to maintain a positive carbon balance in the face of such low photosynthetic rates (Boardman 1977). Wallace and Dunn (1980) found no differences in respiration rates of the leaves of plants grown in high- or low-light regimes in the laboratory. However, they did not examine twig respiration. McLaughlin et al. (1978) examined twig and branch respiration of *L. tulipifera* and found that highest respiration rates correlated with time periods of highest photosynthetic rates and twig elongation. Cunningham and Syvertsen (1977) examined the hypothesis that rates of respiration depended on levels of total nonstructural carbohydrates (TNC). They found this to be true, particularly when TNC levels were low. In shade-tolerant plants, TNC levels may be sufficiently low to preclude a rapid increase in respiration rate; hence growth rate following gap opening would be slow. Further work is needed on carbohydrate levels and respiration rates of field-grown trees before this hypothesis can be verified.

4.   Canopy Arthropods and Herbivory

14. Foliage Consumption and Nutrient Dynamics in Canopy Insects

D.A. Crossley, Jr., C.S. Gist, W.W. Hargrove, L.S. Risley, T.D. Schowalter, and T.R. Seastedt

Coweeta watersheds contain a varied and abundant fauna of insects, spiders, mites, and other invertebrates. Arthropods are usually inconspicuous, except when population excursions of one or more species produce noticeable defoliation in forest canopies. Outbreaks of defoliating or wood-boring insect species clearly have an impact on the ecology of forested watersheds and affect decisions about forest management. Considerable information has been developed on the biology and ecology of these economically important insect species (Coulson and Witter 1984). Much less is known about the ecology of economically unimportant insects, or even on the nonoutbreak phases of the important ones.

Our research on canopy arthropods at Coweeta has centered on their importance in forest ecosystem processes. We have measured arthropod abundance and biomass in functional groups, in an attempt to characterize entire guilds of species. The use of functional guilds has allowed us to devote efforts to interpreting the impact of arthropods on forest processes. Shifts in functional guilds were found to accompany succession or disturbance (Chapter 15). We have measured leaf area removed by insect feeding activities, and performed some experimental work on the effects of feeding on throughfall chemistry.

Twelve years ago Mattson and Addy (1975) proposed that insects may act as regulators of forest primary production, perhaps by stabilizing or optimizing this process. The idea that consumers play a regulatory role in various ecosystems has gained acceptance (see Woodmansee 1978; Kitchell et al. 1979; Zlotin and Kodashova 1980; Seastedt and Crossley 1984). Our major organizing paradigm in our NSF-sponsored research has been one of nutrient cycling as a measure of forest ecosystem function. We have organized the research around the impact of canopy arthropods on forest nutrient cycling. Much of the research has studied arthropods and their effects at low or "background" population sizes. We have been presented with two opportunities to document the impact of insect outbreaks on forest nutrient dynamics: one the effect of a defoliator, the other the effect of a wood-borer. Our studies in nonoutbreak conditions have provided baseline data against which outbreak situations can be evaluated.

A conceptual view of the influence of arthropods on forest nutrient cycling processes is presented in Figure 14.1. Arthropods may influence the timing of litterfall as well as the character and nutrient content of canopy throughfall. Insect frass, secretions, excretions, and eventually their bodies constitute a small but steady rain of nutrients to the forest floor. Leaching of nutrients from grazed foliage may also be important. Alteration of nutrient inputs to the forest floor may catalyze nutrient transfers there, where soil invertebrates have significant influences. Seastedt and Crossley (1984) suggested that canopy arthropods may have the greatest impact on the more mobile elements. These elements are most likely to be affect by enhanced leaching and are more likely to be reassimilated by roots (MacNaughton 1983). Detritivores may have more influence in transfers of less mobile nutrients.

In the Coweeta basin, a partial defoliation by the fall cankerworm (*Alsophila pometaria*) resulted in marked changes in nutrient cycling within the affected watersheds. Stream concentrations of nitrate increased measurably during the defoliation and declined to baseline levels when cankerworm populations declined (Swank

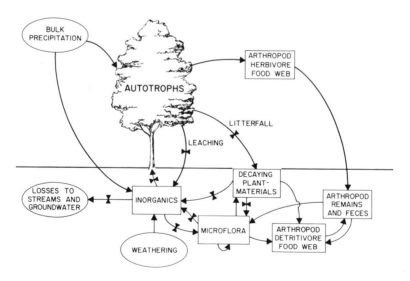

Figure 14.1. A simplified conceptual model of elemental cycling in a terrestrial ecosystem emphasizing the presence and activities of arthropod consumers. Indirect regulation of elemental flows by arthropods is indicated by valves on the flows. Virtually all fluxes within ecosystems are known or believed to respond to varying levels of consumer activity. (After Seastedt TR and Crossley DA Jr. BioScience 34:157–161, 1984. Copyright © 1984 by the American Institute of Biological Sciences.)

et al. 1981). This series of events is presented in Chapter 25. It is reviewed here because it is a clear illustration of the significance of herbivory for nutrient cycling processes at the ecosystem level. There was a net increase in net primary production, large increases occurred in litterfall, nutrient inputs via frass and canopy throughfall increased dramatically, and soil nitrogen pools and associated microflora increased as well (Swank et al. 1981). In another instance (Chapter 25) increased nitrogen concentrations in streamwater draining a successional watershed (WS 6) coincided with high infestations of black locust by the locust stem borer, (*Megacyllene robiniae*). Death or injury to many black locust trees and increased foliage consumption have accompanied increased export of nitrate from this watershed. Studies now in progress are addressing a number of hypotheses to explain the relationship between stem borer attack and stream export of nitrates.

In this chapter we summarize two topics. The characterization of feeding guilds of arthropods in forest canopies is described, along with the use of guilds in describing between-tree and between-watershed variations in arthropod biomasses and standing crops of nutrients. The feeding guilds are used to evaluate nutrient accumulation by the arthropod biomass. The second topic is the analysis of leaf area removed (LAR) by insect feeding as a means of estimating herbivory.

Canopy Arthropod Biomass and Nutrient Element Contents

Samples of canopy arthropods were not sorted to species, but were segregated into trophic units of a preconceived compartment model (Crossley et al. 1976). These trophic units correspond approximately to the guild concept proposed by Root (1967), and we have used the term "guild" to describe them, although the major use of the guild concept has been to consider interactions between species (Yodis 1982). Table 14.1 presents definitions of the guilds, using trophic level, food source, feeding type, size, reproductive rate, and mobility as descriptors. We have used this compartment arrangement as a basis for data organization. The set of compartments includes major feeding types of herbivores and predators, as well as specialized grazers on microflora, forest floor migrants, and emergent stream insects. While the latter may not feed in the canopy, they certainly serve as food for predators there. In situ sampling has been performed with several modifications of an insect net fitted with a plastic bag liner (Crossley et al. 1976; Kaczmerek and Wasilewski 1977; Schowalter et al. 1981c). The plastic bag is closed with a drawstring after being slipped rapidly over a foliage supporting branch or twig. Arthropods are thus trapped along with a sample of foliage. Conversion to biomass was made possible by weighing the contained foliage and comparing the mass to that of the foliage species on the watershed (i.e. Day and Monk 1974). Samples were netted using a long-handled net or by climbing trees. Lower and middle canopies were adequately sampled, but upper canopies on some watersheds could not be reached using the net method.

Weight Intensities

The weight intensities of arthropods in foliage (mg arthropods per 100 g foliage) were useful in comparing the arthropod loading of foliage between different tree species.

Table 14.1. Compartment Definitions for Canopy Arthropod Guilds

Compartment	"Type"	Trophic Level	Biological Characteristics
1	Tree crickets	H	Leaf feeders, large size, low reproduction, chewing, mobile
2	Aphids	H	Leaf feeders, small size, rapid reproduction, sucking, immobile
3	Leafhoppers	H	Leaf feeders, small size, moderate reproduction, sucking, relatively mobile
4	Caterpillars	H	Leaf feeders, large size, low reproduction, chewing, immobile
5	Moths	H	Specialized feeders, large size, low reproduction, sucking, mobile
6	Ants	O	(Social insects with specialized ecology)
7	Some flies, beetles	H	Inflorescence feeders, variable size, moderate reproduction, variable feeding, mobile
8	Chrysomelid beetles	H	Leaf feeders, moderate size, moderate reproduction, chewing, mobile
9	Leaf miners	H	Specialized leaf feeders, small size, moderate reproduction, immobile
10	Elaterids	H	Bark and wood borers, small to moderate size, moderate reproduction, chewing
11	Collembolans	—	(Specialized canopy grazers on fungi or epiphytes)
12	Cockroaches	—	(Forest floor migrants not feeding in canopy)
13	Braconids	P	Parasitoids with specialized ecology
14	Spiders	P	General predators, variable size, moderate reproduction, specialized feeding, immobile
15	Ladybird beetles	P	General predators, small size, rapid reproduction, chewing, mobile
16	Lacewings	P	General predators, small size, low reproduction, sucking, relatively immobile
17	Pentatomid bugs	P	General predators, moderate to large size, low reproduction, sucking, mobile
18	Some flies	P	General predators, small to moderate size, moderate reproduction, specialized feeding, mobile
19	Phalangids	P	General predators, small size, low reproduction, chewing, relatively mobile
20	Stoneflies	—	(Emergent stream insects not feeding in canopy)

Table 14.2. Arthropod Weight Intensities in Foliage of Major Tree Species, WS 18, Coweeta Hydrologic Laboratory, North Carolina, Summer 1972

Compartment Number	Tree Species					
	Dogwood	Tulip-Poplar	Scarlet Oak	Red Maple	Chestnut Oak	Hickory
1	—[a]	0.36	–	0.06	0.90	7.42
2	0.10	1.57	0.39	0.30	0.17	0.08
3	13.84	8.62	4.59	5.30	9.73	6.32
4	82.37	4.47	19.53	13.63	21.28	3.46
5	0.06	0.18	–	1.37	1.09	0.23
6	2.29	3.17	0.11	0.32	0.22	1.39
7	2.09	1.42	0.38	–	2.90	0.94
8	6.08	23.70	7.32	10.56	16.49	4.56
9	0.12	1.29	–	0.64	0.05	0.21
10	6.11	10.34	–	6.61	8.16	3.82
11	0.50	0.07	0.05	0.20	0.08	0.89
12	0.29	0.30	0.02	0.64	–	3.29
13	0.06	0.91	0.29	0.49	0.16	0.09
14	4.45	11.01	1.78	11.21	2.61	3.51
15	–	–	0.17	0.11	0.03	0.14
16	–	–	–	–	–	0.44
17	0.14	0.42	2.99	–	2.29	0.09
18	–	0.09	–	–	–	–
19	8.57	0.90	–	5.39	4.17	1.02
20	3.63	1.12	–	0.74	0.76	0.06
Total Weight Intensity (mg per 100 g)	130.70	69.94	37.62	57.57	71.09	37.97
Weight of Leaves (g)	206.1	481.8	319.9	845.8	585.4	609.3
Number of net-bag samples	50	102	41	130	88	106

[a] None detected.
Variates are mg (dry wt) arthropods per 100 g (dry wt) foliage.

Table 14.3. Weight Intensities for Two Trophic Levels in Six Canopy Arthropod Communities, WS 18, Summer, 1972

Weight Intensity (mg/100 g)	Dogwood	Tulip-Poplar	Scarlet Oak	Red Maple	Chestnut Oak	Hickory
Herbivores	110.8	51.9	32.2	38.5	60.8	27.0
Carnivores	13.2	13.3	5.2	17.2	9.3	5.3
Others	6.7	4.7	0.2	1.9	1.1	5.6
Totals	130.7	69.9	37.6	57.6	71.2	37.9

Table 14.2 presents weight intensities for foliage from six major canopy and subcanopy trees on WS 18, averaged over the summer of 1972. The dogwood fauna was largely dominated by caterpillars (Compartment 4), whereas at the other extreme, the hickory fauna does not appear dominated by any one compartment. Dogwood foliage supported the highest arthropod biomass (130.7 mg per 100 g foliage). Compartment 4 ("caterpillars") was the most abundant compartment across all tree species. Tulip poplar supported large masses of chrysomelid beetles (Compartment 8), elaterids (10), and spiders (11). Table 14.2 shows that every compartment was represented in the sampling, although not for every tree species.

Major differences occurred between tree species in the allocation of weight intensities between compartments. About 75% of arthropod weight could be classified as herbivore and 20% as carnivore (Table 14.3). The weight intensity of carnivores was not correlated with the weight intensity of herbivores across the six tree species ($r = 0.399$, n.s.). Spiders (Compartment 14) accounted for 50% of the weight intensity of carnivores.

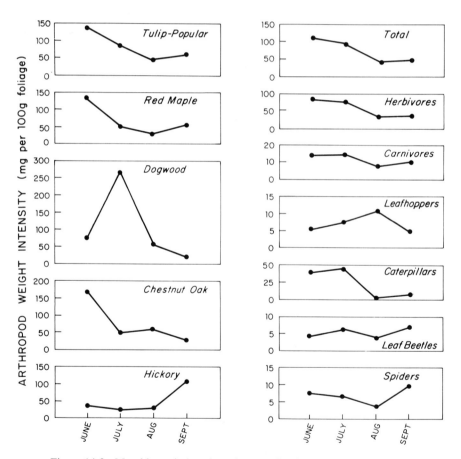

Figure 14.2. Monthly variations in estimates of arthropod weight intensity.

Monthly variation in weight intensities (Figure 14.2) shows that arthropod mass in foliage generally declined during the summer, with a slight increase in September. Exceptions were dogwood (Figure 14.2, left column) with high caterpillar biomass in July, and hickory, which supported large numbers of tree crickets in September. Individual compartments did not follow the general pattern of decrease during the summer (Figure 14.2, right column), but total weights of herbivores and carnivores did follow the pattern.

Nutrient Concentrations in Canopy Arthropods

Nutrient analyses of canopy arthropods required lumping of samples by trophic level, because individual masses were too low for separate analysis. Table 14.4 presents cation concentrations (K, Na, Ca, Mg) for herbivores, carnivores, and additional compartments. These values are consistent with others reported in the literature (Crossley et al. 1975; Gist and Crossley 1975a; Webster and Patten 1979). (See also Chapter 15.) An obvious trophic level effect was found for Na, which averaged almost five times higher concentration in carnivores than in herbivores ($t = 16.88, p < 0.01$). Calcium concentrations in carnivores were also significantly higher ($t = 4.73, p < 0.05$). Such increases in Na concentrations in carnivorous arthropods have been reported previously (Reichle and Crossley 1969; Crossley et al. 1975; Schowalter et al. 1981c). Non-feeding forest floor migrants (Compartment 12) had K concentrations lower than other canopy dwellers, but similar to values reported by Gist and Crossley (1975a) for forest floor arthropods. Emergent stream insects (Compartment 2) had nutrient concentra-

Table 14.4. Nutrient Concentrations in Samples of Canopy Arthropods, WS 18, Summer 1972

Trophic Level		Dogwood	Tulip-Poplar	Scarlet Oak	Red Maple	Chestnut Oak	Hickory	Mean
Herbivores	K	11.30	10.02	12.93	12.01	11.38	11.49	11.52
	Na	0.31	0.88	0.39	0.42	0.46	0.75	0.53
	Ca	0.85	0.84	$-a$	0.98	0.98	1.06	0.94
	Mg	1.27	1.63	1.82	1.07	1.04	1.30	1.35
Carnivores	K	9.88	11.43	9.45	10.11	9.67	13.09	10.61
	Na	2.70	3.11	2.41	2.33	2.01	2.67	2.54
	Ca	1.93	1.79	1.21	$-a$	1.33	1.43	1.54
	Mg	1.17	1.70	1.14	1.63	1.05	1.35	1.34

Additional Compartments

	Forest Floor Migrants	Emergent Stream Insects
K	6.99	12.90
Na	1.39	3.42
Ca	1.10	1.58
Mg	1.83	1.39

a No measurement.
Values are mg per g.

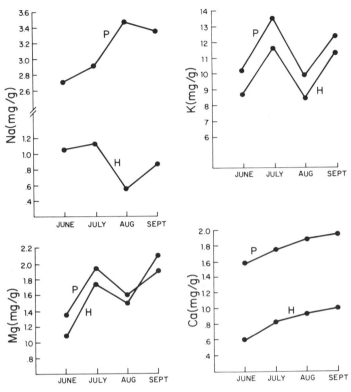

Figure 14.3. Monthly variations in nutrient element concentrations in herbivores (H) and predators (P) for the canopy arthropod fauna of tulip poplar.

tions resembling carnivore trophic levels of canopy arthropods, and they fell within the
ranges of Ca and K concentrations reported by Webster and Patten (1979) for stream
insects from WS 18.

Within-season variation in arthropod nutrient concentration was examined for tulip
poplar faunas (Figure 14.3). Results showed that nutrient concentrations in arthropods
varied seasonally, and followed closely nutrient concentrations in tulip poplar foliage
reported by Day and Monk (1977b) for WS 18 during the same year (1972). Potassium
concentrations in herbivores were strongly correlated with those in carnivores ($r =$
$0.9764, p < 0.05$). High values occurred in July and again in September; the same pattern appears in analyses of foliage as reported by Day and Monk (1977b). No other tree
species showed such variation in K concentration in foliage after the middle of May
(Day and Monk 1977b). Sodium concentrations in carnivores and herbivores appeared
to vary at random and independently, with carnivores averaging about four times the
concentrations in herbivores. Calcium concentrations in both herbivores and carnivores increased regularly throughout the summer and were closely correlated ($r =$
$0.9921, p > 0.05$). In foliage, Ca concentrations increased also (Day and Monk 1977b)
and arthropod values closely track the foliage values ($r = 0.9395, p < 0.05$). Our

values for Mg in arthropods show a dip in concentrations during August. Again, Monk and Day (1977b) illustrated a similar dip (23 samples in July) for Mg concentrations in tulip poplar foliage; no other tree species showed such a variation in foliar Mg concentrations after May.

These results suggest that nutrient element turnover by canopy arthropods must be rapid. Nutrient concentrations in arthropods (except for Na) tracked fairly closely nutrient changes in the foliage. Either nutrient element excretion by canopy arthropods must be rapid, or biomass turnover is rapid, or both occur rapidly. In any case foliage, consumption and nutrient removal by arthropods greatly exceed the standing crop of insects in the canopy.

Arthropod Biomass and Nutrient Standing Crops

Arthropod biomasses were calculated from the weight intensities on tree species and adjusted for the abundance of that species on the watershed, as follows:

$$B = W \cdot L/100$$

where B = arthropod biomass (mg/m²), W = weight intensity (Table 14.2 and other data), and L = leaf biomass (Table 14.5). The results (Table 14.5) reflect the dominance of chestnut oak on WS 18. Although the weight of insects per gram foliage was not high on chestnut oak, the preponderance of foliage for that species results in a large arthropod biomass. Each of the herbivore compartments exceeding 10 mg per

Table 14.5. Foliage and Arthropod Biomasses by Trophic Levels, WS 18, Summer 1972

Tree Species	Foliage[a] Biomass (g/m²)	Herbivores (mg/m²)	Carnivores (mg/m²)	Others (mg/m²)	Total (mg/m²)
Dogwood	17.5	19.38	2.31	1.18	22.87
Tulip-poplar	25.1	13.03	3.34	1.16	17.53
Scarlet oak	30.0	9.67	1.57	0.05	11.29
Red maple	40.4	15.54	6.95	0.77	23.26
Chestnut oak	84.6	51.39	7.83	0.88	60.10
Hickory	21.2	5.73	1.12	1.19	8.04
Red oak	32.1	5.46	7.00	0.96	13.42
White oak	7.0	1.55	1.67	0.05	3.27
Black oak	18.7	4.55	1.42	0.06	6.03
Sourwood	20.1	4.85	2.10	0.61	7.56
Black gum	16.2	2.05	0.46	0.00	2.51
Totals	312.9	133.20	35.77	6.91	175.88
Unsampled	68.5	–	–	–	–
Watershed Totals	381.4	162.36[b]	43.60[b]	8.42[b]	214.38[b]

[a] After Day and Monk 1977, Day 1971.
[b] Calculated as Total/0.8203, see text.

m² had its largest biomass in chestnut oak foliage. To extrapolate to the entire canopy on WS 18, we took Day's (1971) estimate of foliage biomass, subtracted the estimates for those trees we did not sample, and calculated the proportion sampled for arthropods (312.9/381.4 = 0.820). Totals for arthropod biomasses were then divided by this number to account for unsampled foliage. (We did not include *Rhododendron* in our calculations). Our estimate of 214 mg per m² is similar to other estimates for the eastern deciduous forest (e.g. Reichle and Crossley 1969) and in the Coweeta basin (see Chapter 15). It probably should be considered as a baseline value during nonoutbreak conditions; leaf area removed was less than 10% during the summer of 1972.

Table 14.6 shows nutrient standing crops in canopy arthropods, calculated as a product of biomass (Table 14.5) times nutrient concentration (Table 14.3 and other data). Potassium standing crops exceed other nutrients by an order of magnitude. A definite food chain effect appears for Na, in that the standing crop of Na in carnivores exceeds the standing crop in herbivores. The higher concentrations of Na more than compensate for lower biomass in carnivores.

The standing crop of nutrients in canopy arthropods is, of course, much smaller than foliage nutrients, but nutrient flow through the arthropod biomass greatly exceeds its standing crop. Canopy arthropods have been estimated to consume about 50 to 150% of their biomass per day (Kaczmarek 1967; Reichle et al. 1973; Schowalter et al. 1981c).

Analysis and Interpretation of Leaf Area Removed

Canopy foliage is damaged when arthropods feed upon it. In early work, Bray (1961) interpreted the leaf area removed (LAR) by insects as a measure of feeding rate. Bray used graph paper methods, but mechanical planimeters and a variety of photoplanimeters (Gist and Swank 1974) have been used. Interpretation of results has lead to a variety of feeding estimates for canopy insects (Table 14.7).

For graphic estimation of LAR the original leaf outline must be reconstructed, but with photoplanimeters another method must be found. Schowalter et al. (1981c) used aluminum foil cutouts to restore the original leaf shape. Risley (1983) matched partially consumed leaves with identical unconsumed ones; the difference in transmission represented LAR. Waide, Crossley and Todd (unpublished) used a random sampling method: leaves were randomly removed from a large garbage bag containing foliage to be measured until a sample of 100 completely uneaten leaves was obtained. The areas

Table 14.6. Standing Crops of Nutrients in Canopy Arthropods, WS 18, Coweeta Hydrologic Laboratory, Summer 1972

	Herbivore	Carnivore	Others	Total
K	1.541	0.380	0.069	1.990
Na	0.081	0.091	0.017	0.189
Ca	0.136	0.055	0.009	0.200
Mg	0.185	0.048	0.011	0.244

Variates are mg per m².

Table 14.7. Estimates of Leaf Area Removed (LAR) for Coweeta Watersheds (WS)

WS	Year	LAR	Method (See Text)	Reference
2	1977	1.90%	Light box	Schowalter et al. 1981a
6	1984	13.40%	Light box	Risley 1983
7	1977	2.74%	Light box	Schowalter et al. 1981a
7	1980	9.67% (*Robinia*) 3.00% (*Acer*)	Digitizer	Seastedt et al. 1983
18	1972	7.10%	Light box	(unpubl.)
18	1981	12.10%	Light box	Risley 1983
27	1974	32.00%[a]	Random sample	(unpubl.)
27	1975	16.00%[a]	Random sample	(unpubl.)

[a] During period of defoliation by fall cankerworm, *Alsophila pometaria* (see Swank et al. 1981)

of the uneaten leaves were measured photometrically, and all leaves were returned to the bag and thoroughly mixed with the contents. A second 100 leaf sample was taken at random, containing both consumed and unconsumed leaves. The leaf area for the second sample was then measured; the difference in transmission between the first and second samples was taken to represent LAR. With this method reconstruction of leaf shapes is unnecessary, but estimation of variability depends upon regression models. Brown (1982) used regressions of individual leaf length on leaf area to monitor LAR in tropical systems.

Our most recent method (Seastedt et al. 1983; Hargrove 1983) utilizes photocopies of leaves which have been flattened on a plant press. Outlines of leaves on the photocopies are then completed with a pencil, and a computer system used to digitize the areas. These procedures yield extensive data sets. It seems likely that the next development in these techniques will be the use of image analyzers or image processors. Also, leaf damage can be classified into a number of categories (Coulson and Witter 1984) and related to the compartment model of feeding guilds.

Estimates of leaf area removed by canopy insects in Coweeta watersheds are well within the range of results reported by others (Risley 1983). Table 14.7 presents values ranging from 1.9% LAR in the more xeric WS 2 to 32% during a caterpillar outbreak. Data suggest that LAR is generally higher in more mesic, north-facing watersheds. Also, clearcutting of WS 7 may have led to increased LAR after 4 years of regrowth.

The Linear Consumption Model

Hargrove (1983) digitized thousands of *Robinia pseudo-acacia* leaflets and examined the frequency distribution of damage classes. The distribution of LAR yielded a negative exponential model (Figure 14.4). Good fits to negative exponential distributions were obtained for insecticided and fertilized *Robinia* trees (Seastedt et al. 1983; Hargrove 1983) as well as untreated ones. Other models may be equally appropriate, but the linear negative exponential model has the advantage of simplicity since it is described by a single constant.

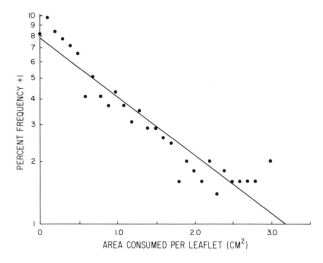

Figure 14.4. Log-transformed frequency histogram of damage to *Robinia pseudo-acacia* L. leaflets.

The slope of the LAR distribution, the consumption constant, was examined in detail for regrowth black locust on WS 7 (Hargrove 1983). LAR was found to accrue during the summer months in untreated foliage, from 5% in June to 14% in September. Insecticide treatment (with Sevin) reduced consumption to 2.5% in June and lesser values in later summer months; the consumption constant was reduced correspondingly. Fertilization of black locust produced mixed results for LAR. In June, applications of fertilizer (8-8-8 NPK) resulted in increased consumption, but consumption was reduced in the remaining summer months (Hargrove et al. 1984).

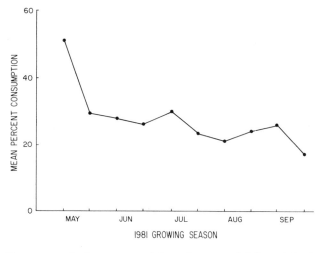

Figure 14.5. Mean percent leaf area removed from chestnut oak foliage, WS 18, summer 1981.

These results show that analysis of the frequency distribution of LAR can improve estimates of canopy herbivory. Changes in herbivore pressure or nutrient status in black locust caused detectable changes in LAR distribution in experimental trees. Current research is investigating the frequency distribution of leaf area removed in different successional status, and through the growing season. The distributions will be examined for deviation from the linear model. One interpretation of the linear, negative exponential model is that all foliage is equally available for consumption. Previous feeding upon a given leaf should not change its palatability for further feeding. Recent experimental work (Schultz and Baldwin 1982; Ryan 1983) has shown that leaf chemistry for some plants can change rapidly under herbivore pressure. Even undamaged trees may respond to herbivory elsewhere in the same stand (Rhoades 1983). Such plant responses should result in significant deviations from linearity in the LAR frequency distributions.

Seasonal Distribution of Leaf Area Removed

Estimates of leaf area removed vary seasonally. Our measurements on black locust showed that LAR accumulated during the growing season. Data from the undisturbed hardwood WS 18 showed that LAR decreased on some tree species, including chestnut oak, during the summer (Figure 14.5). Earlier, Reichle et al. (1973) demonstrated that holes punched in tulip poplar leaves would expand during the growing season, resulting in an overestimate of consumption. There is an increasing body of literature which suggests that damaged leaves are abcissed prematurely. The high measurements of LAR obtained in May (ca. 50%) declined to around 10% during the summer months (Figure 14.5). Since leaf expansion is complete in chestnut oak by July, dilutions of percentage LAR could only come from production of new, undamaged leaves or abscission of heavily consumed leaves. Current research is attempting to follow seasonal development of LAR, including tracking of individual leaves, hole punching, and premature abcission of damaged vs. undamaged leaves.

15. Canopy Arthropods and Their Response to Forest Disturbance

T.D. Schowalter and D.A. Crossley, Jr.

Most ecological studies of forest canopy arthropods have focused on population dynamics. These studies have contributed much to our understanding of population responses to changes in environmental conditions. Moreover, advances in population theory have indicated the importance of understanding the dynamics of species assemblages. Population irruptions do not result simply from changes in abiotic or host conditions, but also reflect competitive interactions within guilds and predisposing changes in host or predator conditions affected by other guilds (Schowalter 1985). Bark beetle population irruptions, for example, may result from host stress aggravated by defoliator populations (Berryman and Wright 1978).

The nature of interactions within and between arthropod guilds and the effects of those interactions on primary productivity have suggested regulatory roles (Mattson and Addy 1975; Odum 1969; Patten and Odum 1981). Schowalter (1981, 1985) advanced the cybernetic view of ecosystems by proposing that the state of predator–prey interactions varies as a function of ecosystem carrying capacity. Resource subsidy promotes prey population growth while resource stress promotes predator population growth. This view also casts succession as a consequence of regulatory responses to delayed disturbance, i.e., as a succession of associations adapted to decreasing disturbance frequency, rather than as a goal-oriented community strategy. However, few studies have documented changes in arthropod assemblages resulting from changes in environmental conditions. Forest canopy arthropods have been particularly difficult to study because of their taxonomic complexity and difficulty of access to them.

Studies of canopy arthropod assemblages at Coweeta have produced an extensive data set documenting assemblage structure across a wide range of ecosystem states. In this chapter, these data are used to assess the magnitude of canopy arthropod responses to changing environmental conditions, to evaluate the relative importances of factors influencing arthropod responses, and to identify mechanisms of regulation.

Canopy Arthropod Responses to Disturbance

We have compiled data from four watershed-level studies representing 8 watershed-years. Included are four studies of canopy arthropods in mature canopies (Chapter 14 this volume, H. Peturssen and Crossley unpubl. data; Risley 1983; Schowalter et al 1981c), one study of arthropods in first and second-year regrowth (Schowalter et al. 1981c), and one study of arthropods in 13-year-old regrowth (Risley 1983). Arthropod mass intensities (mg/kg foliage) on three late successional dominants (chestnut oak (*Quercus prinus* L.), red maple (*Acer rubrum* L.), hickory (*Carya* spp.), and three early successional dominants (dogwood (*Cornus florida* L.), tulip poplar (*Liriodendron tulipifera* L.), and black locust (*Robinia pseudoacacia* L.) across this successional gradient are presented in Tables 15.1 through 15.6. Canopy arthropod guild structure

Table 15.1. Arthropod Mass Intensities (mg/kg Foliage) on Chestnut Oak (*Quercus prinus*) at Coweeta Hydrologic Laboratory, North Carolina

Guild	Watershed Age (yrs)			
	1^a (mg/kg)	2^a (mg/kg)	13^b (mg/kg)	$>60^c$ (mg/kg)
Phytophages				
Aphids and aleyrodids	10	3	19	3 (3)
Other sap-suckers	89	41	27	111 (29)
Caterpillars	21	34	397	160 (70)
Tree crickets	268	6	60	110 (130)
Beetles	146	14	41	140 (131)
Leaf miners	1	0	6	4 (4)
Flower-feeders	0	0	61	23 (15)
Bark and wood borers	1	2	0	35 (34)
Omnivores				
Ants	81	60	91	9 (8)
Predators				
Beetles	0	0	3	9 (19)
Lacewings	3	0	2	1 (1)
Flies and wasps	4	6	5	5 (4)
Assassin bugs	7	0	246	29 (23)
Spiders and phalangids	9	45	84	90 (70)
Others	16	22	26	25 (33)
Total	656	233	1068	723 (130)

[a] Data from Schowalter et al. (1981c).
[b] Data from Risley (1983).
[c] $\bar{x} \pm$ SD. Data from Crossley et al. (this volume), H. Peturssen and Crossley (unpublished data), Risley (1983), and Schowalter et al. (1981c).

Table 15.2. Arthropod Mass Intensities (mg/kg Foliage) on Hickories (*Carya* spp.) at Coweeta Hydrologic Laboratory, North Carolina

	Watershed Age (yrs)			
Guild	1[a] (mg/kg)	2[a] (mg/kg)	13[b] (mg/kg)	>60[c] (mg/kg)
Phytophages				
Aphids and aleyrodids	25	7	2	5 (4)
Other sap-suckers	159	66	133	82 (49)
Caterpillars	2	47	196	117 (85)
Tree crickets	0	333	251	27 (32)
Beetles	35	3	72	48 (46)
Leaf miners	0	0	1	0
Flower-feeders	0	1	14	18 (9)
Bark and wood borers	0	89	0	19 (9)
Omnivores				
Ants	36	43	22	13 (8)
Predators				
Beetles	0	0	2	3 (2)
Lacewings	0	0	3	4 (5)
Flies and wasps	1	1	2	3 (2)
Assassin bugs	50	11	49	11 (15)
Spiders and phalangids	40	3	90	58 (25)
Others	18	4	73	32 (30)
Total	366	605	910	453 (147)

[a] Data from Schowalter et al. (1981c).
[b] Data from Risley (1983).
[c] $\bar{x} \pm$ SD. Data from Crossley et al. (this volume), H. Peturssen and Crossley (unpublished data), Risley (1983), and Schowalter et al. (1981c).

Table 15.3. Arthropod Mass Intensities (mg/kg Foliage) on Red Maple (*Acer rubrum*) at Coweeta Hydrologic Laboratory, North Carolina

	Watershed Age (yrs)			
Guild	1[a] (mg/kg)	2[a] (mg/kg)	13[b] (mg/kg)	>60[c] (mg/kg)
Phytophages				
Aphids and aleyrodids	33	13	6	5 (5)
Other sap-suckers	94	61	36	52 (24)
Caterpillars	15	14	550	80 (54)
Tree crickets	21	1	0	71 (133)
Beetles	96	15	47	65 (59)
Leaf miners	1	0	4	7 (8)
Flower-feeders	31	13	21	18 (15)
Bark and wood borers	5	23	0	32 (34)
Omnivores				
Ants	273	26	2	5 (3)

(Continued)

Table 15.3. *(Continued)*

Guild	Watershed Age (yrs)			
	1[a] (mg/kg)	2[a] (mg/kg)	13[b] (mg/kg)	>60[c] (mg/kg)
Predators				
Beetles	22	18	7	1 (1)
Lacewings	0	1	0	2 (2)
Flies and wasps	9	61	7	13 (9)
Assassin bugs	6	0	2	18 (21)
Spiders and phalangids	29	19	183	76 (61)
Others	15	27	14	13 (6)
Total	650	292	879	468 (100)

[a] Data from Schowalter et al. (1981c).
[b] Data from Risley (1983).
[c] $\bar{x} \pm$ SD. Data from Crossley et al. (this volume), H. Peturssen and Crossley (unpublished data), Risley (1983), and Schowalter et al. (1981c).

Table 15.4. Arthropod Mass Intensities (mg/kg Foliage) on Dogwood (*Cornus florida*) at Coweeta Hydrologic Laboratory, North Carolina

Guild	Watershed Age (yrs)			
	1[a] (mg/kg)	2[a] (mg/kg)	13[b] (mg/kg)	>60[c] (mg/kg)
Phytophages				
Aphids and aleyrodids	0	1	1	1 (0)
Other sap-suckers	110	162	49	77 (48)
Caterpillars	37	10	1022	319 (381)
Tree crickets	47	6	39	2 (3)
Beetles	3	12	27	100 (122)
Leaf miners	2	0	0	1 (1)
Flower-feeders	4	6	55	31 (42)
Bark and wood borers	0	113	1	66 (93)
Omnivores				
Ants	7	1	4	6 (10)
Predators				
Beetles	2	5	5	20 (30)
Lacewings	0	0	1	4 (6)
Flies and wasps	3	31	5	65 (110)
Assassin bugs	0	21	44	55 (73)
Spiders and phalangids	78	117	76	90 (51)
Others	8	4	11	51 (21)
Total	301	489	1390	831 (325)

[a] Data from Schowalter et al. (1981c).
[b] Data from Risley (1983).
[c] $\bar{x} \pm$ SD. Data from Crossley et al. (this volume), H. Peturssen and Crossley (unpublished data), Risley (1983), and Schowalter et al. (1981c).

Table 15.5. Arthropod Mass Intensities (mg/kg Foliage) on Tulip Poplar (*Liriodendron tulipifera*) at Coweeta Hydrologic Laboratory, North Carolina

	Watershed Age (yrs)			
Guild	1[a] (mg/kg)	2[a] (mg/kg)	13[b] (mg/kg)	>60[c] (mg/kg)
Phytophages				
Aphids and aleyrodids	226	106	105	19 (15)
Other sap-suckers	135	93	37	50 (26)
Caterpillars	469	395	90	81 (93)
Tree crickets	80	0	392	1 (2)
Beetles	7	2	25	100 (99)
Leaf miners	2	6	0	3 (6)
Flower-feeders	2	10	26	51 (84)
Bark and wood borers	0	108	0	31 (49)
Omnivores				
Ants	14	12	14	17 (10)
Predators				
Beetles	27	0	7	15 (19)
Lacewings	0	2	1	4 (7)
Flies and wasps	4	33	14	12 (6)
Assassin bugs	0	0	42	22 (23)
Spiders and phalangids	33	25	276	93 (47)
Others	16	22	11	14 (8)
Total	1042	814	1040	520 (234)

[a] Data from Schowalter et al. (1981c).
[b] Data from Risley (1983).
[c] $\bar{x}$ ± SD. Data from Crossley et al. (this volume), H. Peturssen and Crossley (unpublished data), Risley (1983), and Schowalter et al. (1981c).

showed a surprising degree of stability upon clearcutting at Coweeta. Mass intensities of most guilds showed no more change as a result of clearcutting than as a result of climatic and edaphic variation between years and watersheds.

The most pronounced responses to clearcutting appeared to be decreased caterpillar and spider biomass and significantly increased aphid and ant biomass on most tree species. Changes in mass intensities following clearcutting were relatively small and apparently compensatory (i.e., little change in total mass) on the late successional tree species (Tables 15.1 through 15.3) but were dramatic and variable on the early successional tree species (Tables 15.4 through 15.6). Aphid and ant mass intensities were significantly and positively correlated following clearcutting, but not in mature canopies (Schowalter et al. 1981c). Following clearcutting, aphid mass intensities increased one to two orders of magnitude on tulip poplar, due to population increases of *Macrosiphum liriodendri* Monell, and on black locust, due to population increases of *Aphis craccivora* Koch (Tables 15.5 and 15.6, Figure 15.1). Ant response was most conspicuous on black locust, where we observed ants feeding on extrafloral nectaries and on aphid and membracid honeydew. Aphid and ant responses were conspicuously absent on dogwood (Table 15.4) despite the abundance and apparency of this species, which codominated the clearcut with black locust (Boring et al. 1981).

Table 15.6. Arthropod Mass Intensities (mg/kg Foliage) on Black Locust (*Robinia pseudoacacia*) at Coweeta Hydrologic Laboratory, North Carolina

	Watershed Age (yrs)		
Guild	1^a (mg/kg)	2^a (mg/kg)	$>60^b$ (mg/kg)
Phytophages			
Aphids and aleyrodids	999	1548	14 (15)
Other sap-suckers	108	205	180 (95)
Caterpillars	822	196	190 (80)
Tree crickets	22	0	340 (480)
Beetles	79	34	79 (62)
Leaf miners	4	4	4 (3)
Flower-feeders	4	54	19 (12)
Bark and wood borers	9	58	14 (19)
Omnivores			
Ants	235	247	38 (36)
Predators			
Beetles	36	10	7 (10)
Lacewings	4	0	13 (11)
Flies and wasps	6	5	15 (2)
Assassin bugs	21	0	1 (1)
Spiders and phalangids	35	41	40 (34)
Others	20	81	46 (62)
Total	2404	2483	1108 (259)

[a] Data from Schowalter et al. (1981c).
[b] $\bar{x} \pm$ SD. Data from Crossley et al. (this volume), H. Peterssen and Crossley (unpublished data), and Schowalter et al. (1981c).

Data from a 13 year old watershed undergoing thinning of black locust by the locust borer, *Megacyllene robiniae* (Forster), suggest a sharp increase in abundance of caterpillars and spiders on other tree species (Tables 15.1 through 15.5). Unfortunately, data for black locust during this thinning period are not available.

The changes in arthropod mass intensities on individual tree species following clearcutting were reflected in arthropod biomass at the ecosystem level (Table 15.7). The increases in aphid and ant biomass following clearcutting were statistically significant ($p < 0.02$) (Schowalter et al. 1981c); decreases in caterpillar and spider biomass were not ($p > 0.05$). Notice that differences between the first and second years were much larger as biomass than as weight intensity.

Changes in population dynamics also were observed following clearcutting. Aphid biomass in mature deciduous forests typically peaks in the spring and fall during periods of active nitrogen translocation to and from foliage sinks (Figure 15.2; see also Van Hook et al. 1980). By contrast, aphid biomass continued to increase during the summer on the clearcut watershed (Figure 15.2), presumably because of continued nitrogen translocation in rapidly growing shoots, especially of black locust (Schowalter 1985), and declined earlier in the fall, suggesting that regrowth foliage translocated less N prior to leaf fall.

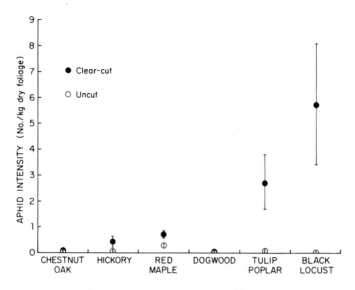

Figure 15.1. Aphid intensities ($\bar{x}$ + SE) in 1977 on chestnut oak (*Quercus prinus*), hickory (*Carya* spp.), red maple (*Acer rubrum*), dogwood (*Cornus florida*), tulip poplar (*Liriodendron tulipifera*), and black locust (*Robinia pseudoacacia*), in a mature forest (WS 2) and an adjacent first-year clearcut (WS 7) at Coweeta Hydrologic Laboratory, North Carolina.

Table 15.7. Canopy Arthropod Biomass (g/ha) in Clearcut and Mature Deciduous Forests at Coweeta Hydrologic Laboratory, North Carolina

	Watershed Age (yrs)		
Guild	1[a] (g/ha)	2[a] (g/ha)	>60[b] (g/ha)
Phytophages			
Aphids and aleyrodids	121	702	24 (6)
Other sap-suckers	141	482	253 (64)
Caterpillars	105	276	322 (153)
Tree crickets	30	24	207 (166)
Beetles	63	62	186 (167)
Leaf miners	19	5	11 (5)
Flower-feeders	77	201	54 (37)
Bark and wood borers	13	126	133 (84)
Omnivores			
Ants	49	153	23 (7)
Predators			
Beetles	14	78	19 (24)
Lacewings	1	2	20 (20)
Flies and wasps	72	45	33 (19)
Assassin bugs	9	175	58 (15)
Spiders and phalangids	43	332	189 (56)
Others	12	82	43 (13)
Total	769	2745	1577 (218)

[a] Data from Schowalter et al. (1981c).
[b] $\bar{x}$ ± SD. Data from Crossley et al. (this volume) and Schowalter et al. 1981c.

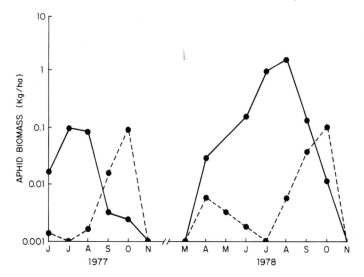

Figure 15.2. Seasonal trends in aphid biomass during 1977 and 1978 in a mature forest (WS 2, ---) and an adjacent clearcut (WS 7, —) at Coweeta Hydrologic Laboratory, North Carolina. [Reproduced from Schowalter (1985) with permission of Academic Press, Inc.]

Factors Influencing Arthropod Responses to Disturbance

The canopy arthropod data from Coweeta can be used to evaluate the relative importances of canopy microclimate, architecture, and nutritional quality as factors influencing arthropod responses to disturbance. These factors have been shown to significantly influence arthropod population dynamics (Mattson 1980; MacMahon 1981; Schowalter et al. 1981b). Clearcutting dramatically altered these environmental variables. Canopy depth and height above the soil surface were greatly reduced, placing the canopy within the influence of the hot, dry conditions prevailing at the soil surface (Seastedt and Crossley 1981). Canopy architecture was altered in at least one additional way as habitat structure changed from woody material with dispersed, growing shoots to clumped, succulent shoots. The relative abundances of host species shifted from oaks and hickories to dogwood and black locust. Nutritional quality was changed as nutrients became relatively more available to the reduced plant biomass and were actively translocated to foliage sinks over an extended period, as shoot elongation and foliage production continued through the summer. Although the influence of these factors on canopy arthropods was not assessed directly, indirect approaches can be used to evaluate relative importances.

Canopy microclimate and architecture appeared to have relatively little effect on arthropod assemblage structure. If either of these factors had a primary effect, we would expect similar arthropod assemblages among the regenerating tree species, which had similar phenologies and architectures and, presumably, similar microclimates. Instead, dramatic differences in arthropod assemblages were observed among the tree species. For example, compare aphids and ants on dogwoods and black locust which assumed canopy dominance following clearcutting (Tables 15.4 and 15.6). Furthermore, arthropod mass intensities in mature forests showed considerable variation between years and watersheds (Tables 15.1 through 15.6) despite relatively constant canopy microclimate and architecture.

This assessment is not intended to suggest that individual arthropod species were not affected by changes in canopy microclimate and architecture. Rather, our data indicate that, at the guild level of resolution, any effect of these factors was masked by compensatory responses among arthropod species or by responses to other factors.

The observed responses of canopy arthropods to clearcutting at Coweeta appeared to be explained best by changes in nutritional factors. This explanation is supported by several lines of evidence. First, the most dramatic arthropod response occurred on black locust, a nitrogen-fixing legume (Table 15.6). Nitrogen availability is known to influence aphid populations (Mattson 1980; Van Hook et al. 1980) and, thereby, to influence ant foraging behavior indirectly (Messina 1981; Schowalter et al. 1981c). Nitrogen availability also directly influenced ant foraging behavior at Coweeta. Black locust shoots on the clearcut produced extrafloral nectaries, which attracted ants in early spring prior to aphid population growth (Schowalter, Crossley and Seastedt, personal observation). Aphid and ant biomass subsequently increased on black locust through positive feedback. Second, fertilization experiments involving 4-year-old black locust (Hargrove et al. 1984) resulted in arthropod responses similar to those observed as a result of clearcutting. Addition of nitrogen–phosphorus–potassium fertilizer resulted in increased foliar nitrogen and phosphorus (but not potassium) concentrations. Hargrove et al. (1984) observed an initial increase in foliage loss to defoliators, but a subsequent decrease due to a combination of a tolerance response (increased foliage production) and a resistance response (reduced foliage consumption). These responses resembled the increased defoliator biomass the first year after cutting, followed by greatly reduced defoliator biomass the second year, especially on black locust (Tables 15.1 through 15.6). Third, an apparent increase in caterpillar biomass coincided with a locust borer population irruption and increased nitrate export on the 13-year-old successional stand (Tables 15.1 through 15.6), suggesting that changes in the nitrogen regime resulting from locust borer activity increased foliage nutritional quality for caterpillars.

These data indicate an important influence of nutritional factors on canopy arthropods and also explain some variations in the pattern of arthropod responses on different tree species. Different nutrient uptake rates between tree species could account for different arthropod responses. The most dramatic and variable arthropod reponses occurred on the early successional tree species, Cornus and Robinia (Tables 15.4 through 15.6), which showed the most rapid responses to clearcutting. Species specific responses of trees to the altered availability of nutrients following clearcutting altered foliar nutrient pools (Chapter 12) in ways which could elicit differing responses from the arthropod guilds. For example, increased nitrogen can affect sap-suckers positively and defoliators negatively, while increased potassium can have the opposite effect (Bogenschütz and König 1976), depending on the nutrient allocation pattern of the plant (Schowalter 1985).

The conspicuous absence of aphid and ant responses on dogwood is particularly intriguing (Table 15.4). The failure of aphids to respond suggests that dogwood's nitrogen budget was not affected by clearcutting. The failure of ants to respond suggests non-random foraging influenced by aphid abundance. However, the apparent absence of ant response on tulip poplar, which did support increased aphid populations, and also ant attraction to extrafloral nectaries on black locust suggest that nutritional factors were involved as well. Thus, indirect evidence indicates that nutritional factors were the primary stimuli influencing canopy arthropod responses to clearcutting at Coweeta.

Regulatory Effects of Canopy Arthropod Responses

Schowalter (1981, 1985) developed a cybernetic model of ecosystems in which changes in host allocation of available resources trigger consumer responses which tend to regulate net primary productivity (NPP) and protect ecosystem carrying capacity (Figure 15.3). This model advances ecosystem theory beyond earlier cybernetic models (Mattson and Addy 1975; Odum 1969; Patten and Odum 1981) in several ways.

First, the individual is clearly the unit of selection. Ecosystem stability is not a community goal, but rather a consequence of increased fitness accruing to individuals which, in addition to other adaptive attributes, interact in ways that contribute to persistence in a temporally-variable environment.

Second, changes in resource allocation pattern, resulting from changes in resource availability at each trophic level, are identified as the mechanism which communicates the current state of the system throughout the food web and triggers feedback responses. Underexploitation and consequent reduction of ecosystem carrying capacity can be remedied by rapid growth and reproduction of vigorous individuals aided by mutualistic interactions; disruptive overexploitation can be controlled by selective cropping, often by the same species functioning as mutualists under more favorable conditions.

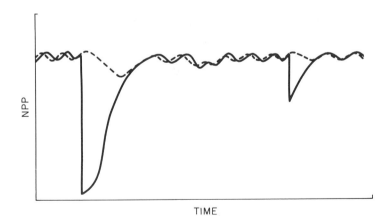

Figure 15.3. Trends in ecosystem net primary production (—) and ecosystem carrying capacity (---) through time. Reduced carrying capacity immediately after disturbance is due to dissipative forces (e.g., leaching and erosion) uncontrolled by biota.

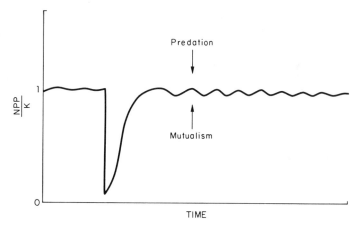

Figure 15.4. Hypothesized mechanism by which species interactions dampen deviation in the ratio between ecosystem net primary production (NPP) and ecosystem carrying capacity (K). Mutualistic interactions stimulate production and accelerate recovery when NPP falls below carrying capacity (NPP/K < 1). Predation in response to increasing host competitive stress reduces NPP at NPP/K > 1. The efficiency of species interactions increases as community development increases the strength of species interactions.

Finally, the most significant advance in ecosystem theory is provided by casting succession not as a goal-oriented strategy, but as a consequence of delayed disturbance. Disturbance frequency selects for species adaptive strategies which also influence rates of resource accretion or export (Grime 1977). Hence, ruderal species characterizing frequently-disturbed ecosystems are highly productive and tend to minimize resource export, often through mutualistic interactions (Figure 15.4). Stress-tolerant species characterizing infrequently-disturbed ecosystems are less productive and tend to prevent overexploitation of resources through predatory interactions (Figure 15.4). The spatial-temporal framework of disturbance maintains species associations representing the range of resource utilization strategies within the ecosystem. Succession occurs because ruderal associations adapted to frequent disturbance colonize disturbed sites. If disturbance does not recur in time to prevent overexploitation and rejuvenate the disturbance-adapted, competition-intolerant association, the plants become stressed, vulnerable to herbivores, and replaced by competition- or stress-tolerant associations adapted to less-frequent disturbance. Although competitive interactions could eventually accomplish species replacement, consumer responses to host stress are viewed as providing more efficient regulation by accelerating replacement and adjustment of NPP to carrying capacity.

Data from canopy arthropod studies at Coweeta support this model. The apparent sensitivity of key guilds to changes in host nutritional quality reflecting changes in ecosystem carrying capacity (Schowalter et al. 1981c) indicates a mechanism for regulating NPP. A clearly disturbance-adapted species assemblage with strong mutualistic interactions became predominant after clearcutting. A relatively high ecosystem carrying capacity (K > NPP) inherited from the preceding community promoted rapid growth. Productivity was enhanced by stimulation of plant growth (especially of black locust) by nitrogen-fixing microorganisms (Boring et al. 1981), stimulation of nitrogen-fixation by aphids (Petelle 1980), and stimulation of aphid productivity, and perhaps plant protection, by ants (Schowalter et al. 1981c; Tilman 1978; but see Fritz 1983). Cycling of nitrogen, potassium, sodium, and perhaps phosphorus, was accelerated by canopy arthropod feeding (Hargrove et al. 1984; Schowalter et al. 1981c; Seastedt et al. 1983).

Rapid growth by the colonizing association eventually should result in overexploitation of resources (K < NPP) despite soil nutrient retention and nitrogen accretion. Such a situation could explain the irruptions of locust borer and defoliator populations which reduced NPP, increased nitrate and potassium mobilization (Schowalter and Crossley 1983; Schowalter et al. 1981a), and which appear to be accelerating a successional transition from the ruderal community to a more stress-tolerant community (sensu Grime 1977).

Finally, periodic overexploitation of resources is likely to continue in mature forests as a result of fluctuating environmental conditions (Figure 15.3). We have observed fall cankerworm [*Alsophila pometaria* (Harris)] population irruptions initially reducing NPP, but subsequently increasing mobilization and availability of nutrients (especially nitrogen), and, ultimately, increasing NPP (Swank et al. 1981). Such regulation might be enhanced by the differing sensitivities of different arthropod guilds to different types or intensities of environmental change (Bogenschütz and König 1976; Schowalter 1985).

Conclusions

Studies of canopy arthropod assemblages at Coweeta support a hypothesis that host resource allocation pattern and consumer response constitute the mechanism for cybernetic control of NPP in temporally-variable environments. We recognize the need for increased resolution of canopy arthropod assemblage structural and functional interactions. If the functional interactions between species represent mechanisms for cybernetic control of NPP, then increased species diversity and community connectivity should contribute to ecosystem stability only to the extent that redundancy ensures an adequate response. Furthermore, the relative importances of mutualistic, competitive, and predatory interactions at any point in time should indicate the instantaneous direction and amplitude of deviation of NPP from ecosystem carrying capacity (Figure 15.4). Consequently, the instantaneous abundance of particular taxa does not necessarily indicate ecological importance. Regulation can be accomplished most efficiently by small, inexpensive (in terms of resource needs) biomass, which can be rapidly multiplied (e.g., as in insects) and amplified through indirect effects (e.g., stimulation of nitrogen-fixation). Increased understanding of canopy arthropod assemblages and responses to changes in environmental conditions will contribute to ecosystem management strategies.

5. Forest Floor Processes

16. Changes in Soil Nitrogen Pools and Transformations Following Forest Clearcutting

J.B. Waide, W.H. Caskey, R.L. Todd, and L.R. Boring

Factors that regulate the cycling of nitrogen are important determinants of the metabolism and organization of forest ecosystems (Gosz 1981; Melillo 1981). The nitrogen cycle includes a number of important gaseous components and is strongly regulated by biological processes, particularly those mediated via microbial populations. The cycle of this element is closely coupled to the cycles of other essential elements (e.g., C, S). Thus, the dynamics of nitrogen may largely regulate the productivity of forests and their response to intensive management and other exogenous disturbance (Kimmins 1977; Swank and Waide 1980; Vitousek 1981; Johnson 1984). Because nitrogen transformations are so tightly coupled to other biogeochemical processes, it is especially difficult to predict a priori impacts of human disturbance on forest metabolism and nitrogen cycling.

The storage and turnover of nitrogen in soil pools are essential components of the forest nitrogen cycle. In most temperate forests, 80 to 90% of the total nitrogen capital may be localized in labile and recalcitrant forms of soil organic matter (Rodin and Bazilevich 1967; Cole and Rapp 1980; Swank and Waide 1980; Melillo 1981). Factors which regulate rates at which nitrogen is stored and subsequently mineralized from these soil pools determine both the availability of nitrogen for uptake by plants and microorganisms (Keeney 1980) and the losses of nitrogen from the forest ecosystem to the atmosphere and to drainage waters (Vitousek and Melillo 1979). Thus, information on these factors is prerequisite to understanding and managing forests. Yet it is those processes which regulate soil nitrogen storage and turnover which are the most poorly understood portions of the forest nitrogen cycle, and about which improved knowledge is urgently needed.

Since the initiation of Coweeta research on forest productivity and nutrient cycling in the late 1960s (Douglass and Hoover, Chapter 2), major emphasis has been placed on the nitrogen cycle. Early work focused on the solution flux of nitrogen through forested watersheds (Swank and Waide, Chapter 4), the cycling of nitrogen through vegetation pools (Monk and Day, Chapter 11), nitrogen inputs to and release from forest litter layers (Cromack and Monk 1975), and biological transformations of nitrogen in forest litter layers and soils, particularly microbial gaseous transformations (Cornaby and Waide 1973; Todd et al. 1975b, 1978). This work was synthesized in the form of a compartment model by Mitchell et al. (1975). This model included estimates of nitrogen standing crops in and transfer rates among 15 storage pools, and rates of nitrogen exchange with the surrounding environment.

This model of nitrogen cycling in an aggrading mixed hardwood forest was implemented to predict and assess likely consequences of intensive management on long-term forest productivity and nitrogen cycling (Waide and Swank 1976, 1977). Swank and Waide (1980) subsequently provided comparative analyses of nitrogen cycling at Coweeta and three other well-studied forest ecosystems, and identified components of the forest nitrogen cycle which appeared to be most susceptible to management impacts. These components were the mineralization of nitrogen from decaying litter and stable soil organic matter, nitrogen inputs via biological dinitrogen fixation, and losses via denitrification. Nitrogen losses due to removal of forest products and leaching to drainage waters were quantitatively less important than these processes in relation to long-term changes in site productivity.

At the time this earlier research was completed, data on changes in nitrogen dynamics immediately following forest disturbance were not available for any forested watershed at Coweeta. Such data were important to verify model predictions and to document both the extent of disruption of forest nitrogen cycling immediately following disturbance and the patterns, rates, and mechanisms of forest recovery. To satisfy this need, a forest clearcutting experiment was implemented. Detailed studies of the treated watershed began 2 years prior to cutting and are still in progress. In this chapter we summarize select results of this ecosystem experiment. Specifically, we focus on changes in soil nitrogen storages and transformations following forest clearcutting. We emphasize major changes at the watershed scale of resolution. Detailed analyses of spatial and temporal variability in nitrogen standing crops and transformations, and of causal relations among measured variables, will appear in subsequent papers. Related data on this experiment are discussed in the chapters by Boring et al. (Chapter 12) and Swank (Chapter 25). Other studies on nitrogen cycling in disturbed watersheds at Coweeta have recently been completed by Montagnini (1985) and White (1986).

Methods

Site Description and Treatment History

Research described here was conducted on Watershed 2 (WS 2) and WS 7. WS 2 served as the control and is located adjacent to WS 7. Vegetation on this 12.1 ha south-facing watershed consists of a mixed hardwood forest (cove hardwoods at low elevations and along ravines, oak–hickory on side slopes) which has been undisturbed since 1924 except for the chestnut blight. Soils are predominantly Fannin fine sandy loams (fine-loamy, micaceous, mesic Typic Hapludults), with Chandler gravelly loams (coarse-loamy, micaceous, mesic Typic Dystrochrepts) predominating on upper slopes and ridges.

WS 7 is a south-facing, 58.7 ha catchment. Prior to logging, vegetation consisted of three major forest types (see Figure 12.1 of Boring et al., Chapter 12): (1) cove hardwoods at lower elevations and along ravines at intermediate elevations; (2) mesic (southeast-facing) and xeric (southwest-facing) oak–hickory forests on intermediate-elevation side slopes; and (3) pine-hardwood forests at upper elevations and on ridges. Soils are similar to those on WS 2.

Treatment of WS 7 can be separated into three components: road construction and stabilization, logging, and site preparation. Three roads were constructed during mid-April through mid-June 1976, covering about 5% of total watershed area. Road cuts and fills were stabilized by seeding grass and applying fertilizer and lime both following construction and again in July 1977 and June 1978. Logging began in January 1977 and continued through June 1977. Approximately 28% of the watershed was not logged due to the poor quality of timber, 56% was logged with a mobile cable system, and 16% (slopes <20%) was logged with tractor skidding. Site preparation was completed in October 1977, and consisted of clearfelling all remaining stems ≥ 2.5 cm dbh.

Sample Collection

Prior to logging, vegetation was inventoried on 142 0.08 ha plots systematically located over WS 7. Sixteen of these were selected randomly (stratified by vegetation zone) as intensive study plots for prelogging sampling. After logging, 12 of 142 plots were selected in a stratified random fashion. Sampling was conducted on WS 2 only following treatment of WS 7, on four randomly located plots. Sampling commenced in March 1975 on a biweekly schedule through February 1979, when the frequency changed to a monthly basis. No samples were collected during the period of August 1976 to July 1977. Results are summarized here through the October 1980 sampling period.

On each sampling date, bulk soil samples were collected from 0 to 10 cm and 10 to 30 cm soil depths. These two depths generally correspond to the A and AB-Blt (Hapludults) or A and AB–Bw (Dystrochrepts) soil horizons, respectively. On each watershed, one-half of the total number of plots were sampled on alternate dates. Samples were placed on ice and returned to the laboratory within 1 day of collection. Samples were then refrigerated (4°C) until analysis, which occurred within 1 week of collection. Chemical analyses (except NO_3-NH_4) were performed on air-dried samples; biological (and NO_3-NH_4) analyses, on field-moist samples.

Resulting data were subjected to analysis-of-variance procedures to detect significant differences between sampling depths and following forest removal. A significance level of $\alpha = 0.05$ was employed in these analyses.

Changes in Soil Microenvironments

Many of the changes in nitrogen cycling processes to be discussed are largely due to altered soil microenvironments resulting from removal and successional recovery of the forest canopy. To evaluate impacts of these changes, direct measures were made of soil moisture and temperature on each sampling date (Table 16.1). Moisture was determined gravimetrically in the laboratory. Temperature was measured directly in the field (0–10 cm depth only) at the time of sampling.

Removal of the forest canopy resulted in large increases in radiation reaching the forest floor. As a consequence, soil temperatures were elevated during the first postcut year. During the first summer of postcut sampling, temperature readings in the range of 40 to 60°C were common. Rapid recovery of the canopy (leaf area index was 68% of control values by the third year (Chapter 12)) provided shade and moderated soil temperatures.

Removal of the canopy during clearcutting also reduced evapotranspiration, and resulted in elevated soil moisture levels during the first year. However, increased solar radiation inputs and soil temperatures also accelerated the drying of surface soil horizons following precipitation events. Canopy recovery provided shade as well as partial recovery of evapotranspiration. Thus, soil moisture had returned to precutting levels by the third year of sampling.

These postcutting changes in soil microenvironments had adverse impacts on overall microbial biomass and metabolism, as revealed by measures of soil ATP concentrations and CO_2 evolution. For example, average soil ATP pools declined about 60% (from about 71 μg m^{-2} to about 29 μg m^{-2}) in the first year following forest removal. Similarly, total soil CO_2 evolution declined about 32% in the first post-treatment year, from 8.3 g m^{-2} d^{-1} to 5.7 g m^{-2} d^{-1}. The harsh soil microenvironments in the first year following treatment also negatively impacted soil microarthropod populations and surface litter decomposition processes (Abbott et al. 1980; Seastedt and Crossley 1981; Abbott and Crossley 1982). Even in this first postcut year, however, short bursts of high microbial activity were observed in fall and spring during brief periods of warm, moist soil conditions. Measures of CO_2 evolution and ATP concentrations suggested that general microbial growth processes had recovered to near pretreatment levels after three years, coincident with canopy recovery and the moderation of soil microenvironments.

Changes in Soil Organic Matter and Nitrogen Pools

Organic Matter

Soil organic matter (SOM) was determined using Walkley-Black titrations (Nelson and Sommers 1982). For both watersheds, SOM concentrations were significantly higher in the upper sampling depth (Table 16.2). Following clearcutting of WS 7, SOM values increased substantially (74%) in the 0 to 10 cm depth and to a lesser extent (17%) in the 10 to 30 cm depth. Thus, postcutting differences between the two depths were greater than precutting differences. Both the differences between depths and following clearcutting were statistically significant. SOM concentrations increased immediately following treatment, reaching peak values after about 1 year. Thereafter, concentrations declined slightly. However, even after 3 years, SOM values were still significantly greater than prior to cutting at both depths.

The exact source of these post-treatment SOM increases is not known, particularly the large increases in the upper sampling depth. These increases presumably include

Table 16.1. Changes in Soil Microenvironments Following Forest Clearcutting at Coweeta

Variable	Units	Depth (cm)	WS 2	WS 7 Precut	WS 7 Postcut, Yr 1[a]	WS 7 Postcut, Yrs 1–3[a]
Soil moisture	g g^{-1}	0–10	0.35 (0.05)[b]	0.35 (0.02)	0.56 (0.04)	0.34 (0.04)
		10–30	0.29 (0.04)	0.30 (0.01)	0.36 (0.01)	0.26 (0.02)
Soil temperature	°C	0–10	13.3 (0.6)	13.6 (0.8)	16.9 (0.7)	12.8 (1.1)
		10–30	ND[c]	ND	ND	ND

[a] To indicate temporal trends in the variables of interest, resulting from successional recovery of vegetation, postcutting data for WS 7 are summarized separately for Year 1 (August 1977–July 1978) and Years 1–3 (August 1977–October 1980). Note that the first year of soil sampling began near the end of the first summer of vegetation regrowth.
[b] Data displayed as $\bar{x}$ (SE) over the period of interest.
[c] Not determined.

organic matter inputs from mortality and turnover of fine roots, and from the decay and leaching of surface litter layers following forest removal. Fine root turnover has been shown to contribute large amounts of organic matter to soil pools in temperate deciduous forests (Harris et al. 1980; McClaugherty et al. 1982; Aber et al. 1985). Dominski (1971) and Covington (1981) documented reductions in surface litter mass following forest removal at Hubbard Brook and other sites in the White Mountains of New Hampshire. Rapid decomposition of logging residues in the first 7 years following watershed treatment, particularly in the smaller size fractions (≤ 5 cm diameter), contributed to SOM increases (Mattson 1986), though not immediately following site preparation. The more modest SOM increase at the lower depth indicated that some C was transported downward, probably in water-soluble form, but also that most of the C inputs were metabolized and immobilized or respired in the upper 10 cm of soil.

Increases were not observed in SOM values following the whole-tree harvesting of another Coweeta watershed, WS 48 (Waide et al. 1985), suggesting that the decay of logging slash (removed from WS 48) played some role in the large SOM increases on WS 7. However, Boring and Swank (1984a) documented SOM increases in surface soil horizons of successional forests on several other disturbed watersheds at Coweeta.

Table 16.2. Changes in Soil Organic Matter and Nitrogen Pools Following Forest Clearcutting at Coweeta[a]

Variable	Units	Depth (cm)	WS 2	WS 7 Precut	WS 7 Postcut, Yr 1	WS 7 Postcut, Yrs 1–3
Organic matter	%	0–10	7.4 (0.3)	6.5 (0.2)	11.3 (0.6)	11.1 (0.5)
		10–30	4.8 (0.4)	4.6 (0.1)	5.4 (0.3)	5.5 (0.2)
Total Kjeldahl N	%	0–10	0.19 (0.01)	0.17 (0.01)	0.28 (0.02)	0.25 (0.06)
		10–30	0.14 (0.01)	0.13 (0.01)	0.16 (0.01)	0.16 (0.01)
NO_3^-	µg g^{-1}	0–10	1.6 (0.4)	1.2 (0.2)	4.3 (0.8)	3.2 (0.6)
		10–30	2.1 (0.7)	1.0 (0.1)	3.0 (0.7)	1.9 (0.3)
NH_4^+	µg g^{-1}	0–10	3.1 (0.3)	4.6 (1.1)	9.2 (1.6)	8.9 (1.2)
		10–30	3.0 (0.4)	3.6 (0.7)	4.2 (0.5)	3.8 (0.3)

[a] Refer to footnotes to Table 16.1.

Edwards and Ross-Todd (1983) observed no changes in SOM levels following forest harvesting in Tennessee. Precut SOM values at that site were 2 to 7 times lower than values measured on WS 7. Also, in contrast to WS 7 results, these authors reported increased soil respiration rates (1.5 to 2×) following cutting which may have precluded SOM increases.

Total Nitrogen

The above discussions of SOM changes following cutting are important in terms of their influence on soil organic N pools. Total N (TKN) was determined as NH_4^+, using the cyanurate–salicylate reaction with autoanalysis following micro-Kjeldahl digestion (Bremner and Mulvaney 1982; Reynolds et al. 1986). Patterns in TKN values paralleled those for SOM (Table 16.2), i.e., statistically higher concentrations in the 0 to 10 cm depth and statistically significant increases following logging, particularly in the upper sampling depths. TKN values peaked shortly after completion of site preparation and have declined slowly since. After 3 years, TKN values were still significantly elevated above precutting measurements.

Much of the postcutting increase in TKN in the upper 10 cm of soil is due to comparable increases in SOM. TKN increases in this upper sampling depth (65%) were slightly less than SOM increases, possibly due to the wide C:N ratios of logging slash left on site. Also, some of this increase must have resulted from microbial immobilization of nitrogen mineralized following cutting. Microbial uptake of N was shown to be an important mechanism retaining N in disturbed forests, particularly in response to large standing crops of C in decaying logging residues (Vitousek and Matson 1984). Increases in TKN values in the lower sampling depth (23%) were slightly greater than SOM increases. These TKN increases at the 10 to 30 cm depth thus resulted partly from the leaching of water-soluble SOM and low-molecular-weight organic N compounds from the upper 10 cm of soil, and also from the immobilization of mineral N percolating downward through the soil profile. For example, NO_3-concentrations in porous cup lysimeter collections were much higher on WS 7 than on WS 2 at both 30 cm (0.50 vs 0.04 mg L^{-1}) and 100 cm (0.25 vs 0.01 mg L^{-1}) depths (Waide et al. 1985).

Mineral Nitrogen

Soil concentrations of NH_4^+ and NO_3^- were determined using the Berthelot and diazotization reactions, respectively, with autoanalysis following extraction of field-moist samples with 2 M KCl (Technicon 1971a, 1971b; Keeney and Nelson 1982; Reynolds et al. 1986). Soil NH_4^+ concentrations were highly variable among sampling dates and sites (Table 16.2). In spite of this high variability, statistically significant increases (100%) in NH_4^+ were observed in the 0 to 10 cm depth following cutting. Although NH_4^+ concentrations did not differ between the two sampling depths prior to forest removal, post-harvest concentrations were significantly higher in the upper 10 cm than in the 10 to 30 cm depth. These patterns in NH_4^+ are expected in relation to the large increases in SOM and TKN in the 0 to 10 cm depth and to the higher mineralization rates measured there (see next section). Soil NH_4^+ concentrations in this upper depth peaked midway through the first post-treatment year and then declined slightly. However, concentrations were still significantly elevated after 3 years. Increases in NH_4^+ in the 10 to 30 cm depth were not significant.

Much less variability was observed in soil NO_3^- concentrations (Table 16.2). Differences between the two sampling depths were not statistically significant before or after logging. However, postcutting concentrations were statistically higher at both 0 to 10 cm (25%) and 10 to 30 cm (200%) depths. At both depths, soil NO_3^- concentrations peaked midway through the first post-harvest year and then declined, but were still significantly above precut values after 3 years. Declines in NO_3^- were slightly greater in the lower sampling depth. These patterns are explained by the higher NH_4^+ concentrations and mineralization–nitrification rates in the upper 10 cm of soil (see next section), and by the high mobility of NO_3^- in the moist forest soils on WS 7 following forest removal.

Changes in Soil Nitrogen Transformations

Nitrogen Mineralization

To assess the regulation of observed mineral N dynamics by microbial processes, mineralization and nitrification potentials were measured in the laboratory. Mineralization and nitrification rates were quantified as the changes in total mineral N and NO_3^-, respectively, over a 33-day aerobic incubation at standard temperature (25°C) and moisture (33% of dry weight) conditions (Keeney 1982; Reynolds et al. 1986). For these two processes only, samples were collected on a monthly basis from the intensive study sites previously described beginning in November 1979. Sampling frequency decreased to a quarterly basis in November 1980; data are summarized here through May 1982. Thus, sampling for these two processes began 2 years after completion of site preparation on WS 7, and continued during the time period when soil organic matter and nitrogen concentrations were declining. Because of the mobility of mineral N, data are summarized here for the O2 litter layer as well as 0 to 10 and 10 to 30 cm soil depths.

For both watersheds, mineralization potentials declined consistently and significantly from the O2 litter layer through the 10 to 30 cm soil depth (Table 16.3). Rates in the O2 layer were slightly higher on WS 2, whereas rates at both soil depths were slightly greater on WS 7. None of these watershed differences were statistically significant, however.

Because of the time lag involved in these data, it is difficult to relate mineralization results to the previous data on mineral N concentrations. Measurements of mineralization potentials made at the same time on WS 48 immediately following whole-tree harvesting provide some basis for interpreting the data from WS 7. On this watershed, mineralization potentials in the O2 litter layer declined significantly (75%) in the first year following harvest, and then increased slightly (30%) over the next 2 years. At the 0 to 10 cm depth, rates remained unchanged in the first post-harvest year and then increased (45%), whereas at the 10 to 30 cm depth rates increased (100%) in the first year after cutting and then remained constant. Because of different post-harvest SOM dynamics in the two watersheds, these trends can be extrapolated to WS 7 only with caution. Also, overall soil mineralization potentials were higher on WS 48, perhaps due to the removal of logging residues. Nonetheless, results from WS 7 appear to be consistent with the temporal trends observed on WS 48.

Table 16.3. Changes in Potential Rates of Nitrogen Mineralization and Nitrification Following Forest Clearcutting at Coweeta[a]

Process	Units	Depth	WS 2	WS 7 Postcut
Potential nitrogen mineralization rate	$\mu g\ g^{-1}\ 33d^{-1}$	$O2^{b}$	415 (53)[c]	359 (32)
		0–10 cm	17.8 (3.2)	21.7 (3.1)
		10–30 cm	7.6 (1.8)	10.1 (1.5)
Potential nitrification rate	$\mu g\ g^{-1}\ 33d^{-1}$	O2	5.8 (2.2) (1.4%)[d]	88.4 (18.1) (25%)
		0–10 cm	4.5 (1.7) (25%)	15.4 (3.1) (71%)
		10–30 cm	2.9 (1.2) (38%)	8.2 (1.7) (81%)

[a] Mineralization and nitrification potentials measured from November 1979–May 1982.
[b] Refers to O2 litter layer.
[c] Data displayed as $\bar{x}$ (SE).
[d] Values represent the percentage of mineralized N which is subsequently nitrified.

Thus, when mineralization potentials are converted to total fluxes (e.g., g m^{-2} yr^{-1}), integrated over the profile, and corrected for actual field temperature and moisture conditions, data in Table 16.3 suggest modest increases in total nitrogen mineralization on WS 7 on the order of 1 to 3 g N m^{-2} yr^{-1} in the first 1 to 3 years following forest removal. These increased rates contributed to the increased concentrations of mineral N discussed above, particularly considering overall reductions in soil heterotrophic activity (reduced ATP levels and CO_2 efflux), and reduced net uptake of N into plant biomass (Boring et al., Chapter 12) in the first few years of successional recovery.

Nitrification

Nitrification potentials tended to decline consistently from the O2 litter layer through the 10 to 30 cm soil depth on both watersheds (Table 16.3). Because of high variability, these differences among sampling depths were not statistically significant except for the O2 layer on WS 7. However, nitrification potentials did increase significantly (3 to 15×) following clearcutting at all sampling depths. Expressed as a percentage of mineralized N, nitrification potentials increased with depth and following logging of WS 7 (Table 16.3). When integrated over the profile and corrected for actual field temperature and moisture conditions, these nitrification potentials suggest increases in nitrogen fluxes on the order of 3 to 5 g N m^{-2} yr^{-1}. These results are consistent with observed increases in soil NO_3^- concentrations and total NO_3^- export from WS 7 (Chapter 25). Comparable increases in nitrification potentials have been measured on other disturbed watersheds, at Coweeta (Montagnini 1985) and elsewhere (Vitousek et al. 1982).

These nitrification potential assays were also consistent with MPN assays of nitrifying bacteria using the microtiter procedure of Rowe et al. (1976). Numbers of both NH_4^+- and NO_2^--oxidizing bacteria were highly variable and did not differ statistically between the two soil depths (Table 16.4). However, numbers of both groups of nitrifiers increased substantially following clearcutting, although the increases in NO_2^--

Table 16.4. Changes in Numbers of Nitrifying Bacteria in Soils Following Forest Clearcutting at Coweeta[a]

Component	Units	Depth (cm)	WS 2	WS 7 Precut	WS 7 Postcut, Yr 1	WS 7 Postcut, Yrs 1–3
NH_4^+ oxidizing bacteria	Bacteria g^{-1}	0–10	46 (14)	25 (6)	270 (100)	470 (130)
		10–30	65 (33)	18 (4)	170 (50)	320 (110)
NO_2^- oxidizing bacteria	Bacteria g^{-1}	0–10	49 (16)	ND	220 (80)	310 (110)
		10–30	53 (16)	ND	83 (23)	130 (30)

[a] Refer to footnotes to Table 16.1.

oxidizers were not significant at the 10 to 30 cm depth. Numbers of nitrifying bacteria continued to increase into the second year following forest removal. This pattern is consistent with nitrification potential measurements on WS 48 following whole-tree harvesting. The magnitudes of increases in nitrifying bacterial populations are similar to values reported by Likens et al. (1968) following forest clearcutting at Hubbard Brook, and by Todd et al. (1975a) and Montagnini (1985) for other disturbed watersheds at Coweeta.

A third measure of nitrification rates on WS 2 and WS 7 employed the chlorate-inhibition procedure (Belser and Mays 1980) during the time period July 1980 through December 1981. These data also showed increases (50 to 100%) in rates following logging of WS 7 at both soil depths. However, large variances and small sample sizes precluded detection of statistical differences between depths or associated with forest removal.

Denitrification

Denitrification potentials were measured on WS 2 and WS 7 as the rate of N_2O production in acetylene-inhibited soil slurries (Swank and Caskey 1982) during Phase I of denitrification (Smith and Tiedje 1979). This assay measures the maximum activity of denitrifying enzymes present in the soil at the time of sampling. As with several other assays, sampling for this process began 2 years after completion of site preparation, and spanned the period October 1979 through February 1982.

Measured rates were lowest in the upper sampling depth on WS 2, and higher and nearly equal for the other watershed-depth combinations (Table 16.5). Because of the extremely large variability in measured rates, no statistical differences were detected between sampling depths or watersheds (i.e., associated with clearcutting).

In order to place these rates into some perspective, estimates of total denitrification fluxes on the two watersheds were calculated by correcting Phase I rates for field temperatures and NO_3^- concentrations, and then relating corrected rates to the occurrence of discrete precipitation events. Temperature corrections were based on the equation of Rickman et al. (1975), as described by Caskey and Schepers (1985). Correction for measured NO_3^- values assumed Michaelis–Menton kinetics (Bowman and Focht 1974; Caskey and Schepers 1985), with an experimentally determined Michaelis constant of

Table 16.5. Changes in Rates of Denitrification and Nitrogen Fixation in Soils Following Forest Clearcutting at Coweeta[a]

Process	Units	Depth (cm)	WS 2	WS 7 Precut	WS 7 Postcut
Denitrification[b]	nl N_2O g^{-1} hr^{-1}	0–10	10.0 (3.5)	ND	61.0 (27.7)
		10–30	54.1 (34.0)	ND	55.9 (31.7)
Nitrogen fixation[c] (acetylene	nl C_2H_2 g^{-1} d^{-1}	0–10	2.44 (0.42)	12.8 (3.0)	97.3 (73.7)
reduction)		10–30	0.92 (0.17)	7.9 (2.1)	30.8 (22.5)

[a] Refer to footnotes to Table 16.1.
[b] Denitrification rates measured from October 1979–February 1982.
[c] Nitrogen fixation rates measured over the same time period as shown in Table 16.1; postcut refers to years 1–3 following treatment, August 1977–October 1980.

10.74 mM. Estimated fluxes were about 5 and 11 kg N ha^{-1} yr^{-1} for WS 2 and WS 7, respectively. These estimates are similar in magnitude to values reported by Robertson and Tiedje (1984) for successional oak–hickory forests in Michigan. Earlier, Swank and Waide (1980) simulated denitrification fluxes for uncut and clearcut hardwood forests at Coweeta as about 11 and 16 kg N ha^{-1} yr^{-1}, respectively. Estimated fluxes cited above are slightly lower than these simulation predictions, but the increase attributable to cutting is comparable.

More recent studies of denitrification on other Coweeta watersheds call into question the magnitude of the above flux estimates. But these studies also confirm the expectation of some increase in denitrification losses following disturbance, largely due to increased NO_3^- transport into the riparian zone of disturbed watersheds. Davidson (1986) conducted intensive field studies of denitrification rates on a control (WS 18) and a disturbed watershed (WS 6), as well as factorial laboratory experiments on major regulatory variables (O_2 tension, NO_3^- concentrations, available C). Results of these studies suggest that soil denitrification fluxes from undisturbed forests are no more than 1 kg N ha^{-1} yr^{-1}, and are perhaps much less. Increased N losses are expected following forest disturbance, but annual flux estimates at a watershed scale are not possible based on available information. Interpretation of these results in regard to clearcutting responses will require synthesis of field measurements of soil moisture, temperature, and NO_3^- concentrations; climatological data on the occurrence of precipitation events; and laboratory studies on controlling variables with process-oriented simulation models of hydrologic and nutrient transport in forest soils.

Nitrogen Fixation

Nitrogen inputs to temperate deciduous forests via biological dinitrogen fixation occur predominantly through the activity of symbiotic associations (*Frankia*, *Rhizobium*) in the roots of early-successional tree species, free-living bacteria in litter and soil layers, and epiphytic lichens living on the external surfaces of trees and decaying wood (Melillo 1981; Waughman et al. 1981; Boring et al. 1986). Nitrogen fixation potentials have previously been measured on other Coweeta watersheds for all the above components except tree symbioses (Cornaby and Waide 1973; Todd et al. 1975b, 1978). Based on these results, significant nitrogen inputs via epiphytic lichens or associated with decaying logging residues would not be expected during the initial phase of recovery (e.g., 0 to 5 years) following forest removal. Thus, WS 7 research initially focused on N inputs via tree symbioses and free-living bacteria.

A species which predominates in early successional forests in the Southern Appalachians following major disturbance such as clearcutting is black locust (*Robinia pseudoacacia*), a woody legume nodulated with *Rhizobium* species. The role of *Robinia* in forest succession and the amount of N fixed by this symbiosis have been extensively studied, both on WS 7 and on older clearcuts within the Coweeta Basin (Boring and Swank 1984a, 1984b; Chapter 12). Locust regeneration occurs predominantly via sprouting; early sprout growth is rapid. These sprouts represent potentially significant inputs to forest N cycles. Biomass and nitrogen accretion of locust stands, as well as biomass and C_2H_2-reducing activity of root nodules, have been measured. Total stand N increased 48, 75, and 33 kg N ha^{-1} yr^{-1}, and nodule biomass was 8, 106, and 4 kg ha^{-1}, in 4-, 17-, and 38-year old stands, respectively. Peak rates of N fixation thus appear to occur in early to intermediate stages of forest succession, declining thereafter. In the youngest stand studied (on WS 7), N-fixing activity was estimated as 30 kg N ha^{-1} yr^{-1} based on extensive measurements of nodule biomass and activity of freshly excised nodules. Assuming that N-fixing activity is proportional to the density of *Robinia* stems, a weighted average of 10 kg N ha^{-1} yr^{-1} for all of WS 7 was calculated. Although dense locust stands at Coweeta may experience extensive mortality after 15 to 25 yrs, the nitrogen fixed by this symbiosis could impact forest N dynamics for 50 years or longer.

Nitrogen fixation potentials of free-living bacteria in litter and soil layers of WS 7 were measured in the laboratory (22°C) on field-moist samples with the C_2H_2-reduction assay (Hardy et al. 1973; Todd et al. 1978; Knowles 1982; Reynolds et al. 1986). Rates of free-living N fixation in the soil were highly variable, particularly after forest removal (Table 16.5). Differences among sites varied over 1 to 2 orders of magnitude. Highest rates tended to occur in mesic, fertile sites within the mid-elevation chestnut oak vegetation type. These were also the sites of highest symbiotic activity. When the logarithmic distribution of these data was taken into account, N-fixing potentials were shown to be significantly higher in the 0 to 10 cm soil depth, and on WS 7 prior to logging than on WS 2. Rates also increased substantially (3 to 8×) following clearcutting on WS 7. Highest rates tended to occur into the second postcutting year, declining slightly thereafter.

This extensive data base on free-living N fixation provides the basis for estimating N inputs to the forest ecosystem via this process. To facilitate this goal, recent studies have evaluated the temperature dependence of measured C_2H_2-reduction rates (Reynolds et al. 1986). When these results are combined with field-measured soil temperatures, the following estimates of free-living inputs to soil N pools are obtained: for WS 2, 0.3 kg N ha^{-1} yr^{-1}; WS 7 precut, 1.7 kg N ha^{-1} yr^{-1}; WS 7 postcut (years 1 to 3), 9.7 kg N ha^{-1} yr^{-1}. These values revise downward earlier estimates of potential N fixation inputs to Coweeta forests in relation to actual field conditions. Increases in free-living N inputs following cutting are thus estimated at about 8 kg N ha^{-1} yr^{-1}. The major unknown in these estimates of free-living inputs, as with estimates of symbiotic inputs, involves ^{15}N verification of C_2H_2:N_2 ratios. Nonetheless, although symbiotic N inputs are substantially higher within dense locust stands, free-living N inputs are of comparable magnitude averaged over WS 7. The duration of elevated free-living N-fixing activity in soils on WS 7 is unknown, but may be on the order of 6 to 12 yrs, the major determinant probably being successional declines in soil C levels.

As the successional recovery of the forest ecosystem on WS 7 proceeds, decaying logging residue may become an important site of N fixation. Because of the large amounts of decaying wood remaining on WS 7 after treatment (ca. 12 kg m^{-2} in the >5 cm diameter fraction), even modest rates of N fixation could result in substantial N inputs to the forest ecosystem. Thus, complete estimates of N inputs will require detailed future research on this component.

Summary and Conclusions

Together with data on N export in drainage waters (Chapter 25) and N cycling through vegetation pools (Chapter 12), results reported here provide an integrated picture of changes in forest N cycling processes on WS 7 following clearcutting. Soil organic matter and nitrogen pools increased (20 to 70%) immediately following forest removal. Proportionately larger increases (20 to 250%) in mineral N pools were also observed. These increases in available mineral N may be attributed to slight increases in soil N mineralization (ca. 25%, or 1 to 3 g N m^{-2} yr^{-1}), substantial increases in nitrification (ca. 200%, or 3 to 5 g N m^{-2} yr^{-1}), and reductions in general soil heterotrophic activity and plant N uptake in the first few years after logging. Increases in both symbiotic (ca. 1 to 3 g N m^{-2} yr^{-1}) and free-living (ca. 1 g m^{-2} yr^{-1}) fixation also added additional N to soil pools.

Only small fractions of these increases in available soil N were exported from WS 7 in drainage waters (ca. 0.1 g N m^{-2} yr^{-1}). Small increases of unknown magnitude in N losses via denitrification also occurred. But, the majority of these increased soil mineral N supplies were retained on site and recycled through rapidly regrowing early successional vegetation pools. Larger fractions of this vegetation uptake of N cycled through labile leaf tissues (rather than being stored in wood) than was the case in control forests (Chapter 12; see also White 1986 for comparable results on disturbed WS 6). Subsequent increases in soil heterotroph activity, associated with the recovery of the forest canopy and the moderation of harsh soil microenvironments, probably provided a secondary sink of N immobilization, stimulated by large C pools in decaying logging residues.

In subsequent papers resulting from this ecosystem experiment, detailed analyses of spatial and temporal variability in soil N pools and transformations, as well as of causal relationships among measured variables and processes, will provide additional insights into the mechanisms responsible for patterns reported here. Moreover, detailed studies on WS 7 will continue to document patterns, rates, and mechanisms of recovery in forest N cycling processes. Nonetheless, several dominant themes emerge from this analysis of early successional changes in forest N cycling processes: substantial acceleration of rates of N turnover in soil pools and of N recycling through labile vegetation components; rapid recovery of vegetation and microbial processes which foster conservation of N supplies on site; and coupled soil processes which make mineral N available for rapid uptake by early successional plant species, thus contributing to the resilience of the forest ecosystem following logging disturbance.

17. Soil Arthropods and Their Role in Decomposition and Mineralization Processes

T.R. Seastedt and D.A. Crossley, Jr.

About 95% of the annual net primary production of mature Coweeta forests is directly transferred to the detrital food web as foliar litter and woody debris. The decomposition of these plant substrates and the release of elements contained within the litter are necessary for the continued productivity of the forests. Most of the chemical energy released during decomposition is processed by bacteria and fungi; however, interactions with a host of invertebrates (primarily protozoans, nematodes, annelids, and arthropods) are responsible for the patterns of nutrient immobilization and mineralization observed in litter and soil.

Three sites in North America have extensive data sets on soil invertebrates and their effects on decomposition and mineralization processes. These sites include the Jornada desert studied by Whitford and his colleagues (Santos and Whitford 1981; Steinberger et al. 1984), the shortgrass prairie ecosystem studied by Coleman and his associates at Colorado State University (Anderson et al. 1981; Coleman et al. 1983; Ingham et al. 1985), and the Southeastern deciduous forests of Coweeta studied by Crossley and his students (Crossley 1977a,b). The Coweeta site has focused almost entirely on arthropods, in part because of investigator interests, but also because the arthropod fauna at Coweeta is particularly diverse and abundant (Gist and Crossley 1975a,b; Cornaby et al. 1975; Reynolds 1976; Abbott et al. 1980; Seastedt and Crossley 1981) (see Table 17.1). The relatively low pH of the weathered soils has favored arthropods over fauna such as earthworms that require a higher soil pH. The moderate levels of acid precipitation presently occurring at Coweeta dictate an even larger role for certain arthropod species in the future (Hagvar and Kjondal 1981).

Table 17.1. Soil Macroinvertebrate Numbers and Biomass in the Forest Floor of a Defoliated Hardwood Forest (WS 27) and an Undisturbed Forest (WS 7)[a]

Arthropod Taxa	WS 27 (n = 171)		WS 7 (n = 211)		F Value	
	Mean Number per m²	Biomass (mg×m⁻²)	Mean Number per m²	Biomass (mg×m⁻²)	Number	Biomass
Isopoda	0.1	0.1	0.5	2.4	16.00****	15.77****
Chilopoda	25.6	105.5	12.9	78.8	6.86**	5.92*
Diplopoda	31.6	92.8	10.0	383.9	1.47	8.22**
Araneida	11.2	15.3	13.6	28.4	11.06***	29.83****
Orthoptera	0.4	3.6	0.5	21.4	24.06****	11.41***
Coleoptera	0.6	12.5	0.5	7.2	3.41	0.08
Total macro-arthropods[b]	70.4	230.8	45.0	528.6	0.72	5.14*
Mollusca	7.2	278.9	13.4	315.2	12.77***	0.05
Annelids	3.6	10.6	0.6	2.5	19.51****	14.20**
Total inver-tebrates[b]	81.2	520.5	60.0	845.5	0.16	1.61

F values based on tests of log-transformed values.
$*p < 0.05$, $**p < 0.01$, $***p < 0.001$, $****p < 0.0001$
[a] WS 7 was subsequently clearcut. Results presented here were obtained prior to logging.
[b] Includes taxa not listed separately.

The studies of Gist (1972), Cornaby (1973), and Cromack (1973) at Coweeta provided some of the first empirical evidence supporting the hypothesis that arthropods have large effects on nutrient cycling processes; far larger impacts than are apparent by measurements of the arthropods' contribution to community respiration. Their work preceded the large number of consumer papers that began to appear in the mid 1970s and that have remained abundant in the literature. Their work also provided the conceptual framework for many of the studies that followed at the site. In this chapter we attempt to summarize the important findings of the Coweeta research by reporting on soil arthropod abundance and response to substrate quantity, substrate quality, and perturbations. We then discuss how arthropods influence litter chemistry and decay rates. These different subjects demonstrate the interdependence between the biotic and abiotic components of the forest floor, and show that arthropods are both regulated by and regulators of forest floor nutrients and organic matter.

The Effects of Substrate Quantity, Quality, and Perturbations on the Arthropod Community

Global patterns of arthropod densities and composition in relation to substrate quantity have been addressed by Harding and Studdard (1974) and Swift et al. (1979). The general pattern is one of macroarthropods such as millipedes diminishing in importance and abundance as one goes from the tropics to the poles. Conversely, microarthropods such as mites and collembolans tend to be much more numerous in northerly environments. Macroarthropods are believed to be more important in the direct breakdown of litter than are microarthropods (Anderson 1973; Herlitzius 1983). The response of the macroarthropod group to disturbances affecting amounts of litter is therefore likely to be larger than the response of the microarthropods. Conversely, microarthropods are believed to be much more sensitive than are macroarthropods to changes in litter quality via their interactions with microorganisms. Thus, microarthropod densities should fluctuate greatly with substrate quality changes, but macroarthropod biomass should be inversely correlated with litter standing crops. Here, we examine the relationship between arthropod standing crops and detritus variables measured within several Coweeta watersheds.

In the early 1970s an outbreak of a defoliator, the fall cankerworm (*Alsophila pometaria*), was discovered on one of the Coweeta watersheds. Several ecosystem processes were subsequently measured for this and a reference watershed. Among the variables were litterfall, litter standing crops, and macroarthropod densities and biomass along several transects within several distinct plant community types. Comparisons between the defoliated watershed (WS 27) and the reference hardwood watershed (WS 7) yielded a number of differences (Table 17.1). Annelid and centipede numbers and biomass were greater on WS 27, while isopteran, spider, and orthopteran numbers and biomass were greater on WS 7. Millipede biomass was greater on WS 7, as was total arthropod biomass. *Narceus annularis*, a large spirobolid millipede, was abundant on portions of WS 7 but was not found on WS 27. When *Narceus* was excluded from the analysis, no differences were observed between watersheds for either millipede or total arthropod biomass. Of course, only site differences are indicated by these statistics; treatment differences cannot be statistically evaluated (Hurlburt 1984).

We examined the relationship between invertebrate biomass to litter standing crops using the three transects on WS 27 and the four plant community types on WS 7 (Table 17.2). The mesic oak–hickory area on WS 7 had significantly more macroinvertebrate biomass than did all other areas. The pine-hardwood and xeric oak–hickory associations had less macroinvertebrate biomass than the two mesic strata on WS 7, but did not differ from the biomass estimated for the transects on WS 27. The ratio of standing crops of invertebrate biomass to litter standing crop varied almost five fold, with WS 27 values intermediate among those observed for WS 7. Macroinvertebrate biomass appeared inversely proportional to litter standing crops on WS 7, but not on WS 27. Overall, there was a nonsignificant negative relationship between macroinvertebrate biomass and litter standing crops ($r= -0.44$).

Partial defoliation by the cankerworms on WS 27 resulted in many measurable changes in nutrient export and internal elemental dynamics (Swank et al. 1981). The arthropod response, based on comparisons with WS 7, remains difficult to interpret. Differences in macroarthropod densities between WS 27 and WS 7 were observed, but these could not necessarily be attributed to the defoliation. An examination of within-watershed differences on WS 7 does, however, allow for some generalizations. Based on litterfall and litter standing crop data (Table 17.2), WS 7 can be divided into two areas: A productive zone with relatively high litterfall and low litter standing crops (the cove hardwood and mesic oak–hickory associations), and a low productive zone with low litterfall and high litter standing crops (the pine-hardwood and xeric oak–hickory associations). The productive zone had high standing crops of invertebrates, while the

Table 17.2. Estimates of Litter, Litterfall in 1974, and Soil Invertebrates in Summer 1975 on Defoliated and Control Watersheds.

Parameter	Defoliated (WS 27)			Control (WS 7)			
	Lower	Middle	Upper	Pine Hard-wood	Oak–Hickory Xeric	Oak–Hickory Mesic	Cove Hard-wood
Litterfall (g m^{-2})	424.2	360.9	536.0	326.8	346.0	426.9	346.2
(sample size)	(4)	(4)	(4)	(8)	(5)	(4)	(3)
Litter standing crop (g m^{-2})	686.4 AB	680.4 AB	807.0 AB	879.6 A	801.9 AB	737.4 AB	626.6
(sample size)	(32)	(32)	(32)	(32)	(32)	(32)	(32)
Total arthropods (mg m^{-2})	273.4 BC	230.1 C	189.3 C	297.4 C	193.4 C	912.3 A	699.6
Total macroinvertebrates	533.8 BC	607.9 BC	399.7 BC	499.0 C	309.7 C	1421.6 A	1178.4
Ratio of macroinvertebrates biomass to litter standing crop ($\times$ 10^{-3})	0.69	0.89	0.58	0.51	0.39	1.93	1.88
Litterfall/litter standing crop	0.62	0.53	0.66	0.37	0.43	0.57	0.55

Means followed by different letters are significantly different at $p < 0.05$ (Duncan's Multiple Range Test).

unproductive area had lower standing crops (Table 17.2). This correlation between forest productivity, low standing crops of litter, and large standing crops of forest mac- roarthropods is similar to relationships observed in European forests (e.g. Wallwork 1976; Swift et al. 1979). The absence of such a relationship on WS 27 probably resulted from the disturbance of defoliation. A time lag likely occurred between increased litter inputs and the subsequent increase in the detritivore standing crop (Seastedt, unpubl. data); thus, equilibrium conditions implicit to the relationship observed on WS 7 were not present on WS 27.

Microarthropod densities range from about $1.25–2.25 \times 10^5$ individuals per m² in the top 5 cm of litter and soil at Coweeta (Gist and Crossley 1975a; Crossley 1977b; Seastedt and Crossley 1981). A high percentage of microarthropods are fungivores, and these organisms are thought to have only a small direct effect on litter comminution (e.g. Anderson 1973; Swift et al. 1979). Microarthropod densities are not always posi- tively correlated with litter decomposition rates (Seastedt 1979). Nonetheless, their contributions to decomposition and mineralization are usually measurable and often substantial (Crossley 1977a; Seastedt 1984). At Coweeta as in other ecosystems, densi- ties appear roughly correlated with substrate quality, and, we believe, microbial activity (Seastedt and Crossley 1981; Abbott and Crossley 1982). Densities of microarthropods at Coweeta have been increased either by adding additional substrate (Webb 1976) or by enriching litter by adding artificial canopy leachates (Seastedt and Crossley 1983). The highest naturally occurring densities of microarthropods that we have observed on 1-year-old foliar litter (in litterbag studies) were about 100 individuals per gram (Seastedt et al. 1983). However, by fertilizing this litter with an N, P, K, Ca, Mg mixture we significantly increased densities to over 160 individuals per gram of litter (Seastedt and Crossley 1983). In spite of this increase in microarthro- pod densities, we did not observe a significant increase in the decomposition rate of the fertilized litter over that of the unfertilized controls.

The Effects of Microarthropods on Substrate Chemistry

The first microarthropod exclusion study to be conducted at Coweeta was reported by Cromack (1973). His results, along with subsequent findings, indicated that the presence of microarthropods always increases the nitrogen content of the litter (Figure 17.1). There is little doubt that this result occurs from microarthropod stimulation of microbial respiration (Hanlon and Anderson 1979. Either the microorganisms them- selves or their waste products (which are assumed to be intimately mixed with the decaying litter) contain the nitrogen. These findings need to be evaluated in respect to the reports of Aber and Mellilo (1980) and Mellilo et al (1982) that initial substrate quality (either nitrogen, lignin or nitrogen:lignin ratios) predict subsequent changes in the nitrogen and mass of the litter. Aber and Mellilo (1980) reported an inverse linear relationship between litter decay and the initial nitrogen content of the litter, with an average correlation coefficient of -0.93 based on the analysis of 30 data sets. A graphi- cal analysis of Cromack's data (see Mellilo et al. 1982) indicates a modest lack of fit to their linear models. High nitrogen–low lignin substrates such as dogwood foliage decayed too rapidly relative to other litter types. Indeed, it is difficult to understand

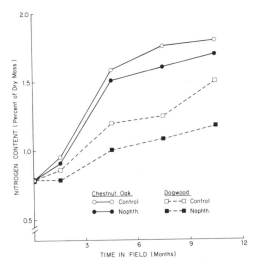

Figure 17.1. Effects of microarthropods on the nitrogen concentrations of decomposing dogwood (*Cornus florida*) and chestnut oak (*Quercus prinus*) foliage. Data are from Cromack (1973). Results shown are from naphthaline-treated and untreated plots.

why initial lignin or nitrogen concentrations should be such a strong predictor of amounts of litter remaining after one year in the field. If decay is evaluated as a function of biotic activity, perhaps most simply expressed by the use of a correlate such as evapotranspiration (Meentemeyer 1978), then an explanation for the increased variability and lack of fit of the Coweeta data is apparent (Figure 17.2). Decay in a single year is much more rapid, and substrate quality much more altered, in the warmer and wetter environment of the Southeast. Microhabitat and microenvironmental influences on decomposition tend to be cumulative (Seastedt et al. 1983). Thus, initial substrate quality is of lesser importance to decay rates during the second half of the year in the Southeast than it is in the Northeast. Faunal effects are not shown in Figure 17.2, but they are implicit. Furthermore, the seasonality of temperature and moisture regimes may influence decay rates (e.g., Seastedt et al. 1983). Differences in nitrogen content of substrates with or without fauna may be large or small, but this faunal effect also contributes a source of variation that diminishes the usefulness of using initial substrate characteristics as predictors of decomposition and mineralization rates. Of course, if initial substrate quality was correlated with subsequent microbial and microarthropod activity, a strong relationship would be expected. Meentemeyer (1978) combined initial lignin content of litter with actual evapotranspiration values to predict decomposition rates. However, this relationship remains useful only if the actual evapotranspiration values are correlated with biotic activity (Whitford et al. 1981). If the biota circumvent temperature or moisture constraints, then actual evapotranspiration measurements lose much of their usefulness as predictors of litter decay rates.

We have continued to analyze the influence of microarthropods on the concentrations of nitrogen and other elements in foliar litter. Feeding activities of microarthropods appear to consistently increase the concentrations of most elements (Table 17.3).

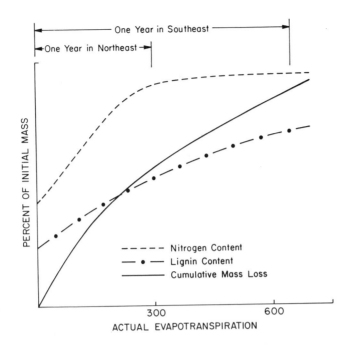

Figure 17.2. Cumulative mass loss, nitrogen content, and lignin content of a hypothetical sub-strate on the forest floor in the Northeastern U.S. versus the Coweeta region. Actual evapotran-spiration values are lower in the Northeast, producing a slower decay rate and, in this example, a linear relationship with nitrogen concentrations of the decaying foliage. In the Southeast, decay is more rapid, and the relationship between decay rates and nitrogen concentrations is nonlinear. The effects of the detritivore fauna are implicitly represented by the relationship between actual evapotranspiration and invertebrate feeding activities.

Concentrations of elements increase if the absolute amounts of the measured elements remain constant in litter while the absolute amounts of carbon, hydrogen, and oxygen of the substrate diminish. Concentrations may also increase by inputs of elements in canopy leachates or by the recruitment of nutrients from the soil by fungal hyphae. Previously (e.g., Seastedt and Crossley 1980, 1983), we argued that the increased nutrient concentrations observed in litter with fauna are a consequence of stimulation of microbial respiration by fauna. In the absence of leaching of the litter substrate, an increase in the concentration of all mineral elements is predicted. With leaching, however, only those elements concentrated or immobilized by the microorganisms should remain within the litter-microbial system. This interaction between fauna, microorganisms, and microclimatic conditions was established by Witkamp and his colleagues (e.g., Patten and Witkamp 1967; Witkamp 1969; Witkamp and Frank 1970), but the consequences of these interactions for biogeochemical cycles remains some-what controversial (Seastedt 1984). All else being equal, enhanced nutrient concentra-tions in litter indicates that elements are being immobilized, and this result has been observed in field studies (e.g., Seastedt and Crossley 1980). Effects of fauna on

Table 17.3. Concentrations of Nutrients in Decaying Dogwood (*Cornus florida*) Litter

| | | Concentration (% of Dry Mass) | | | |
| | | Day 180 | | Day 364 | |
Element	Day 0	With Fauna	Without Fauna	With Fauna	Without Fauna
		No Canopy Leachates			
N	1.25	1.61	1.27	1.83	1.50
P	0.09	0.11	0.08	0.12	0.10
K	0.92	0.71	0.68	0.47	0.33
Ca	2.40	2.72	2.74	3.35	3.06
Mg	0.29	0.26	0.25	0.38	0.28
		With Canopy Leachates			
N	1.25	1.58	1.29	1.91	1.78
P	0.09	0.11	0.09	0.12	0.12
K	0.92	0.67	0.90	0.65	0.66
Ca	2.40	2.48	2.56	3.52	3.38
Mg	0.29	0.27	0.35	0.47	0.43

Unpublished data from a study reported by Seastedt and Crossley (1983). Fauna were excluded by application of Naphthalene (100 g m^{-2}).

elemental budgets cannot be assessed on the basis of nutrient concentrations alone. Nonetheless, comparisons of litter nutrient concentrations with and without fauna and with and without throughfall still provide useful insights into biotic processes occurring on the forest floor.

Results shown in Table 17.3 illustrate the effects of throughfall (in this example artificial canopy leachate additions) on litter nutrient concentrations. Occasionally, throughfall additions appear to mask a faunal effect (i.e., K, Mg in Table 17.3), and also show that a portion of the elements in litter originate from throughfall rather than from the litter itself. Also, the effect of fauna on the seasonality of nutrient immobilization and mineralization patterns appears to differ for the different elements. For example, increased N and P concentrations are very evident after 180 days of litter decomposition, while similar effects of fauna on Ca and Mg concentrations are not apparent until later. Such results may reflect a fauna-mediated change in either the successional sequence or composition of the microflora of litter (Parkinson 1980). The effect of microarthropods on enhanced P concentrations in litter appears to diminish later in the year. Such changes generally indicate a pulse of mineralization (e.g., Seastedt and Crossley 1984).

The Effects of Arthropods on Nutrient Cycling

We have conducted a number of studies of microarthropod effects on nutrient dynamics of litter, but have regrettably devoted little time doing similar studies on the effects of macroarthropods. The modeling efforts conducted using Coweeta data all suggest a substantial macroarthropod effect (e.g., Gist and Crossley 1975a; Cornaby et al. 1975; Webb 1976, 1977). More recently, Anderson and Ineson (1983) and J. M. Anderson (personal communication) have suggested that macroarthropod effects on the nutrient flux of such elements as nitrogen are equal to the entire contribution of the microflora and mesofauna. The subsequent discussion, then, may be regarded as a conservative evaluation of arthropod effects on nutrient budgets.

A generalized nutrient cycling model used by Seastedt and Crossley (1984) (Figure 17.3) illustrates that the arthropods may directly influence elemental cycling by ingestion rates, storage within living tissue, and storage within dead tissues and feces. The works of Gist (1972) and Cornaby (1973) suggested that arthropods function as a modest sink for elements such as calcium. Subsequent work by Seastedt and Tate (1981) and Seastedt et al. (1981) indicated that soil fauna exoskeletons composed a slightly larger elemental sink than that found in living fauna, but that this amount, when elevated on the basis of total system standing crops, remained relatively small. Only a few species of detritivores appeared to specialize on invertebrate carrion and exoskeleton fragments; most of the species appeared to be generalist litter feeders. Exoskeletons of millipedes and oribatid mites represent concentrated calcium sources (e.g., Cromack et al. 1977; Crossley 1977b), and may be fed upon by a number of organisms.

Webb (1976, 1977) studied the effects of feces from a large millipede species on decomposition and cation mineralization processes of litter. His results were similar to those reported by Nicholson et al. (1967b): feces did not decompose faster than the

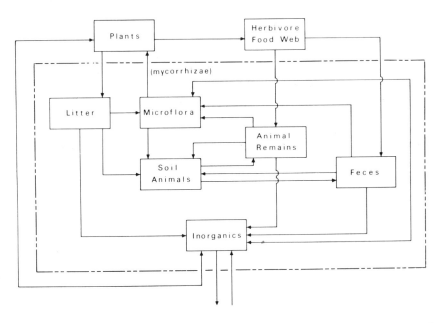

Figure 17.3. Conceptual model of an ecosystem emphasizing the presence and activities of the forest floor fauna. Direct effects of fauna are represented by the flows of material through the faunal component. A potential nutrient and energy sink is created by feces production. For microarthropods, at least, the dominant effect of the fauna is expressed by stimulation of microbial respiration and mineralization activity.

initial litter substrates, and the addition of feces to the forest floor did not stimulate decomposition or mineralization processes of other litter. The relatively small surface area of the fecal pellets, which supported a relatively small microbial population, was suggested as a reason for these negative results. Webb's findings appear very different from those reported by Anderson and Ineson (1983), who demonstrated large increases in ammonium–nitrogen leaching from litter containing feces of a millipede. The source of the nitrogen was believed to have originated from heterotrophic bacteria on the feces themselves. These bacteria produced a pulse of N mineralization in spite of a relatively high carbon to nitrogen ratio of the substrate. No similar studies have been conducted using microarthropod feces; however, both Anderson and Ineson (1983) and Petersen and Luxton (1982) cite studies suggesting lower rates of decomposition of the feces than that measured for the original litter or microbial substrates.

As previously stated, microarthropods are primarily microbivores and therefore the dominant effect of microarthropods on nutrient budgets results from direct and indirect effects of their feeding activities on microflora. Net mineralization is largely decided by whether microbial immobilization accompanies the increase in microbial respiration that usually results from faunal activities (e.g., Hanlon and Anderson 1979; Anderson and Ineson 1983). Seastedt (1984) developed a procedure to quantify this faunal effect, i.e.:

$$\text{Net faunal effect} = \text{CONC}_f/\text{CONC}_w \times \text{MASS}_f/\text{MASS}_w$$

where CONC_f = concentration of a nutrient in litter containing fauna; CONC_w = concentration of a nutrient in litter without fauna; MASS_f = mass remaining of litter containing fauna; and MASS_w = mass remaining of litter without fauna. This procedure combines both the effect of the arthropods on the concentrations of elements within the detritus and its effect on mass loss. Ironically, fauna may occasionally increase nutrient concentrations of the litter and decrease the mass of the substrate such that their effect on net mineralization is negligible. Of course, such substrates that have been grazed by microarthropods have different chemical constituents (generally higher nutrient content per unit of mass) than litter without fauna (e.g., Figure 17.1), and can be expected to subsequently mineralize at a more rapid rate. A summary of the Coweeta data (Seastedt 1984; Seastedt and Crossley 1984) indicates that the microarthropod effect on microbial respiration and therefore mass loss usually exceeds the increase in nutrient concentration in litter by stimulation of microorganisms. Thus, net mineralization is the usual result. However, the patterns observed for all elements are by no means the same, and, to date, no consistent pattern has been observed for some of the cations.

Conclusions

The Coweeta studies, along with many others conducted elsewhere, suggest that litter and soil arthropods are directly and indirectly responsible for a 20% to 40% increase in the cycling rates of most elements. The mechanisms responsible for these increased mineralization rates remain poorly known. We do not know if the net faunal effect is due to more efficient processing of nutrients by a continually cropped (hence, younger and more physiologically active) microflora, or if the response is primarily caused by a shift in the species composition of the microflora (e.g., Parkinson 1980; Newell 1984a,b). Our current conclusions are drawn from studies of relatively short duration. An untested assumption is made that the faunal effects measured in the short-term studies can be extrapolated using standard exponential decay models. Recently, Mellilo (pers. comm.) suggested that decomposition apparently ceases when the ratio of lignin to lignin plus cellulose approaches a given value. He suggested that physical mixing or some other process that reduces overall lignin content of a substrate must occur. His findings concur with data previously presented by Howard and Howard (1974). Soil fauna may not only speed decomposition, but perform a function that is ultimately responsible for the continued decay and mineralization of plant substrates.

Current research at the ecosystem level has indicated that the ratios of important elements such as C:N:P ratios may result in certain predictable patterns of energy allocation by plants. These patterns have subsequent consequences to consumers. Herbivores appear to both respond to plant C/N ratios (Mattson 1980), and also cause changes that influence other consumers (e.g., Hutchinson and King 1980; Smolik and Dodd 1983). A similar process also occurs within the detrital food web. Feeding activities of soil fauna produce measurable changes in the chemical content of the detritus (e.g., Table 17.3), and this chemically modified material influences the subsequent activities of detritivores and microbivores. At any given time, a system or subsystem such as the forest floor may therefore be represented by a nutrient and energy matrix that, free from extrinsic perturbation, will interact with the microclimate and generate a subsequent matrix. The soil arthropod fauna at Coweeta clearly represents a major factor in the regulation of this nutrient and energy reservoir. Continued studies of soil invertebrates at Coweeta will undoubtedly contribute to our knowledge of these emerging patterns.

18. Sulfur Pools and Transformations in Litter and Surface Soil of a Hardwood Forest

J.W. Fitzgerald, W.T. Swank, T.C. Strickland, J.T. Ash,
D.D. Hale, T.L. Andrew, and M.E. Watwood

Work on sulfur transformations was initiated at Coweeta in 1980 in an effort to determine the relevance of the plant sulfolipid as a source of sulfate in forest soil. It soon became apparent from the results of this study (Strickland and Fitzgerald 1983) that A-horizon soil from several watersheds exhibited not only rapid S-mineralization rates for this compound, but samples also rapidly incorporated a substantial portion of the released sulfate into a fraction which could only be recovered by acid extraction. This work was followed closely by a study of the fate of sulfate in soils of several ecosystems at Coweeta (Fitzgerald et al. 1982). The results confirmed initial observations made with the sulfolipid and suggested that the acid extractable fraction might be comprised of soil organic sulfur.

Although work at Coweeta by Johnson et al. (1980) indicated that soil adsorption is an important sulfate retention mechanism, it was also apparent that microbial metabolism was also important in sulfur accumulating properties of watersheds located in the Coweeta basin (Chapter 4). Sulfate arising from atmospheric deposition is immobilized by adsorption as well as by conversion to organic forms of sulfur. In consequence, subsequent work has concentrated on sulfate adsorption and on providing quantitative evidence for the incorporation of sulfate-S into organic matter (Fitzgerald et al. 1983, 1985). Preliminary estimates of the annual flux of sulfate into the forest floor and all mineral horizons of a control hardwood forest (WS 18, 30 Kg S ha^{-1}; Swank et al. 1984) suggest that organic S formation is a major pathway in the sulfur cycle at Coweeta.

Because the conversion of sulfate-S to organic S will immobilize the anion, this process together with adsorption act as sulfate reservoirs, and should also reduce cation leaching and thus lessen the impact of acid precipitation on forest soils of the Coweeta basin (Swank et al. 1984, 1985). However, organic S formation generates a pool which is subject to mobilization and subsequent mineralization (Strickland et al. 1984). Examination of the mobilized S-pool indicated that most of the sulfur (94%) was present as sulfate, whereas the remainder consisted of soluble organic S. Since the subsequent metabolism of organic S determines the importance of the formation process as a sulfur retention mechanism, considerable effort (Strickland et al. 1984; Strickland and Fitzgerald 1984) has also been made to characterize organic S mobilization in forest soil and litter at Coweeta. The data indicate that organic S mobilization occurs at rates substantially less than those for organic S formation (Swank et al. 1985), thus indicating a net accumulation of insoluble organic S.

The purpose of this chapter is to review sulfur research conducted over the past 5 years on control WS 18. Included in this objective is an attempt to provide an ecosystem level perspective for organic S formation, mobilization, and accumulation for the forest floor and A_1 horizon soil utilizing data obtained from samples collected on a monthly basis over at least a year interval in most cases. Data on the in situ levels of organic and inorganic S present in forest floor and soil will also be presented together with a description of the sulfur linkage groups comprising the organic S pool.

Experimental Site and Sulfur Analyses

Data reported in this chapter are based upon analyses of samples taken along a transect of WS 18 established at mid-elevation on the catchment in May 1982. The transect, consisting of 10 equally spaced 0.01 ha circular plots, transverses the watershed from ridge to stream to ridge. Prior to July 1983, a single sample of the O_1 and O_2 layer and three samples of A_1 horizon were taken at random on a monthly basis from each of the 10 plots. These samples were analyzed separately. In June 1983 and thereafter, samples taken as above were collected on a quarterly basis. Forest floor components were analyzed separately, but triplicate A-horizon samples from each plot were mixed in equal proportions and each mixture was then analyzed. Sample collection dates and frequency are given in the text as a footnote for each particular analysis.

Total S was determined by hydriodic acid reduction after oxidation of each sample in hypobromite (Tabatabai and Bremner 1970). Ester sulfate content was determined by reduction with hydriodic acid (Freney 1961). Since hydriodic acid will not reduce sulfur which is directly bonded to carbon, the amount of this form of sulfur was calculated as the difference between the total S content of a given sample and the amount of hydriodic acid reducible S present in that sample. The amount of carbon bonded S present in each sample as amino acid S was determined by reduction with Raney Ni (Freney et al. 1970). Because this catalyst will not reduce the other major form of carbon bonded S (e.g., sulfonate-S), the difference between the total carbon bonded S content and the Raney Ni reducible S represents the amount of sulfonate-S in a given sample. Soluble and adsorbed S was determined by hydriodic acid reduction of extracts obtained by extraction of the samples with water and basic phosphate, respectively. Inorganic

Figure 18.1. Naturally occuring organic sulfur linkage groups.

ESTER SULFATE $R-C-O-SO_3^-$

CARBON BONDED-S AS:

AMINO ACID-S $R-C-S-CH_3$

                                $R-C-SH$

SULFONATE-S $R-C-SO_3^-$

SULFAMATE-S $R-N-SO_3^-$

SULFATED THIOGLYCOSIDE $R=N-O-SO_3^-$

sulfate in these extracts was determined by anion chromatography (Dick and Tabatabai 1979). Hydriodic acid reduces ester sulfate and inorganic sulfate; therefore, values for soluble and adsorbed ester sulfate were obtained as the difference between hydriodic acid reducible S and sulfate determined by chromatography. Structures for the various sulfur linkage groups detected by these reagents are shown in Figure 18.1.

In Situ Forms of S and Pool Sizes

Analyses of S forms present in the upper horizons of WS 18 are shown in Table 18.1. The extent of analyses is comparable to that previously available for agricultural systems (Fitzgerald 1976) and represents, to our knowledge, the only analyses of its kind for a forest ecosystem. In agreement with observations made for a hardwood forest in the Adirondacks (David et al. 1982), carbon bonded S, as opposed to ester sulfate, is

Table 18.1. Mean Concentrations (μg S g^{-1} Dry Weight), % of Total S[a], and Coefficients of Variation (C) for Litter and Soil from WS 18[b]

	O_1 Layer			O_2 Layer			A_1 Horizon		
Form	$\bar{x}$	%	C	$\bar{x}$	% Total S	C	$\bar{x}$	% Total S	C
Ester sulfate	185	13	0.82	360	27	0.31	118	40	0.07
Carbon bonded S	1298	87	0.72	949	72	0.29	161	54	0.40
Amino acid S[c]	270	18	0.42	452	34	0.25	69	23	0.55
Sulfonate S	1028	69	0.80	497	38	0.33	92	31	0.29
Soluble S[d]	0	—	—	1.43	0.1	3.08	1.6	0.5	3.48
Adsorbed S[d]	0	—	—	13.8	1.0	0.75	17.1	6	1.0
Total S	1482	—	0.74	1325	—	0.31	295	—	0.42

[a] Carbon bonded S = amino acid S + sulfonate S.
[b] Samples collected from plots 1–10 in May 1984; $n = 20$ except as noted.
[c] $n = 30$
[d] $n = 10$

Table 18.2. In Situ Levels of Soluble and Adsorbed Sulfur in Litter and Soil from WS 18[a]

| Horizon | Amount (μg S g^{-1} Dry weight) $\pm$ SE and (% E) as: | |
	Soluble	Adsorbed
O_1	6.9 $\pm$ 2.1 (30.4)	12.2 $\pm$ 2.8 (23.0)
O_2	25.2 $\pm$ 6.5 (25.8)	40.7 $\pm$ 7.7 (18.9)
A_1	4.4 $\pm$ 1.1 (25.0)	35.5 $\pm$ 3.9 (11.0)

[a] Samples collected from plots 1-10 quarterly, November 1983 through August 1984; $n = 40$.

the dominant component in the forest floor representing 87% and 72% of the total S in the O_1 and O_2 layers of WS 18, respectively. However, unlike the Adirondack forest in which the level of carbon bonded S comprised about 74% of the S in the soil, Table 18.1 shows that about 54% of the total S of the A_1-horizon of WS 18 consists of sulfur in this linkage. After correction for the levels of soluble and adsorbed S in the samples, the remainder (40%) of the sulfur in soil from WS 18 is comprised of ester sulfate. An analysis of the linkage groups present in the carbon bonded S pool (Table 18.1) indicates that sulfonate-S is a major component in all horizons, and especially in the O_1 layer in which this form comprised 69% of the total S. This finding was not unexpected, since the plant sulfolipid in which S is present in a sulfonate linkage is considered to represent a major component of leaf sulfur (Harwood and Nicholls 1979). This observation coupled with previous results showing that the A-horizon can mineralize plant sulfolipid and amino acid S (Strickland and Fitzgerald 1983; Fitzgerald and Andrew 1984), suggests that carbon bonded S represents an important source of sulfate for forest soil. Soluble and adsorbed S (Table 18.1) represented minor components ($\leq$6%) of total S

Table 18.3. Nature and Pool Sizes of In Situ Soluble and Adsorbed Sulfur in Litter and Soil from WS 18[a]

| S-Pool and Determination | Amount (μg S g^{-1} Dry Weight) $\pm$ SE in: | | |
	O_1 Layer	O_2 Layer	A_1 Horizon
Soluble S by HI-reduction (SO_4^{2-} + ester SO_4^{2-})	22.9 $\pm$ 3.6	85.2 $\pm$ 14.1	4.1 $\pm$ 1.0
Soluble S by IC (SO_4^{2-} only)	20.8 $\pm$ 2.5	49.8 $\pm$ 8.6	5.8 $\pm$ 0.6
Soluble ester SO_4^{2-} by difference	5.8 $\pm$ 2.5[b]	35.4 $\pm$ 5.9	0.4 $\pm$ 0.2[c]
Adsorbed S by HI-reduction (SO_4^{2-} + ester SO_4^{2-})	0	86.0 $\pm$ 19.3	65.9 $\pm$ 4.9
Adsorbed S by IC (SO_4^{2-} only)	0	28.2 $\pm$ 4.1	22.5 $\pm$ 3.5
Adsorbed ester SO_4^{2-} by difference	0	57.7 $\pm$ 16.4	43.3 $\pm$ 5.5

[a] Samples collected from plots 1-10, August 1984 and extracts analyzed by HI (hydriodic acid) reduction and IC (ion chromatography); $n = 10$.
[b,c] HI > IC value for 4 and 2 of the plots, respectively.

found in the forest floor and soil of WS 18 during May of 1984. These sulfur pools also exhibited the greatest variability in concentration across the sampling transect, with coefficients of variation (CV) ranging from 0.75 to 3.5 compared with the other S forms reported in Table 18.1, where the CV was ≤ 0.75. The O_1 layer exhibited the highest variability for all forms of sulfur considered. Although soluble and adsorbed S was undetectable in the O_1 layer of all plots during May 1984 (Table 18.1), some of the O_1 samples collected between November 1983 and August 1984 did have sulfur in these forms (Table 18.2). Variability was large, especially for levels of soluble S in all three horizons. Since hydriodic acid reduces inorganic, as well as ester-linked S, this reagent was utilized to quantify soluble and adsorbed S so that forms of S, apart from sulfate, could be detected. Previous work on the fate of sulfate in forest soil (Fitzgerald et al. 1982) suggested that ester sulfate formed from this anion was not confined solely to a nonsalt extractable sulfur pool. By analyzing water and salt extracts by hydriodic acid reduction and by anion chromatography, it is evident (Table 18.3) that ester sulfate represents a major component of the soluble and adsorbed sulfur pools of O_2 layer samples collected from the transect during August 1984. A similar observation was made for the adsorbed S pool of the A_1 horizon, whereas, soluble S in this horizon, as well as in the O_1 layer was comprised almost totally of inorganic sulfate (Table 18.3).

Transformations of Amino Acid S

In situ levels of amino acid S have not been documented for other forest ecosystems, and in view of the apparent importance of sulfate input to soil from these sources a study of the metabolic fate of ^{35}S-methionine in forest floor and soil from WS 18 was conducted (Fitzgerald and Andrew 1984, Fitzgerald et al. 1984). Table 18.4 summarizes data on rates of S-mineralization and incorporation of methionine into organic matter. The rates for the O_1, O_2 forest floor layers and A_1 horizon are compared with in situ levels of amino acid S found in each horizon. The relative differences in the in situ amino acid levels between each horizon reflect the relative differences in the methionine mineralization and incorporation rates for each horizon. Table 18.4 shows that the rate for incorporation exceeds the mineralization rate in both the O_1 and O_2 layers and these horizons contained an in situ level of amino acid S which was at least fourfold greater than the A_1 horizon. As expected from mineralization rates of the plant sulfolipid (Strickland and Fitzgerald 1983), the A_1 horizon exhibited a mineralization

Table 18.4. Metabolism of Added Amino Acid Sulfur in Litter and Soil from WS 18[a]

	(μg S g^{-1} 12h^{-1} at 20°C) $\pm$ SE		μg S g^{-1} In Situ Amino Acid S[b]
Horizon	Amount Mineralized	Amount Incorporated into Organic Matter	
O_1	0.05 ± 0.01	0.37 ± 0.04	270
O_2	0.38 ± 0.07	0.69 ± 0.07	452
A_1	0.16 ± 0.01	0.07 ± 0.01	69

[a] Added as ^{35}S-methionine; samples collected from plots 1–10 in August 1982; $n = 10$.
[b] Taken from Table 18.1.

rate for methionine which was about twofold greater than the incorporation rate, and this difference may account for the low in situ amino acid S level in this horizon relative to the forest floor.

Transformations of Sulfate and Organic S

In view of atmospheric inputs of sulfate to forests of the Coweeta basin, a concentrated effort has been made over the past 4 years to document the metabolic fate of this anion in litter and soil from WS 18. In brief, ^{35}S-labelled sulfate at a concentration similar to the average annual input concentration from throughfall and soil solutions is incubated with each horizon for 48 hr at 20° C. The samples are then washed with water, extracted with salt, and finally extracted with a strong acid and base. The extracts are analyzed separately to determine the level of ^{35}S which remains nonmetabolized (water extract), which has been adsorbed (salt extract) and which is only recovered by acid-base extraction. Analysis of these extracts by electrophoresis revealed that sulfate was the major ^{35}S-labelled component. Details of the procedure have been published (Fitzgerald et al. 1982, Strickland and Fitzgerald 1984). Results of these analyses on samples collected on a monthly basis for 1 year are summarized in Table 18.5. Total recoveries of the label approached 90% for all analyses and in terms of a percentage of ^{35}S which could be recovered, between 9 and 13% of the label remained unmetabolized in all horizons. Table 18.5 also shows that the potential for adsorption was greatest in the A_1 horizon and least in the O_1 horizon (72% and 12% of the ^{35}S present in salt extracts, respectively). On a dry weight basis, the O_1 layer exhibited the greatest potential for incorporating the label into acid-base extractable organic matter, whereas the A_1 horizon possessed the lowest capacity for incorporation. Nevertheless, approximately 20% of the ^{35}S was found in the combined acid-base extract of this latter horizon (Table 18.5). Moreover, when the quantities of substrate comprising the forest floor and A_1 horizon are taken into consideration (Table 18.6), estimates of the annual potential flux of sulfate into organic matter were greatest for the A_1 horizon. This horizon exhibits a poten-

Table 18.5. Potential Fates of Added Sulfate in Litter and Soil from WS 18[a]

Fraction	Amount (nmole SO_4^{2-} g^{-1} 48 h^{-1} at 20°C) ± SE in:		
	O_1 Layer	O_2 Layer	A_1 Horizon
Water extract	2.48 ± 0.40	2.48 ± 6.1	0.74 ± 0.07
	(8.8)[b]	(12.5)	(8.8)
Salt extract	3.29 ± 0.41	6.34 ± 0.44	6.03 ± 0.22
	(11.7)	(32.0)	(71.5)
Acid-base	22.4 ± 0.91	11.0 ± 0.66	1.67 ± 0.04
	(79.5)	(55.5)	(19.8)
Total percent recovery	88.3 ± 0.80	89.3 ± 0.68	89.6 ± 0.40
n	120	120	360

[a] Samples collected from plots 1–10 monthly May 1982 through June 1983, inclusive.
[b] Percent of total ^{35}S recovered is given in parentheses.

Table 18.6. Estimates of Annual Potential Flux of Sulfate into Salt Extractable (Adsorbed S) and Acid plus Base Extractable Organic S in Litter and Soil from WS 18[a]

| Horizon | Flux (Kg SO_4^{2-} − S ha^{-1} y^{-1}) into: | |
	Adsorption	Organic Matter
O_1	0.12	0.78
O_2	0.22	0.38
A_1	41.38	11.46

[a] Calculations made using means shown in Table 18.5.

tial at least 10-fold greater than that for the forest floor. Similar considerations also apply to the potential capacity for sulfate adsorption (Table 18.6) but, in this case, the potential for the A_1 horizon was more than 100-fold greater than the forest floor. When data are placed within an ecosystem perspective, it is clear that sulfate adsorption and incorporation into organic matter are important soil processes in the sulfur cycle of this forest. The importance of the incorporation process is not confined solely to the A_1 horizon. Based upon a soil profile study conducted during August 1982, Swank and co-workers (1984, 1985) estimated a flux of sulfate into organic matter in the B_W horizon which was equivalent to that estimated for the A_1 horizon for this sampling date.

Characterization of Acid-Base Extractable ^{35}S

Since the incorporation of sulfate into organic matter is based on incorporated S that can only be recovered by acid and base extraction, some attention has been given to characterizing this conversion in an effort to prove that the sulfur is incorporated into organic matter by covalent bond formation. Recovery of ^{35}S under conditions that extract organic matter does not, of necessity, mean that ^{35}S was originally present as organic S. An alternative possibility is that the ^{35}S was simply adsorbed to organic matter and could not be released by salt extraction. Several lines of evidence rule out this alternative, and these will be reviewed briefly. The incorporation of ^{35}S into the acid-base extractable pool was shown to be time- and temperature-dependent (Fitzgerald and Johnson 1982; Fitzgerald et al. 1983), and subject to stimulation by increased energy availability (Strickland and Fitzgerald 1984). While these characteristics do not unequivocally rule out adsorption of S in favor of covalent bond formation, they do show that the process is microbially mediated with characteristics unlikely for a purely physical phenomenon such as adsorption. We were subsequently able to isolate ^{35}S-labelled organic matter by pyrophosphate extraction at pH 8 (Fitzgerald et al. 1985). Unlike acid-base extractions (Fitzgerald et al. 1983), organic S can be extracted with this reagent with minimal destruction of the sulfur linkage groups. Unequivocal proof for the incorporation of sulfate-S into organic matter was obtained when ^{35}S was retained after dialysis of the pyrophosphate extract under conditions which would completely release adsorbed sulfate. Moreover, the dialyzed extract reacted with reagents specific for organic S linkage groups (Fitzgerald et al. 1985).

Table 18.7. Mobilization of Organic Sulfur Formed from Sulfate in Litter and Soil from WS 18[a]

Horizon	Organic S Mobilized 24 h^{-1} (%) $\pm$ SE[a]	Organic S Formed 48 h^{-1} (n mole S g^{-1})[b]	Organic S Retained 24 h^{-1} (n mole S g^{-1})
O$_1$	13.2 $\pm$ 0.71	22.4	19.4
O$_2$	33.6 $\pm$ 2.3	11.0	7.3
A$_1$	41.3 $\pm$ 2.0	1.7	1.0

[a] Samples collected March through November 1983; $n = 42$.
[b] Taken from Table 18.5.

Mobilization and Accumulation of Organic S

Studies of the capacity of soil to mobilize organic S were initially conducted on samples collected from Coweeta control WS 2 (Strickland et al. 1984). The methodology for these determinations is complex and will not be described here. Detailed descriptions have been published elsewhere (Strickland et al. 1984; Strickland and Fitzgerald 1984). Although most of the mobilized S consists of sulfate, some of this sulfur is released as soluble organic S. Based upon these findings, Strickland and co-workers (1984) suggested that the mobilization process involves depolymerization of the insoluble organic S matrix to yield soluble forms which then undergo desulfation (mineralization).

Potential capacities to mobilize organic S are now available for the forest floor and A$_1$ horizon of WS 18. Means of data for samples collected over a 9-month period from the entire transect are shown in Table 18.7. The A$_1$ horizon and the O$_2$ forest floor layer mobilized between 34 and 41% of the available organic S during a 24-hr time interval, and only 13% was mobilized in the O$_1$ layer. Variability was low in all cases with SE of estimates <7%; this small variation corresponds with the low variation of sulfate-S incorporation observed in the A$_1$ horizon (Table 18.5). Moreover, the low variability in the two processes, irrespective of the season in which samples are taken, provides some justification for an attempt to estimate potential organic sulfur accumulation in the forest floor and A$_1$ horizon of WS 18. By utilizing the amount of organic S formed within a 48-hr interval (Table 18.5) and by applying the percentage of this sulfur which could be mobilized during a subsequent 24-hr interval, it can be seen (Table 18.7) that during this latter period substantial quantities of organic S may be retained within each horizon. These calculations are based on a number of assumptions which will be examined in future studies and are presented to provide a perspective of the relative importance of incorporation and mobilization processes.

Conclusions and Future Considerations

Results derived from the initial study of the capacity to mineralize plant sulfolipid S conducted some 5 years ago have opened many previously unexpected avenues of sulfur research at Coweeta. The most prominent of these may be transformations involving inorganic sulfate. Clearly, a long term study of organic S formation and mobilization will be absolutely essential for better understanding of soil nutrient dynamics and the true effects of acidic deposition on forests. These processes together with sulfate adsorption dominate the sulfur cycle at Coweeta. The extraordinary lack of spatial variability of the respective activities in forest soil indicates that when other factors which regulate these processes are quantified, a realistic and predictable model for these transformations at the ecosystem level will be tractable. In view of the long-term data base on nutrient and water budgets which already exists for the Coweeta basin, the influence of sulfur transformations on net accumulation of sulfate (by adsorption and organic S formation) and stream chemistry can be tested. This combined effort should lead to a more accurate interpretation of atmospheric deposition impacts.

6. Stream Biota and Nutrient Dynamics

19. Aquatic Invertebrate Research

J.B. Wallace

As pointed out by the editors of several recent books (Barnes and Minshall 1983, Resh and Rosenberg 1984; Merritt and Cummins 1984), the freshwater biology literature, especially that concerned with stream ecology and aquatic invertebrates, has expanded enormously in the last 15 years. Research at Coweeta follows this worldwide trend. The purpose of this chapter is to review the aquatic invertebrate work at Coweeta during the past and present and to address future prospects, particularly as they relate to long-term ecological research at Coweeta.

There are approximately 73.4 km of streams within the confines of the Coweeta Hydrologic Laboratory. This distance is composed primarily of small, low-order streams: first, 41.7 km (57%); second, 15.8 km (21.5%); third, 11.1 km (15.1%); fourth 4.3 km (5.8%); and, fifth, 0.6 km (0.9%) (Table 19.1). These estimates were made from digitized stream lengths on a 1:7200 map, so a number of small first-order reaches were not measured; They are thus conservative for first-order streams. More than three-quarters of the total stream length at Coweeta is composed of smaller first- and second-order streams.

Most small streams draining undisturbed catchments are heavily shaded by the surrounding forest. Rhododendron is especially dense along most stream margins. On undisturbed catchments, litter fall and lateral inputs of coarse particulate organic matter (CPOM) into Coweeta streams range from ca. 350 to 568 g DM m^{-2} yr^{-1} (Webster and Patten 1979; Webster and Waide 1982; Webster et al. 1983; Webster 1983). The few available measurements of primary production in Coweeta streams indicate a low

Table 19.1. Stream Lengths in Various Orders Within the Confines of the Coweeta Experimental Forest

Drainage Basin	Total Stream Length (km)	Length (km) in Various Orders				
		1st	2nd	3rd	4th	5th
Dryman Fork	15.2	8.5	2.4	2.6	1.8	–
Ball Creek	27.7	15.6	6.4	4.8	1.0	–
Shope Fork	27.9	16.3	6.4	3.7	1.5	–
Coweeta Creek	2.7	1.4	0.6	–	–	0.6
Coweeta Creek Basin[a]	58.2	33.2	13.4	8.5	2.5	0.6
Sum of all[b]	73.4	41.7	15.8	11.1	4.3	0.6

Based on digitized stream lengths by J. O'Hop from 1:7200 U.S. Forest Service Coweeta Map.
[a] Includes Ball Creek, Shope Fork, and Coweeta Creek.
[b] Includes Dryman Fork, Ball Creek, Shope Fork, and Coweeta Creek.

level of net primary production in an undisturbed second-order stream, i.e., 2.6 g ash free dry mass (AFDM) m^{-2} yr^{-1} (Hains 1981; Webster et al. 1983). This represents <1% of the allochthonous inputs (Webster et al. 1983; Webster 1983). Thus, the small streams that represent >78% of stream length at Coweeta can be characterized as primarily heterotrophic systems.

Response of Invertebrates to Ecosystem Disturbance

Aquatic invertebrate studies at Coweeta have traditionally focused on the effects of forest disturbance on stream ecosystems. Tebo (1955) studied the influence of sediments discharged from a logged catchment [Watershed (WS) 10] on the downstream fauna in Shope Fork. Invertebrate densities in Shope Fork were reduced significantly by accumulated sediment which altered the streambed. The sedimentation problem was ameliorated to some extent by high water during the spring months, which reworked the streambed and exposed the original substrate. Tebo's study was among the first in North America to document the influence of sediments on invertebrates, and it additionally demonstrated the importance of physical factors (i.e., current velocity and discharge) on invertebrate fauna.

Gurtz (1981) and Gurtz and Wallace (1984) studied the response of aquatic macroinvertebrates to a major catchment disturbance, the clearcutting of WS 7, and found that substrate type was an important factor in determining the direction and magnitude of the aquatic invertebrate response. Following road building and logging in WS 7, inorganic and organic seston increased in Big Hurricane Branch (BHB), which drains WS 7 (Gurtz et al. 1980), and silt deposition caused a redistribution of stream fauna among substrate types. In BHB, more taxa increased in the moss covered rock face habitat followed by the cobble riffle, pebble riffle, and sand substrates; whereas, more taxa decreased in abundance in sand followed by pebble riffle, cobble riffle and rock face (boulder) substrates. Larger substrates require more stream power to move them, occur where current velocity is high, and are less susceptible to deposition of fine particles

(Gurtz and Wallace 1984). The response of various invertebrate taxa to disturbance suggests that biological stability is closely associated with physical stability of the habitat. Moss associated with large substrates apparently facilitates the colonization by invertebrates of these otherwise exposed surfaces.

Georgian (1982) showed that several species of grazers in Shope Fork were associated with different substrate particle sizes and current velocities. Haefner and Wallace (1981a,b) found that certain benthic taxa displayed different distributions with regard to substrate preference. Malas and Wallace (1977), Wallace et al. (1977), and Ross and Wallace (1982) also found that the distribution of some species of net-spinning caddisflies were closely associated with larger substrate sizes and hence higher velocities.

Woodall and Wallace (1972) investigated the benthic fauna in four Coweeta streams draining catchments with different treatment histories: initial stages of old field succession (WS 6), undisturbed hardwood forest (WS 18), coppice hardwood forest (WS 13), and a catchment converted to white pine forest (WS 17). Significant differences in invertebrate densities and biomass existed among streams. The old field catchment stream had the highest faunal densities. The coppice catchment stream supported the highest biomass. The pine plantation catchment had significantly lower densities and biomass than the other three streams. The old field catchment stream also had higher densities of taxa belonging to the scraper functional group, while the streams in hardwood and coppice hardwood catchments had higher densities of leaf-shredding organisms (e.g., Merritt and Cummins 1984).

Haefner and Wallace (1981a) found that there were significant changes in invertebrate densities in the stream draining WS 6 between 1968 and 1969 (Woodall and Wallace 1972) and 1978 to 1979. The major changes in invertebrate fauna were a decrease in insect scrapers and an increase in shredder densities. However, stream shredder taxa densities in WS 6 remained lower than those of the hardwood forest stream (WS 18). The shift in functional feeding group densities coincided with 10 years of terrestrial secondary succession which had resulted in more shading of the old field stream, which, coupled with increases in allochthonous detritus inputs, altered the energy base of the stream. Changes in taxa and functional groups can occur in fairly short time periods following catchment disturbance. Gurtz (1981) and Gurtz and Wallace (1984) found stream collector-gatherers and scraper taxa increased while the dominant shredder species declined during the first year of logging of the BHB catchment.

Webster and Patten (1979) studied three of the streams investigated by Woodall and Wallace (1972) (WS 6, WS 17, and WS 18). Webster and Patten found that although the rates of organic matter processing differed among streams, the pathways were similar in all streams. They concluded that stream ecosystems have low resistance to perturbation, but high resilience following disturbance. A major aspect of stream resilience in headwater streams is replacement of organic matter by allochthonous inputs (Webster and Patten 1979; Gurtz et al. 1980). Consequently, recovery of streams from catchment clearcutting is closely linked to restoration of surrounding terrestrial vegetation and therefore recovery of allochthonous inputs (Gurtz et al. 1980; Haefner and Wallace 1981a; Webster et al. 1983).

There is another aspect of stream recovery that extends beyond the simple restoration of allochthonous leaf litter inputs. This concerns woody litter, which has a very slow rate of decomposition in streams (Triska and Cromack 1980). Most (> 77%) leaf litter inputs occur in the 3 month autumn season (Webster and Patten 1979). Stream flow at Coweeta is generally highest in winter and early spring (Chapter 3). Thus, there must be some mechanism for retention of CPOM inputs within stream reaches if the organic matter is to be fully utilized by microbial flora and animal fauna within these streams. Low stream power (Leopold et al. 1964), high roughness (Chow 1959), and shallow narrow channels which are readily subject to obstruction by woody debris, enhance debris dam formation and consequently CPOM and FPOM retention (Sedell et al. 1978; Naiman and Sedell 1979; Bilby and Likens 1980; Bilby 1981; Wallace et al. 1982a; Cuffney et al. 1984).

Recent measurements of wood in various Coweeta streams indicate that Sawmill Branch (WS 6) has <10% the volume of wood per meter squared measured for any other Coweeta stream considered to date (Wallace and Huryn, unpublished data). All woody debris was removed or burned on WS 6 prior to its conversion to grass in 1958. In Sawmill Branch (WS 6), early successional species such as black locust have not yet entered the stream channel in quantities sufficient to restore significant woody debris structure. As retention of both coarse particulate organic matter (CPOM) and fine particulate organic matter (FPOM) is enhanced by woody debris (Bilby and Likens 1980), this suggests that simple restoration of leaf litter inputs alone will not restore the stream to its previous state. Interestingly, Webster and Golladay (1984) recently reported that both organic and inorganic seston in Sawmill Branch still remains extremely high 27 years since disturbance. Likens and Biley (1982) and Swanson et al. (1982) suggested that recovery of streams may actually lag behind, or be out of phase with that of the terrestrial forest, since large stable debris dams will be reestablished only after mature trees die and fall into the stream channel. Thus, while total allochthonous inputs of leaf fall may be restored to stream channels quite early in succession, retention of CPOM in the stream channels having little woody debris may be much lower than that of streams with extensive woody debris. Molles (1982) found that for New Mexico streams, differences in woody debris may influence the relative abundance of both shredder and grazer functional groups. Although there was no difference in taxa present, Molles found relative abundance of shredders and benthic organic matter were much higher in streams with abundant debris dams.

Influence of Invertebrates on Stream Ecosystem Processes

Most, but not all, of the aquatic research at Coweeta has focused on the effect of watershed disturbance on stream biota and stream ecosystem function. These studies have involved changes in energy inputs to streams including alteration of the surrounding forest and changes in physical characteristics of streams such as increased sediment loads during and after catchment disturbance. With several exceptions, studies have not focused on examining what role macroinvertebrates have in ecosystems. If physical characteristics of Coweeta headwater streams favor retention of organic matter, what is the influence of the invertebrate fauna on ecosystem processes? Benthic invertebrates are small and represent an insignificant portion of total catchment biomass; therefore, the suggestion that they influence ecosystem function would seem remote at first glance. The suggestion that insect shredders can generate considerable quantities of FPOM and DOC by their feeding is supported by indirect evidence such as high ingestion rates, low assimilation efficiencies, and their ability to comminute CPOM to FPOM (McDiffit 1970; Grafius and Anderson 1979; Golladay et al. 1983; Meyer and O'Hop 1983). However, little direct evidence has been obtained to substantiate the existence, magnitude, and importance of macroinvertebrates in producing FPOM from CPOM in streams. Indeed, Winterbourn et al. (1981) suggested for New Zealand streams that macrobenthos, especially insect shredders, are of minor importance in the production of FPOM.

Wallace et al. (1982b) devised an experiment to examine the role of insects in processing CPOM in two small headwater streams at Coweeta. We applied a pesticide to one of two adjacent streams, which resulted in massive invertebrate drift (primarily aquatic insects) from the treated stream. Aquatic insect densities were reduced to < 10% of those in the adjacent reference stream by applications of pesticide. Benthic community structure shifted from a system dominated by insect shredders to one dominated by noninsects such as Oligochaetes, Turbellaria, and various other noninsect groups. Chironomids and a few predaceous dragonfly nymphs were the only insects remaining in this stream in any abundance (Wallace et al. 1982b; Cuffney et al. 1984).

Leaf species breakdown rates were studied concurrently with the pesticide treatment and exhibited the same relative sequence of breakdown in both the treated and the reference stream (i.e., dogwood > red maple > white oak > rhododendron). However, reduction of densities of shredders and other insect fauna in leaf bags in the treatment stream significantly reduced leaf breakdown rates well below those of the reference stream. The more refractory the leaf species, the greater the effect insect exclusion had on breakdown rates (Wallace et al. 1982b). Furthermore, suspended POM concentrations in streamwater and transport to downstream areas were significantly lower following treatment, whereas no significant difference existed prior to treatment. Nonstorm FPOM export to downstream reaches in the post-treatment period was 3.9 times higher in the reference stream than that of the treatment stream (Wallace et al. 1982b). Average particle size of transported organic matter was also smaller in the treated stream than in the reference stream (Cuffney et al. 1984).

The above studies (Wallace et al. 1982b; Cuffney et al. 1984) indicated that benthic invertebrates play an active and substantial role in the processing of CPOM to FPOM in headwater streams. Therefore, the conclusion of Winterbourn et al. (1981) that insects have little influence on the generation of FPOM in streams seems unwarranted, at least in the small headwater streams of Coweeta. Through comparison of the physical characteristics of Coweeta and New Zealand streams, a contrast in the amount of woody litter and therefore retentiveness of the streams becomes apparent. The Coweeta streams had more woody litter allowing for the development of high shredder densities; whereas in the New Zealand streams, the paucity of wood reduced the potential for instream processing of litter by shredders and the resulting comminution of available CPOM to FPOM. This conclusion is supported by another study of a New Zealand stream in which the quantity of wood was similar to that of Coweeta streams

(Rounick and Winterbourn 1983). Here these authors acknowledge that in the retentive, wood dominated stream, the general pattern of the invertebrate community did agree with the generalization of the continuum concept of Vannote et al. (1980).

Physical Characteristics of Streams vs. Food Exploitation by the Biota

Undisturbed Coweeta streams tend to retain a large portion of their CPOM inputs. The biota, dominated by shredder biomass, exploit these retentive characteristics by feeding on the retained CPOM and in doing so comminute CPOM to FPOM, which is more easily entrained and transported downstream (Table 19.2). Conversely, in larger downstream reaches which are less easily subjected to channel obstruction, the physical characteristics favor entrainment. The biota of these larger streams are dominated by collector-gatherers (especially filtering collectors) and favor retention. The biota are exploiting the physical characteristics of these systems. In the unidirectional, harsh physical environments of streams, it is difficult to visualize how the biota could function otherwise, i.e., shredders in downstream reaches where there is little retention and passive filter feeders in environments with high retention and little entrainment. In small headwater streams, the biota may exert considerable influence by comminution of CPOM to FPOM, thereby promoting downstream transport of POM. The downstream fauna probably has a progressively smaller role in offsetting stream physical characteristics as a result of higher downstream discharge and large cross sectional areas in proportion to the substrate available. Entrainment and physical transport probably overwhelm the retention capacity of the biota (e.g., Haefner and Wallace 1981b; Ross and Wallace 1983). This suggestion is supported to some extent by Webster's

Table 19.2. Some Characteristics of Streams vs. that of the Accompanying Fauna

Characteristic	Headwater Streams of Forested Regions	Large Rivers
Stream length	ca. 74% of total stream length in North America[a]	Much less
Interface with terrestrial environment	Maximal	Less(?)[b]
Inputs of terrestrial litter	Maximal	Less(?)[b]
Retention devices	Abundant[c]	Few
Organic matter inputs	Retained	Entrained
Invertebrate community structure[d]	Favor entrainment (i.e., shredders)	Favor retention (i.e., filter-feeders)

[a] First- and second-order streams represent ca. 74% of the total stream length in North America (Leopold et al. 1964).
[b] These characteristics are probably not true of large, lowland rivers with extensive flood plains subject to seasonal flooding, which may result in large inputs of allochthonous organic matter.
[c] Influenced by low stream power, high bed roughness, and narrow, shallow channels which are easily obstructed.
[d] Microhabitats with high entrainment, e.g., steep-gradient, rock face substrates, exist within headwater reaches and influence invertebrate community structure over relatively short mesospatial scales (see text).

(1983) computer simulation, which suggested that the importance of macroinvertebrates in overall organic matter processing decreases downstream.

These retention and entrainment characteristics don't necessarily relate to an entire stream or a given stream reach. Even Coweeta headwater streams have localized areas which may favor deposition or transport. For example, excluding filter-feeding chironomid larvae, the average densities of insect filter feeders in Sawmill Branch, Grady Branch, Hugh White Creek, and BHB are about 2200/m² of rock face substrate vs. ca. 320, 260, and 205/m² for cobble, pebble, and sand substrates, respectively (based on data from Haefner 1980 and Gurtz 1981). Conversely, debris dams retain large amounts of CPOM and FPOM and may harbor high densities of shredders such as *Lepidostoma* and *Tipula*. These microhabitats have fauna which reflect local physical characteristics.

Biologists often fail to recognize that very localized physical conditions may exert strong influences on stream biota and microhabitat preferences and even influence food availability to the organism. Rock face substrates are shallow, high velocity habitats (Haefner and Wallace 1981a,b; Gurtz and Wallace 1984). Here, filter feeders such as *Parapsyche cardis* occupy shallow depths, where a great proportion of potential food passes within the height (<1 cm) of their catchnets (Smith-Cuffney and Wallace, in press). Higher stream velocities above these rock face substrates also enhance the rate at which food is delivered to these passive filterers. Furthermore, moss growing on this substrate may facilitate the presence of invertebrates by providing heterogeneous microhabitats with respect to velocity gradients as well as attachment sites for invertebrates.

Moss-covered rock face substrates also provide habitat for some of the fauna found in debris dams, e.g., *Peltoperla*. Moss both traps and retains particulate organic and inorganic material. Francie Smith-Cuffney has been using different densities of "artificial moss" on rock outcrop substrates in Big Hurricane Branch and Hugh White Creek, and has found up to 100 g AFDM/m² of organics may be retained in this "moss" within a 30-day sampling period. About 40% of the trapped organic materials are particles >250 μm diameter. Mosses also serve as major sites of primary production and provide attachment sites for algae in these streams (Hains 1981). Thus, very localized areas of streams possess physical characteristics which are exploited on a mesospatial scale by the biota.

Secondary Production in Coweeta Streams

Several studies have addressed secondary production of aquatic invertebrates in Coweeta streams. The production studies to date are biased toward Trichoptera (Table 19.3). In an adjacent drainage system, Benke and Wallace (1980) estimated net-spinning caddisfly production as 1 g AFDM m^{-2} yr^{-1}. Similar results were obtained by Ross and Wallace (1983) in Coweeta streams. Haefner and Wallace (1981b) found that production of two species of net-spinning caddisflies in Sawmill Branch (WS 6) was 4.9 g AFDM m^{-2} yr^{-1}, or about 3.4 times higher than that of an undisturbed reference stream (Grady Branch, WS 18). All of the above studies, as well as that of Georgian and Wallace (1981) and Ross and Wallace (1982), attempted to quantify the influence of

Table 19.3. Secondary Production of Aquatic Invertebrates in Coweeta and Nearby Streams

Organism	Reference	Production mg AFDM m^{-2} yr^{-1}	Location
Copepoda			
Bryocamptus zschokkei	1	ca. 360	WS 14
Ephemeroptera			
Baetis spp.	2	63–1,112[a]	WS 14, WS 7
Plecoptera			
Peltoperla maria	3	414–560	WS 6, WS 18
Trichoptera			
Rhyacophila spp.	4	2–115[b]	Dryman Fork
Wormaldia moesta	5	67	Dryman Fork
Diplectrona modesta	5–7	31–647	Several locations
Parapsyche apicalis	5	180–188	Dryman Fork
Parapsyche cardis	5–7	161–4,274	Several locations
Arctopsyche irrorata	6	604	Tallulah Headwaters
Hydropsyche spp.	5,6	27–175[c]	Several locations
Cheumatopsyche h. enigma	5	26–151	Dryman Fork
Glossosoma nigrior	8	612	Shope Fork
Agapetus spp.	8	21	Shope Fork
Neophylax consimilis	8	150–176	Shope Fork
Goera fuscula	8	9–16	Shope Fork
Goerita semata	10	238	Rock Face, Ball Ck.
Brachycentrus spinae	9	261	Dryman Fork
Diptera			
Chironomidae	11	224[d]	litterbags WS 54
Chironomidae	12	1,608[e]	Ball Ck. (WS 27)
Blepharicera spp.	8	307–325[f]	Shope Fork
Prosimulium spp.	4	32–167	Dryman Fork
Simulium spp.	4	54–348	Dryman Fork

References 1–11 as follows: 1, O'Doherty (1985); 2, Wallace and Gurtz (1986); 3, O'Hop et al. (1984); 4, D. H. Ross and J. B. Wallace (unpublished data); 5, Ross and Wallace (1983); 6, Benke and Wallace (1980); 7, Haefner and Wallace (1981b); 8, Georgian and Wallace (1983); 9, Ross and Wallace (1981); 10, Huryn and Wallace (1985); 11, Huryn and Wallace (1986); and, 12, Huryn, (unpubl). In addition to the above, production estimates for several other taxa of Ephemeroptera, Plecoptera, Trichoptera, Diptera, and crayfish are completed (Huryn, Wallace, Gurtz, unpublished data).
[a] Weighted stream production for all substrates for 21-month period.
[b] Range for individual species (n = six species).
[c] Range for individual species (n = four species).
[d] Average per litterbag; yearly P/B = 42.
[e] Weighted stream production (WS 27) for all substrates.
[f] Total production for three species.

filter feeders on particulate organic seston, and results suggested that filter feeders ingest only a minute portion of the total transported organic matter.

In a study conducted in Shope Fork, the secondary production of a grazer guild was estimated to be about 1.2 g AFDM m^{-2} yr^{-1} (Georgian and Wallace 1983). Overlap between six grazer species was calculated on the basis of density, biomass, and production. (Overlaps based on production should most closely reflect patterns of resource consumption.) Production of each species was concentrated in short, <6 week to 3 month, time periods, and mean overlap based on production was significantly lower than that based on either biomass or densities. This lends support to Vannote's (1978) hypothesis that various species within a functional group are organized to minimize periods of similar resource use by two or more species. While the assumption underlying the cause of the temporal separation of production (the avoidance of interspecific competition) was not tested, the various species do tend to occupy different microhabitats as well (Georgian 1982).

Secondary production estimates require a knowledge of life cycles or specific growth rates, the standing stock densities, and biomass. Thus, they are very labor intensive. Voltinism and length of immature development have been identified as the two most important factors influencing production (Benke 1979; Waters 1979), yet life history studies are often considered unfashionable. Ecosystem studies are often oriented toward the processing of organic matter and nutrient cycling by various groups of organisms. The integration of production, feeding habits, and bioenergetic data can yield a much better understanding of the role of animal populations in ecosystems (e.g., Benke and Wallace 1980; Fisher and Gray 1983; Benke 1984). O'Hop et al. (1984) showed that, although the average benthic density in one stream was more than twice that of the other, *Peltoperla* production was similar. Standing stock densities alone did not detect differences in growth rates and survivorship between the two streams. In summary, numerical abundances and biomass may lead to erroneous interpretations about the role organisms play in ecosystems.

To date, most Coweeta secondary production studies have focused on bivoltine, univoltine, or semivoltine species with clearly discernible life cycles. However, there are exceptions. O'Doherty (1985) estimated production of an harpacticoid copepod as ca. 360 mg AFDM m^{-2} y^{-1}, and Wallace and Gurtz (1986) estimated *Baetis* production in excess of 2 g AFDM m^{-2} y^{-1} for rock face substrates in Big Hurricane Branch for the first year following logging of WS 7. Based on gut analyses, literature-derived assimilation efficiencies, and production, Wallace and Gurtz (1986) estimated that *Baetis* used about the same proportion (7.4 to 9%) of net primary production in both Big Hurricane Branch and a reference stream, Hugh White Creek. Total diatom consumption by *Baetis* was about 25 times higher in Big Hurricane Branch than in Hugh White Creek, and production of both *Baetis* and diatoms was much higher in the former stream (Hains 1981; Webster et al. 1983; Wallace and Gurtz 1986).

If we are going to understand responses of many species to disturbance and their role in energy flow and nutrient cycling, we must have more data on life cycles and species-specific growth rates, feeding habits, and bioenergetic efficiencies. It is rather appalling that the most abundant insects in Coweeta streams are the chironomids, and that only recently have any data on growth rates been obtained for this group in Coweeta streams (Huryn and Wallace 1986). Sixty-six taxa of chironomids were identified from Hugh White Creek and Big Hurricane Branch from three collections spanning a 13-month period (Gurtz 1981; Gurtz and Wallace 1984). Annual estimates of P/B (production/$\bar{x}$ (= mean) biomass) ratios of chironomids range from <1 in arctic tundra ponds to over several hundred in warm streams. Benke (1984) suggested that annual P/B's may exceed 500 under favorable conditions. Some recently initiated field growth studies for chironomid larvae in Coweeta streams indicate daily growth rates of 7 to 20% of larval AFDM/day at temperatures of about 14°C, with highest growth rates in early instars (Huryn and Wallace 1986). Annual P/B's for chironomids may approach

20 to 50 at Coweeta. Based on a pretreatment chironomid biomass of 86 mg AFDM/m^2 for the pesticide treated stream (Cuffney et al. 1984) and a daily growth rate of ca. 0.125 mg mg AFDM body wt d^{-1}, chironomid production alone may approach 4 g AFDM m^2 yr^{-1} in some Coweeta streams. Chironomids are just one of several groups in need of research if we are going to address to role of invertebrates in energy flow and nutrient cycling in streams.

We have little knowledge of the meiofauna composition of Coweeta streams, and less is known about their feeding habits and potential P/B ratios. In published studies to date, the only attempt to address meiofauna is the work of O'Dougherty (1985), who estimated secondary production of the copepod, *Bryocamptus zschokkei*, at ca. 360 mg AFDM m^{-2} yr^{-1}. This production value is slightly less than production of a dominant shredder in Coweeta streams, but greater than several other species (cf. Table 19.3).

Based on existing studies, Table 19.3 shows that secondary production of invertebrates does not appear to be very high in Coweeta streams. Several factors probably attribute to this rather low level of secondary productivity. These include rather cool annual temperature regimes, generally low stream nutrient concentrations, and shading of most streams (which limits autochthonous production and food quality available to invertebrates). Food quality and temperature are important factors influencing growth and life cycles of invertebrates (see Merritt and Cummins 1984; Resh and Rosenberg 1984).

There is little direct evidence that insect predators exert much impact on other invertebrate populations at Coweeta. Pesticide treatment of a headwater stream reduced large insect predators to less than 8% of the population levels for the adjacent reference stream during an 8-month period following treatment (Wallace et al. 1982; Cuffney et al. 1984) and within this same 8-month period, noninsect populations increased in the treated stream. Dragonfly larvae, primarily *Lanthus vernalis* Carle, were the only significant insect predators remaining in the treated stream. In the reference stream, insects represented the most frequent items in *Lanthus* guts (73% of prey). The food of *Lanthus* in the treated stream reflected the shift in invertebrate community structure, with noninsects constituting 87% of all prey items (Wallace et al., unpublished data). Within two years, insect biomass had recovered in the treated stream. This shift in community structure was reflected in *Lanthus* diets, as insects represented 82% of their prey in the treatment stream vs. 78% in the reference stream during the second year of recovery. These data suggest that generalist predators such as *Lanthus* can shift to alternative prey when confronted by massive changes in community structure. However, more research is needed on prey and predator production and turnover, availability of prey production to predators, and the energetic requirements of predators in order to adequately assess the influence of predation on benthic community structure at Coweeta.

Future Prospects

Coweeta offers unique opportunities for assessing both short- and long-term consequences of ecosystem disturbance on invertebrates, and for addressing long-term recovery of lotic ecosystems. Pesticide treatment of a small headwater stream at Coweeta resulted in massive invertebrate drift, subsequent changes in benthic community structure, lower leaf litter breakdown rates, and significant reductions in the amount of fine organic matter transported to downstream reaches. Can restoration of these functional characteristics occur before full structural (i.e., taxonomic) recovery of various insect species? During 1982 to 83, leaf breakdown rates in the formerly treated stream were not significantly different from those of the reference stream. Although populations of the predominant pretreatment shredder *Peltoperla* remained extremely low in treatment stream, *Lepidostoma* and tipulids were present in large numbers. *Lepidostoma* and *Tipula* were actually much more abundant in litter bags from the treated stream than in the reference stream during 1982 to 1983 (Vogel 1984). Particulate organic seston concentrations had increased several fold in the treated stream compared to the 1980 levels. These results suggest that functional recovery may occur before full structural recovery of the shredder group. This work is being continued, and offers an approach to investigate changes in ecosystem processes caused by benthic organisms without altering energy inputs or the physical structure of the stream.

Webster (1983) developed a series of models to evaluate the role of macroinvertebrates in streams. Overall, the simulations suggested that macroinvertebrates were responsible for only a small portion of the respiration of detritus, and that the major role of macroinvertebrates is in the conversion of benthic detritus to organic seston. This model suggested that although macroinvertebrates were responsible for only 27% of the annual POM transport, their activities may contribute as much as 83% of the POM transported during low flow (late summer). Although stochastic processes, i.e., storm flows, may dominate annual export budgets, there may be long periods of base flow in which biological processes predominate. Webster's efforts in this regard represent an excellent contribution, and such models are highly useful in showing us where problems exist in research efforts.

Further knowledge of a number of important aspects of streams are mandatory for additional refinement of Webster's model. These include aspects such as: (1) What are more accurate secondary production estimates for both macroinvertebrates and meiofauna? (2) What are the rates of microbial respiration in Coweeta streams? (3) How do macroinvertebrates influence community metabolism? (4) What are the relative rates of microbial respiration on detritus in streams in which macroinvertebrate densities have been manipulated vs. unmanipulated streams? (5) How comparable are patterns and quantity of particulate organic matter export during storm flows in macroinvertebrate manipulated and unmanipulated streams? (6) Does macroinvertebrate manipulation influence annual output budgets for particulate and dissolved organic export to downstream areas? (7) How does manipulation influence the timing and magnitude of organic matter export to downstream areas during storm and non-storm periods?

Another area where Coweeta offers excellent possibilities for long-term study concerns the potential influence of acidic precipitation on stream ecosystems. Streams draining high elevation catchments with shallow soils underlain by granitic bedrock are thought to represent areas most sensitive to acidic precipitation (Record et al. 1982). Coweeta has many catchments with these characteristics, e.g., WS 27. Acidification of stream water reduces diversity of aquatic fauna, but the exact mechanisms and consequences for ecosystem level processes remain unclear (Wiederholm 1984). Other than brief experimental acidification studies at Norris Brook (Hall et al. 1980), the sequence of changes during acidification are uncertain. Especially lacking is the documentation of long-term changes in a single stream. Recent efforts have focused on studies of the invertebrates and some system processes on WS 27. This project should result in the best invertebrate documentation available for a single stream at Coweeta.

While there is currently no evidence to demonstrate any acidification effects on WS 27 stream fauna, this does not imply that there need be no long-term concern. One reported effect of acidification that may be especially relevant for Coweeta in the future is its influence on decomposer organisms. Leaf litter decomposition proceeds much slower in acidified (pH 4.3 to 5.6) than in more neutral waters (pH 6.0 to 6.5) (Traaen 1980). Hendrey et al. (1976) attributed an abnormal accumulation of coarse organic detritus on the bottom of acidified Swedish lakes to this phenomenon. This reduction in decomposition rates is probably related to a reduction in heterotrophic microorganism activity on detritus and to reduced invertebrate abundances in acidified waters. Based on the studies of Wallace et al. (1982b) and Cuffney et al. (1984), we know that reduction in densities of insects can drastically alter detritus processing and potentially influence energy and nutrient flow to downstream areas. In acidified waters where both invertebrate and microbial activities are reduced, the potential influence of acidification on stream energy and nutrient flow may equal or exceed the effect of reducing invertebrates alone as in the experiments of Wallace et al. (1982) and Cuffney et al. (1984). While there is no strong current evidence to support acidification of Coweeta streams, the potential for impact certainly exists and requires long term vigil. The availability of baseline data on invertebrate abundance, biomass, and production; combined with heterotrophic respiration, leaf litter processing rates, and long-term data on stream chemistry, place this Laboratory in a unique position for assessing any long-term changes in stream ecosystems.

20. The Trophic Significance of Dissolved Organic Carbon in Streams

J.L. Meyer, C.M. Tate, R.T. Edwards, and M.T. Crocker

Accumulations of particulate organic matter (e.g., decomposing leaves, debris dams) are a conspicuous feature of stream ecosystems at Coweeta and in other forested areas. Stream ecologists have devoted considerable research effort toward examining the role of this particulate detritus in stream food webs (e.g. Cummins 1974). In this paper we draw attention to a less visible, but equally important component of stream food webs, namely dissolved organic carbon (DOC)—its sources and the organisms that utilize it as a food resource.

The DOC concentration of Coweeta streamwater is relatively low (<1.5 mg C/L during baseflow, Meyer and Tate 1983), and is similar to the concentration of particulate organic matter in streamwater at baseflow (Wallace et al. 1982a). DOC export per m^2 of undisturbed stream channel (137 to 145 g C/m^2, Meyer and Tate 1983) is slightly less than leaf litter inputs to Coweeta streams (212 g C/m^2, Webster 1983). About 15 kg C/ha (watershed area) is lost annually from undisturbed Coweeta watersheds as DOC (Meyer and Tate 1983). Hence, DOC export is an important component of organic carbon export from forested watersheds at Coweeta as well as in other regions of the country (e.g. Fisher and Likens 1973; Wetzel and Manny 1977; Naiman 1982).

The DOC lost from terrestrial ecosystems enters streams. In this chapter we examine the fate of this DOC, its uptake, and its role in the food web of the stream. We also discuss DOC sources, particularly those within the stream channel, and examine the manner in which streamwater DOC concentration changes as watersheds recover from a disturbance.

Removal of DOC from Streamwater

DOC can be removed from streamwater by flocculation, biotic uptake, and abiotic sorption by sediments. It is unlikely that flocculation is important in Coweeta streams because of their low ionic strength (Lush and Hynes 1973; Mulholland 1981), and removal by abiotic processes is generally a small fraction of biotic removal by stream sediments (Cummins et al. 1972; Lock and Hynes 1975, 1976; Dahm 1981; Kaplan and Bott 1983; Kuserk et al. 1984). Hence, the major pathway for DOC removal from the water in Coweeta streams is probably uptake by the benthic microbial community.

To determine rates at which DOC could be removed from the water column, we experimentally enriched two Coweeta streams with DOC and measured the DOC uptake rate. We enriched short reaches of Hugh White Creek (HWC) and Big Hurricane Branch (BHB) on three occasions. HWC drains Watershed 14 (WS 14), an undisturbed reference watershed with oak-hickory forest and a dense rhododendron understory along the stream. BHB drains WS 7, a successional watershed which was clearcut 3 years before these experiments were done. We enriched the stream for brief periods (1 to 2 hr) with sucrose, a labile form of DOC, by slowly adding water (100 ml/min) from a carboy filled with a concentrated solution of sucrose and NaCl. The sodium was used as a conservative tracer to determine the amount of dilution occurring. On one date, leaf leachate was used instead of sucrose as the DOC source. Leachate was prepared by soaking leaf litter (collected by placing a large net under a mixed-species canopy on WS 14) in stream water for 24 hr and then filtering through bolting cloth.

During the enrichments, streamwater samples were taken at 15 minute intervals from eight stations 10 to 50 m apart and immediately filtered (precombusted Gelman A/E glass fiber filters). Sodium concentration was determined on a Perkin-Elmer atomic absorption spectrometer. DOC was measured with a Dohrman DC-54 carbon analyzer (UV catalyzed oxidation in the presence of persulfate).

Predicted DOC concentrations were calculated for each time at each station using the measured Na concentrations. The mean predicted DOC concentration at each station was calculated as the average over all sampling times and compared with the mean observed DOC concentration at these sampling times. Stream discharge was calculated from the dilution of Na added to the stream at a known rate. Uptake rate (mg C m^{-2} min^{-1}) was calculated as the difference between mean predicted and mean observed DOC concentration (mg/L) times stream discharge (L/min) divided by stream channel area between carboy and sampling point (m^2). This measure of streambed area is an underestimate of area of substrate in contact with DOC, since channel roughness was not considered. Hence, uptake rates per m^2 substrate would be substantially less than the rates reported here. These uptake rates were plotted vs. mean observed DOC concentration (average of observed DOC concentrations at all sites upstream of the sampling point weighted for channel area between sites) to examine the relationship between uptake rate and average DOC concentration over the reach (Figure 20.1).

Uptake rates increased on all dates as DOC concentrations increased, although the slopes and intercepts of the curves differed between dates (Figure 20.1). There was no consistent difference between the two streams; in August 1979 uptake was more rapid in BHB, whereas in July 1980 uptake of both sucrose and leaf leachate was more rapid

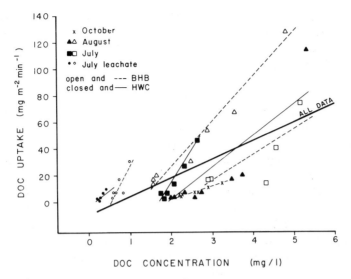

Figure 20.1. DOC uptake rates vs. mean DOC concentration over the reach. These rates were measured in Big Hurricane Branch on three occasions and in Hugh White Creek on two occasions for uptake of added sucrose or leaf leachate. Regression lines for each stream and date are indicated. The regression equation for all data combined is: $Y = -9.85 + 13.61 \, X$ ($r^2 = 0.45$, $p < 0.0001$), where Y = uptake rate (mg C m^{-2} min^{-1}) and X is DOC concentration (mg C/L).

in HWC. Uptake rates for fresh leaf leachate were slightly higher than those observed with sucrose, although data are limited.

When uptake data from all dates are combined, the regression of uptake rate vs. concentration is significant ($r^2 = 0.45$), and the intercept is not significantly different from zero. At a typical DOC concentration of 1 mg C/L, uptake rate is 240 mg C m^{-2} h^{-1}. Kuserk et al. (1984) recently compared DOC uptake rates observed in a wide variety of streams. They reported rates ranging from 5.8 mg C m^{-2} h^{-1} (Kaplan and Bott 1983) to 400 mg C m^{-2} h^{-1} (Lush and Hynes 1978). The uptake rates observed with these labile DOC sources at Coweeta are at the high end of the reported range. This is, in part, a consequence of using a labile DOC source and of measuring these rates in a streambed with high channel roughness, which leads to an overestimate of uptake rate per m^2 of substrate as discussed above. Yet, these experiments clearly demonstrate that the benthos of Coweeta streams is capable of removing DOC from the water column.

Role of DOC in the Stream Food Web

Fungi and bacteria are the organisms most likely to be assimilating DOC from the water column (Dahm 1981; Lock 1981; Kaplan and Bott 1983). We are beginning to examine this in our laboratory by concentrating naturally occurring DOC and examining its ability to stimulate bacterial growth (measured as rate of incorporation of [^{3}H] thymidine into bacterial DNA, Findlay et al. 1984). One source of naturally occurring DOC that we have used in these studies is DOC generated by leaf-shredding insects (Meyer and O'Hop 1983). Dave Lauriano (unpublished data) in our laboratory has demonstrated that bacteria isolated from Coweeta detritus can utilize shredder-generated DOC. He observed removal of DOC from the water and a fourfold increase in bacterial growth rates after 17 hr exposure to this DOC. Streamwater DOC and leaf leachate have also been shown to stimulate bacterial growth in a Pennsylvania stream (Kaplan and Bott 1983).

Little is known of microbial biomass and production in any stream. Fungi are undoubtedly an important component of the microbial community in Coweeta streams, yet we know nothing of their biomass or production. We are beginning to collect information on bacteria using epifluorescent direct counting techniques (Hobbie et al. 1977) to measure abundance. Bacterial production is being estimated by measuring the rate of incorporation of [^{3}H] thymidine into bacterial DNA (Fuhrman and Azam 1982; Findlay et al. 1984). There are few bacteria in stream water at Coweeta (2 to 4 $\times$ 10^4 cells/mL). This is in striking contrast to an annual average bacterial abundance of 1.5 $\times$ 10^7 cells/ml in the Ogeechee River, a sixth-order blackwater river in southern Georgia (Edwards 1985), but within the range reported for streams in the Marmot Creek Basin, Alberta (10^4 to 10^6 cells/ml, Ladd et al. 1982; Geesey et al. 1978). Bacterial density is higher in Coweeta sediments, ranging from 1.6 $\times$ 10^8 cells/cm^3 in sediments with 1.5% organic matter content to 7.8 $\times$ 10^8 cells/cm^3 in sediments with 6.6% organic matter. These values correspond to a bacterial biomass of 64 and 312 mg C/m^2 in the top cm of sediments, which is not significantly different from the bacterial biomass observed in Ogeechee River sediments with similar organic matter content (Findlay et al., 1986).

Sediment bacterial production rates have also been measured, and they appear comparable to those observed in Ogeechee River sediments of similar organic content. We currently have data from Coweeta streams only during March 1984, but at that time bacterial production was 1.1 mg C m^{-2} h^{-1} in sediment with 2.6% organic matter (Crocker 1986). We would expect higher production rates as sediment organic content increases, provided Coweeta streams follow the trends observed in the Ogeechee River (Findlay et al., 1986). The key to determining system-wide measures of bacterial production in Coweeta stream sediments lies in accurate measures of sediment organic content coupled with production measures in sediments of varying organic content.

Bacteria and fungi are also found associated with decomposing leaves in streams, although these microbes may be less reliant on DOC as a carbon source than are water column bacteria. Bacterial biomass in Coweeta leaf detritus varies with leaf species and its decompositional state, but an average value is 20 µg C/g leaf C, which is similar to that observed on leaf detritus in the Ogeechee River (Findlay and Meyer 1984).

Bacterial production rate on decomposing leaves in Coweeta streams is about 0.04 mg C d^{-1} g^{-1} leaf C, similar to that on Ogeechee River detritus (Findlay and Meyer 1984). The population turnover time is 0.5 days (Findlay and Meyer 1984), whereas it is several days in the sediments. In leaf packs, we appear to have a smaller bacterial population that is turning over more rapidly compared to the sediment population. This may be due to greater grazing pressure by invertebrates in the leaf material (Findlay and Meyer 1984). A measure of benthic leaf detritus is necessary to express these production rates per m² of streambed. Standing crop of benthic leaf detritus changes throughout the year, but a reasonable annual average is 86 g DW/m² in an undisturbed Coweeta stream (Webster et al. 1983) or 39 g C/m², assuming carbon is 45% of leaf dry weight. Hence, bacterial production associated with leaf detritus in Coweeta streams is on the order of 1.5 mg C m^{-2} d^{-1}.

When we compare these estimates of bacterial production in Coweeta streams with rates of primary production, the potential significance of bacteria in the trophic structure becomes obvious. Average primary productivity (^{14}C) in HWC is 0.3 mg C m^{-2} h^{-1} (Webster et al. 1983), or 3.6 mg C m^{-2} d^{-1} assuming a 12-hr day. Bacterial production in the top cm of sediments is 1.1 mg C m^{-2} h^{-1}, or 26.4 mg C m^{-2} d^{-1}. Bacterial production in leaf detritus is 1.5 mg C m^{-2} d^{-1}. Clearly, bacteria provide a greater potential food resource than do primary producers in these forested headwater streams.

Bacteria and fungi, some of which are utilizing DOC, can provide food for higher trophic levels in the stream. The following Oregon experiment demonstrates the potential of this microbial link in the food web. In sections of stream continuously enriched with low concentrations of sucrose (0.4 to 1.7 mg C/L), trout production was seven times greater than in untreated sections. The increased production was due to increased trout consumption of a benthic chironomid that was feeding on the bacterial slime growing at the expense of the added DOC (Warren et al. 1964). Benthic meiofauna (those organisms ranging in size from 50 to 500 µm) represent one potentially important link in this microbial portion of the stream food web in Coweeta streams. Very little research has been done on meiofauna in streams, and nothing is known of their functional role. In Coweeta streams, meiofaunal organisms are abundant in decomposing leaf packs (Dan Perlmutter, unpublished data). Harpacticoid copepods, a dominant group in the meiofauna, are capable of feeding on the bacteria associated with leaf packs (Rieper 1978) or on protozoans that graze on bacteria (Rieper and Flotow 1981). Harpacticoids have a mean annual density of 5000 animals/m² in Coweeta streams (Erin O'Doherty unpublished data). Preliminary calculations suggest production rates of 360 g AFDW m^{-2} yr^{-1} for one species (O'Doherty 1982, 1985). This is 87% of the annual production measured for *Peltoperla maria* (O'Hop et al. 1984), an important leaf-shredding stonefly in these streams. Meiofaunal secondary production appears to be consumed by benthic insects, although we are only beginning to systematically collect this information. J. B. Wallace (Chapter 19) reports finding them in dragonfly and caddisfly guts. These preliminary data suggest that meiofauna may be an important trophic link between DOC-utilizing microbes and the well-studied benthic insects of these streams (Chapter 19).

Within-Stream Sources of DOC

As a consequence of DOC removal from stream water at Coweeta, one would predict a decrease in DOC concentration from spring seep to weir. This is not what we observe. During the growing season, there are consistent increases in DOC concentration along the length of Hugh White Creek (Figure 20.2, Meyer and Tate 1983) and in streams in other regions of the country (Kaplan et al. 1980). This increase does not appear to be due to higher DOC concentration in subsurface water or other sources at lower

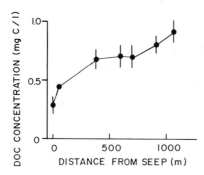

Figure 20.2. Mean (± 95% CI) DOC concentration along the length of Hugh White Creek during the growing season (July–October 1979 and May–June 1980). (Redrawn from Meyer and Tate, 1983.)

elevations, but rather to processes within the stream channel that generate DOC (Kaplan and Bott 1982; Meyer and Tate 1983). A major source of this DOC in Coweeta streams is probably leaching of organic matter in the stream bed. Leaching is a physical/chemical process that is accelerated by biological activity, i.e., microbial metabolism and invertebrate feeding (Meyer and O'Hop 1983).

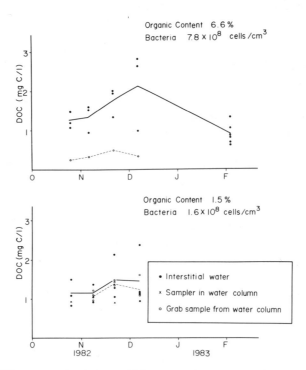

Figure 20.3. DOC concentration in interstitial water and streamwater from two sites differing in sediment organic matter content. The upper graph is from a headwater seep of Dryman Fork and the lower graph is from Hugh White Creek, a second-order stream. Solid lines connect the mean interstitial DOC concentration for each date, and dashed lines connect the mean streamwater DOC concentration.

We are studying this potential source of DOC by monitoring DOC concentrations in interstitial water. Our interstitial water sampler is a 50 ml disposable syringe perforated with holes (2 mm in diameter). It is inserted into the sediments, filled with a dialysis bag (Spectropor, 50,000 MW pore size) containing deionized water, and sealed from the overlying water with the plunger. Dialysis bags are retrieved after 1 week. Based on a preliminary study, DOC concentration in the bags stabilized after 6 days, and appeared to be in equilibrium with the interstitial water (Crocker 1986).

Interstitial DOC concentrations are highly variable (Figure 20.3). In sites with organic rich sediments, interstitial water is consistently higher in DOC than is the overlying streamwater. Since only DOC < 50,000 MW is collected from interstitial water (see above) and all molecular weights of DOC are included in measures of streamwater DOC, these differences are particularly striking. Crocker (1986) observed that interstitial DOC concentrations were consistently higher than baseflow streamwater DOC in a headwater seep of Dryman Fork over a 2-year period. The variability we observe in interstitial DOC may be due, in part, to the sampler's proximity to discrete patches of buried organic matter, because some samplers had consistently high DOC and others consistently low values. Even in the site with low sediment organic matter content (HWC), the highest DOC concentration observed on each date was from an interstitial sampler, suggesting that there are microhabitats of elevated DOC in stream sediments. Experimental additions of organic matter to sediments elevate levels of interstitial DOC (Crocker 1986). Elevated DOC concentrations have been reported in interstitial water of Canadian streams (Wallis et al. 1981, Hynes 1983). These data on interstitial DOC suggest slow leaching of stored organic matter as a potential source of DOC to the stream, while the Canadian data suggest most of the DOC is consumed in the sediments and does not reach the water column (Hynes 1983).

Quality of DOC

DOC has been measured in many streams (reviewed by Moeller et al. 1979), but only rarely has its chemical nature been characterized (e.g. Larson 1978). In undisturbed Coweeta streams, 71 to 91% of DOC in headwater seeps has a molecular weight < 10,000, while only 23 to 33% of DOC is in the < 10,000 molecular weight fraction at the base of the watersheds (Meyer and Tate 1983). Hence, much of the increase in DOC concentration observed along the stream channel (Figure 20.2) is due to increased concentration of higher molecular weight compounds. This is probably the net result of two processes; (1) more rapid uptake of low molecular weight, labile compounds; and (2) leaching of predominantly higher molecular weight compounds from organic matter in the stream channel. One observation supporting the latter process is that shredder-generated DOC consists primarily of high molecular weight compounds (Meyer and O'Hop 1983). The predominance of higher molecular weight DOC in second order Coweeta streams contrasts with observations in a second-order Pennsylvania stream where 68 to 83% of DOC in transport was < 10,000 MW (Kuserk et al. 1984). The Pennsylvania stream flows through grazed pastures, and there are diel fluctuations in DOC associated with algal production (Kaplan and Bott 1982). Clearly, the source of DOC is different in these two streams, and the source affects the nature of DOC (measured here as molecular weight fractions).

We have very limited information on the chemical constituents of DOC in Coweeta streams. Tannin and lignin content of water from HWC and BHB were measured during February and April 1982 (APHA 1975). Concentrations ranged from 0.012 to 0.162 mg C/L, which represented 4 to 20% ($\bar{x} = 13\%$) of total DOC. There was no difference in concentration between streams, although within a stream the seeps tended to have lower concentrations than the base of the watershed. This is in agreement with the molecular weight data discussed above. In a Pennsylvania stream, tannins plus lignins represented a somewhat smaller fraction of total DOC (3 to 7%; Kaplan et al. 1980; Kaplan and Bott 1983).

Total reducing sugars (Mopper 1978a,b) were measured on water collected from BHB and HWC prior to leaf fall in October 1982. Sugar concentrations were slightly higher in HWC than in BHB (0.180 $\pm$ 0.036 mg C/L vs. 0.136 $\pm$ 0.038 mg C/L, $\bar{x} \pm$ 95% CI), although in both streams sugars represented about 15% (range 8 to 28%) of total DOC. Total carbohydrates in a Pennsylvania stream ranged from 11 to 36% of total DOC (Kaplan et al. 1980; Kaplan and Bott 1983). Taken together, the tannins plus lignins and sugars in Coweeta streamwater account for <30% of total DOC. Even in streams where considerable effort has been expended to chemically characterize DOC, a large fraction (80%) remained unidentified (Larson 1978).

The Influence of Watershed Disturbance on DOC in Streams

One important theme in Coweeta research has been the response of ecosystems to disturbance. The most thoroughly investigated recent disturbance has been the clearcutting experiment begun in 1977 (Webster et al. 1983 provide a review of the stream's response). Since July 1979, we have been measuring DOC in streamwater from Big Hurricane Branch draining the clearcut watershed (WS 7), and in water from Hugh White Creek draining a reference watershed of the same size (WS 14). In the first year of the study (1979 to 1980), DOC concentration and export was lower in the disturbed stream, primarily due to lower inputs from subsurface water seeps and tributaries (Meyer and Tate 1983). Differences in water yield between watersheds were insufficient to account for differences observed in concentration or export (Meyer and Tate 1983). We postulated that as Big Hurricane Branch recovered and organic matter inputs increased, DOC concentrations would also increase (Meyer and Tate 1983).

We have continued measuring DOC concentrations in water samples taken weekly from the base of the two watersheds to test this recovery hypothesis. These data are presented as seasonal means in Figure 20.4. There was considerable variability in DOC concentration between years. This annual variability appears to be greater than the variation due to successional change (c.f. Tate and Meyer 1983). Despite this variability, there does seem to be slightly higher DOC concentrations 6 years after disturbance than there were 2 years after disturbance. Mean annual DOC concentration is significantly greater in Big Hurricane Branch 6 years after disturbance than it was 2 years after disturbance (0.689 vs. 0.453 mg C/L, t test, $p < 0.0005$); whereas in Hugh White

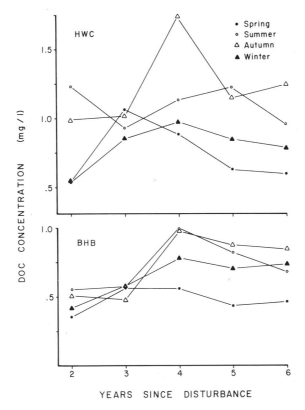

Figure 20.4. Seasonal mean DOC concentration vs. years since Big Hurricane Branch watershed was clearcut. The upper graph is for Hugh White Creek, the reference watershed, and the lower graph is for Big Hurricane Branch. Seasons were designated as follows: summer = June–September; autumn = October–November; winter = December–March; spring = April–May.

Creek there is no significant difference in mean annual DOC concentration for those two years (0.889 vs. 0.786 mg C/L, t test, $p > 0.05$).

This pattern is clearer when we compare DOC concentration in the two streams in each season by examining the ratio of DOC concentration in the reference stream to DOC concentration in the disturbed stream (Figure 20.5). In all seasons, the discrepancy between DOC concentration in the two streams has been decreasing over the 5-year period. The winter data show the least dramatic but most consistent trend, whereas the changes in the other seasons have been greater but more erratic. This suggests that seasonally varying biotic processes are important causal agents for increasing the DOC concentration in streamwater in the disturbed ecosystem. Changes in water yield from the disturbed watershed over this period (Chapter 22) account for only a fraction of observed concentration changes. The seasonal ratio of DOC concentration in the two streams (Figure 20.5) and the seasonal ratio of discharge in the

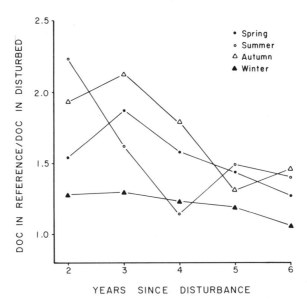

Figure 20.5. Ratio of mean seasonal DOC concentration in the reference stream (Hugh White Creek) over mean seasonal DOC concentration in the disturbed stream (Big Hurricane Branch) plotted vs. years since the Big Hurricane Branch watershed was clearcut.

two streams are significantly related ($p < 0.05$); but the regression explains only 28% of the observed variance in the DOC ratio. Hence, changes in water yield with recovery only partially explain observed changes in DOC concentration. Changing inputs of DOC and leachable particulate organic matter must also be considered.

Summary

DOC is an important energy resource in Coweeta streams. It is removed from the water column by the benthic community and provides a carbon source for benthic bacteria, whose production rates exceed primary production in these forested, headwater streams. The ultimate fate of DOC in the stream trophic structure is an important topic for future research, and benthic meiofauna may prove to be an important link between DOC-utilizing bacteria and benthic macroinvertebrates.

The surrounding watershed is an important source of DOC in these streams. Stream-water DOC changes with watershed disturbance and recovery. Within-channel DOC sources have been less intensively studied, but also appear to be significant. Leaching of buried organic matter in the streambed contributes to elevated DOC concentrations in interstitial water. Interactions between this hyporheic zone and the water column will be an important topic for future research.

21. Effects of Watershed Disturbance on Stream Seston Characteristics

J.R. Webster, E.F. Benfield, S.W. Golladay,
R.F. Kazmierczak, Jr., W.B. Perry, and G.T. Peters

In many ecosystems disturbances are short-lived, and recovery can be followed as a gradual return to predisturbance conditions in the absence of further disturbance. However, the impact of watershed disturbances such as logging, fire, hurricanes, volcanoes, insect outbreaks, etc. on streams is long-term, often lasting as long as recovery of watershed vegetation to predisturbance structure and function (Webster and Patten 1979; Gurtz et al. 1980). Webster et al. (1983) noted that streams are easily disturbed, i.e., exhibit relatively low resistance to disturbance (sensu Webster et al. 1975), but have the potential to recover rapidly following disturbance, i.e., high resilience. However, if the disturbance continues, this potential resilience cannot be realized.

There are many long-term stream disturbances that result from disturbance to the watershed itself. In this chapter, we are concerned primarily with forest logging, although any disturbance that results in vegetation death would have similar results. Studies at Coweeta and other sites have demonstrated that logging increases streamflow (e.g., Likens et al. 1970; Chapter 22). Increased flow has been shown to last 20 to 30 years following logging (Kovner 1956; Hewlett and Hibbert 1963). Dissolved nutrient levels are elevated by logging (e.g., Likens et al. 1970; Chapter 25) and may remain elevated for many years depending on the nature of the forest disturbance. Nutrient levels on Watershed 6 (WS 6) at Coweeta were still elevated 16 years after the watershed was allowed to begin natural succession (Chapter 25). Soil disturbance, primarily due to road-building and skidding methods associated with logging (e.g., Lieberman and Hoover 1948a; Tebo 1955; Brown and Krygier 1971; Chapter 23), increases sediment inputs to streams. Although sediment input may occur only in the first few years following logging and associated activities, redistribution and transport of this material may continue for many years. Coweeta streams draining watersheds logged within the last 20 years still carry elevated sediment levels even during baseflow periods (Webster and Golladay 1984).

Logging opens the canopy, causing increased stream water temperatures (Brown and Krygier 1970; Swift and Messer 1971; Swift 1982). The open canopy and increased illumination, coupled with elevated dissolved nutrient concentrations, result in greatly increased in-stream primary production (Hains 1981). Three years after logging, autochthonous production in Big Hurricane Branch (WS 7) was significantly higher than in a reference stream (Webster et al. 1983). Six years after logging, the two streams showed differences in algal species composition and biomass that appeared to be related to shade tolerance (Lowe et al. 1986).

One of the most significant impacts of forest logging on streams is the reduction in allochthonous inputs. Leaf litter inputs to Big Hurricane Branch 2 years after logging were less than 2% of prelogging levels (Webster and Waide 1982). Seven years following logging, the quantity of litterfall had returned to near original levels (Figure 21.1), however, the quality of inputs was still different. Originally, oak leaves accounted for greater than 32% of the input. In more recent measurements, oak leaves represented < 10% of the input. In contrast, there were much higher inputs of less decay-resistant leaves, such as birch and dogwood (Webster et al. unpublished data).

In addition to the alteration of leaf inputs, logging changes the pattern of woody inputs. In undisturbed Coweeta streams, there are 20 to 30 woody debris dams per 100 m (Table 21.1). Streams draining disturbed watersheds generally have fewer woody debris dams. Similar results have been reported for other Appalachian streams (Silsbee and Larson 1983). According to Swanson et al. (1982) and Likens and Bilby (1982), the

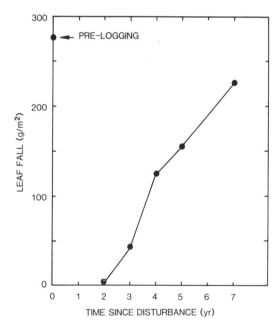

Figure 21.1. Leaffall into Big Hurricane Branch (WS 7) before and after logging. Data from Webster and Waide (1982), J. Meyer (unpublished), and Webster et al. (unpublished).

Table 21.1. Frequency of Woody Debris Dams in Coweeta Streams

Stream	Frequency of Debris Dams, No./100 m	Treatment
WS 6	1.3	Clearcut 1958, slash burned; converted to grass 1959; regrowth since 1967
WS 7	25.4[a]	Commercial clearcut 1977
WS 13	12.0	Clearcut but no products removed, 1939 and 1962
WS 17	11.3	Clearcut but no products removed, 1940; recut annually through 1955; white pine planted 1956
WS 19	20.3	Understory vegetation cut, 1948–1949
WS 22	21.3	Alternate 10-m strips cut without removal, 1955
WS 2	30.0	Reference
WS 14	23.1	Reference
WS 18	28.7	Reference
WS 21	28.7	Reference
WS 34	30.7	Reference

In this survey a debris dam was defined as an aggregation of organic material spanning the stream that included woody material with a diameter greater than 2.5 cm. At least 150 m of each stream were surveyed.
[a] Thirty-one percent of these logs were slash from recent logging

quantity of woody material in streams may be high immediately following logging, depending on the input of logging slash and the extent of channel clearing. During the next 100 or more years of forest regrowth, wood inputs will probably be low since there is little tree death, and the number of woody debris dams in the stream will probably decline as old logs decay. There may be small inputs of wood when successional trees die, such as the death of pin cherry at Hubbard Brook (Likens and Bilby 1982), aspen in western mountains (Molles 1982), or black locust on WS 6 at Coweeta, but this material is generally small, does not span the channel, does not form stable debris dams, and is rapidly washed out of the stream (Molles 1982). Consequently, the number of woody debris dams in streams may remain low for many years (100+) following logging. Since debris dams have been reported to play an important role in stabilizing stream channels (Bilby and Likens 1980; Bilby 1981; Mosley 1981), their absence may result in elevated dissolved nutrient and particle transport especially during storms.

In this chapter we evaluate the effects of long-term disturbances on stream function. We selected 17 Coweeta streams for study, including seven reference streams and 10 streams draining watersheds that had been disturbed 7 to 34 years prior to our study. We made quantitative and qualitative measurements of transported particulate material (seston) in each of these streams. Studies by Wallace et al. (1982b) and Webster (1983) demonstrated experimentally and through computer models that much of the seston in Coweeta streams during baseflow periods results from biological activity. Consequently, we are using seston as an integrative measure of in-stream biological function.

Study Sites

The 17 study streams can be roughly divided into three categories (Table 21.2). Undisturbed streams drain reference watersheds which were selectively logged in the early 1900s. Since Forest Service acquisition in 1929, they have remained undisturbed

Table 21.2. Description of Coweeta Streams Used in this Study

Stream	Watershed Area (ha)	Time Since Disturbance (yr)	Midstream Elevation (m)	Treatment
WS 6	8.9	16	770	Clearcut 1958, slash burned; converted to grass 1959; regrowth since 1967
WS 7	58.7	7	853	Commercial clearcut 1977
WS 10	85.8	27	899	Commercial logging 1942-1956
WS 13	16.2	21	899	Clearcut but no products removed, 1939 and 1962
WS 17	13.4	27	808	Clearcut but no products removed, 1940; recut annually through 1955; white pine planted 1956
WS 19	28.3	34	876	Understory vegetation cut, 1948-1949, 22% of basal area
WS 22	24.3	28	960	Alternate 10 m strips cut without removal, 1955
WS 37	43.7	20	1173	Clearcut but no products removed, 1963
WS 40	20.2	28	990	Selective logging, 1955
WS 41	28.7	28	1036	Selective logging, 1955
WS 2	12.1	60	686	Reference
WS 14	61.1	60	808	Reference
WS 18	12.5	60	754	Reference selective logging prior to
WS 21	24.3	60	899	Reference 1930; chestnut blight
WS 27	38.8	60	1188	Reference 1930s
WS 34	32.8	60	975	Reference
WS 36	48.6	60	1211	Reference

except for chestnut blight in the 1930s. The second category of streams includes six streams draining watersheds that were moderately disturbed by cutting and/or forest species conversions. The third category includes three streams draining watersheds that were more severely disturbed by road construction, logging, and/or species conversion. A complete description of watershed characteristics and treatment histories is given in Chapter 1.

Methods

Water samples were collected in July and November 1983 and March and July 1984 from each of the 17 streams. Collections were made during baseflow periods. Samples for analysis of larger particle sizes were concentrated by pouring measured volumes (40 to 200 L) of stream water through a 20-μm mesh plankton net. Material collected in the net was rinsed into a 1-L nalgene container. In addition, 8-L carboys of unfiltered stream water were collected for analysis of particles smaller than 25 μm. Samples were processed in the laboratory within 12 hr and usually within 2 to 3 hr.

Plankton net samples were separated into four size classes using a wet filtration system consisting of stainless steel sieves (Gurtz et al. 1980): medium large (ML), >280 μm; small (S), 105 to 280 μm; fine (F) 43 to 105 μm; and very fine (VF), 25 to 43 μm.

Material collected on the screens was washed onto glass fiber filters (preweighed and pre-ashed) with distilled water. To collect the ultrafine size fraction (UF, 0.5 to 25 μm), measured volumes (2 to 8 L) of unfiltered stream water were wet sieved and then filtered on glass fiber filters. All samples were oven dried (50°C, 24 hr), desiccated (24 hr), weighed, ashed (500°C, 20 min), rewetted, dried, desiccated, and reweighed. The difference between dried mass and ashed mass provided an estimate of organic seston (ash-free dry mass, AFDM), and the difference between ashed mass and filter mass gave an estimate of inorganic seston (ash). Three replicate samples were collected from each stream for all five size classes.

Samples for density measurements were collected by passing stream water through the 20-μm plankton net until there appeared to be sufficient material for measurement. These samples were subsequently fractioned into ML, S, F, and VF size classes and the material was frozen for later analyses. Density was measured by soaking oven-dried particles in 1-chloronaphthalene (density = 1.19 g/cm³) and adding bromoform (density = 2.89 g/cm³) until the particles became suspended. The density was then determined from a standard curve developed by measuring the refraction of the known mixtures of the two chemicals on a refractometer (Kazmierczak et al. in press). Five replicate measurements were made from each size fraction for each sample collected in July and November 1983 and March 1984.

The fall velocities of each particle size class were measured using a 15-cm diameter settling tube. Wet particles were released into the water column with a Pasteur pipette and were allowed to sink 15 cm before velocity measurements were taken. The time for particles to fall 10 cm was then measured. Ten replicate measurements were made on each size fraction from samples taken from each watershed in July and November 1983 and March 1984.

Median particle size was estimated by regressing the cumulative mass of seston against particle size and then determining the 50% intersection on the particle size axis. In all but 13 of the 204 samples the linear regression was significant ($r > 0.95$, $N = 5$). In each of the other cases the correlation coefficient was >0.90.

Results and Discussion

Our study confirmed the observation by Webster and Golladay (1984) that elevation has a major effect on seston quantity and quality. AFDM, ash, dry mass concentrations, percent ash, and particle size were all correlated with mid-watershed elevation (Table 21.3). Concentrations and percent ash were negatively correlated with elevation; particle size was positively correlated with elevation. Densities of the largest (ML) and smallest (VF) size fractions were also negatively correlated with elevation (Table 21.4) as were the fall velocities of the three smaller particle size fractions (Table 21.5). Webster and Golladay (1984) attributed the correlations of seston concentration and percent ash with elevation to the effect of temperature on biological production of seston. Measurements of stream temperature made in May and July of 1984 on undisturbed watersheds showed that midday stream water temperatures decreased about 5°C/1000 m elevation increase (linear regression, $r^2 = 0.54$, $N = 10$, in May; $r^2 = 0.69$, $N = 13$, in July). Although winter temperature differences may not be as

Table 21.3. Results of Analysis of Covariance for Dependent Variables as a Function of Date, Elevation, and Time Since Disturbance

Dependent Variable	Coefficient of Determination (r^2)	PR > F		
		Date	Elevation	Time Since Disturbance
AFDM	0.61	0.0001	0.0001	0.0001
Ash	0.60	0.0001	0.0001	0.0001
Dry mass	0.62	0.0001	0.0001	0.0001
Percent ash	0.45	0.06	0.0001	0.0001
Median particle size	0.27	0.0001	0.02	0.03

PR > F is the probability that the F value would occur randomly ($n = 204$). The coefficient of determination applies to the multiple regression equation including all three independent variables.

Table 21.4. Results of Analysis of Covariance for Dependent Variables as a Function of Date, Elevation, and Time Since Disturbance

Dependent Variable	Coefficient of Determination (r^2)	PR > F		
		Date	Elevation	Time Since Disturbance
ML density	0.40	0.0001	0.001	0.0002
S density	0.21	0.0001	0.14	0.0001
F density	0.24	0.0001	0.67	0.09
VF density	0.58	0.0001	0.02	0.0001

PR > F is the probability that the F value would occur randomly ($n = 255$). The coefficient of determination applies to the multiple regression equation including all three independent variables.

Table 21.5. Results of Analysis of Covariance for Dependent Variables as a Function of Date, Elevation, and Time Since Disturbance

Dependent Variable	Coefficient of Determination (r^2)	PR > F		
		Date	Elevation	Time Since Disturbance
ML fall velocity	0.25	0.0001	0.37	0.0001
S fall velocity	0.26	0.0001	0.02	0.0001
F fall velocity	0.32	0.0001	0.0001	0.0001
VF fall velocity	0.38	0.0001	0.0001	0.001

PR > F is the probability that the F value would occur randomly ($n = 510$). The coefficient of determination applies to the multiple regression equation including all three independent variables.

great, summer temperature differences of this magnitude would cause measurable differences in the annual biological seston production over the elevational range of Coweeta streams. Also, because seston processing rates are more rapid at lower elevations, particles contain less organic matter (higher percent ash) and tend to be smaller.

This study also supported the observation that nonstorm seston concentrations are highest in summer and lowest in winter (Gurtz et al. 1980; Webster et al. 1983; Webster and Golladay 1984). All measured factors except percent ash were significantly affected by date of sample collection (Tables 21.3 through 21.5). While the seasonal trend in concentrations might be partly explained by winter dilution, Webster and Golladay (1984) showed that transport (concentration × discharge) was highest with warmer temperatures in spring and summer. This further supports the conclusion that much of the nonstorm seston in the streams is the result of biological activity.

Organic seston concentration (AFDM) ranged from 5.06 mg/L to 0.90 mg/L (Table 21.6, Figure 21.2) and was significantly correlated with time since disturbance (Table 21.3). The AFDM concentration was significantly highest on WS 6 and relatively high on four fairly recently disturbed watersheds (Table 21.6). The concentration of ash followed a trend similar to that of AFDM. The correlation of ash concentration with time since disturbance was statistically significant (Table 21.3, Figure 21.2). Ash concentrations were high for WS 6, WS 7, and WS 13, all of which are low-elevation and recently disturbed. Lowest AFDM and ash concentrations were found on high elevation watersheds (Table 21.6).

As noted above, seston percent ash did not change seasonally, but differed significantly among watersheds (Table 21.6). Percent ash was significantly correlated with

Table 21.6. Mean Seston AFDM, Ash, and Percent Ash for the 17 Streams

Stream	Ash Free Dry Mass (mg/L)		Stream	Ash (mg/L)		Stream	Percent Ash	
WS 6	5.06	(0.88)	WS 6	14.88	(2.68)	WS 7	73.5	(1.7)
WS 7	3.48	(0.41)	WS 7	11.23	(1.90)	WS 6	73.5	(1.4)
WS 10	3.37	(0.54)	WS 13	8.80	(2.51)	WS 13	69.0	(1.5)
WS 13	3.33	(0.57)	WS 40	7.29	(1.72)	WS 2	68.7	(1.1)
WS 40	3.27	(0.61)	WS 2	6.90	(1.41)	WS 17	65.7	(3.2)
WS 34	2.96	(0.48)	WS 10	5.56	(0.87)	WS 41	65.4	(0.8)
WS 2	2.81	(0.48)	WS 41	4.68	(0.95)	WS 40	64.7	(1.9)
WS 41	2.34	(0.43)	WS 34	4.14	(0.82)	WS 10	62.9	(0.7)
WS 37	2.31	(0.57)	WS 17	3.84	(1.03)	WS 18	58.7	(1.8)
WS 18	2.08	(0.31)	WS 18	3.32	(0.63)	WS 14	57.8	(0.8)
WS 14	1.89	(0.29)	WS 14	2.56	(0.39)	WS 34	57.2	(2.0)
WS 17	1.51	(0.17)	WS 37	1.55	(0.37)	WS 36	53.1	(1.9)
WS 21	1.06	(0.14)	WS 36	1.28	(0.26)	WS 21	52.3	(2.3)
WS 19	1.03	(0.14)	WS 21	1.15	(0.18)	WS 22	49.4	(2.3)
WS 36	1.01	(0.17)	WS 22	0.93	(0.14)	WS 19	45.1	(1.3)
WS 27	0.94	(0.19)	WS 19	0.83	(0.10)	WS 27	42.4	(1.6)
WS 22	0.90	(0.13)	WS 27	0.65	(0.12)	WS 37	41.0	(0.8)

Standard errors of the means are in parentheses; $n = 12$ for each mean. Vertical bars indicate values that are not significantly different; multiple t tests with a protected α of 0.05.

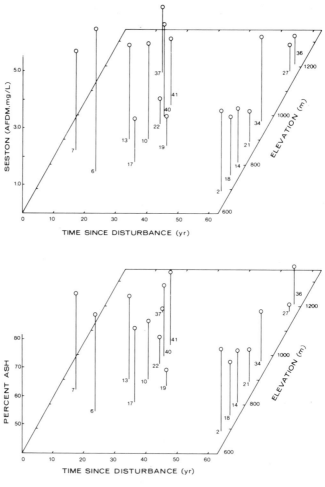

Figure 21.2. Seston concentrations, percent ash, and median particle size vs. time since disturbance and stream elevation. Numbers with each line are watershed numbers. Each point is the
mean of 12 samples.

both elevation and time since disturbance (Table 21.3). The range in percent ash was
large (41% to 73.5%, Table 21.6) and was highest on WS 6 and WS 7, the low-elevation,
recently disturbed watersheds. Lowest percent ash was found on high-elevation
watersheds (Table 21.6).

The similar trends for both AFDM and ash suggest that these measurements do not
represent different particles types. Rather, we propose that most particles are con-
glomerates of organic and inorganic material. This proposition is supported by several
other measurements discussed below and by observations made while measuring densities and fall velocities. In general, all the particles within one size class from an
individual sample exhibited similar densities and fall velocities. The exception was the
largest size class (ML), where we sometimes observed high-density sand grains

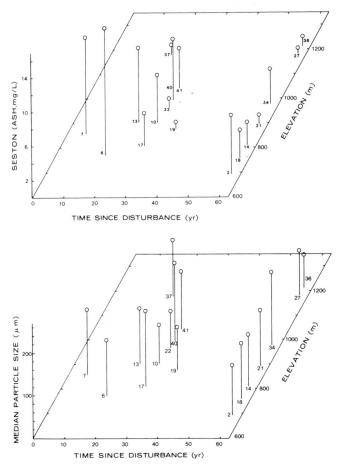

Figure 21.2. *(Continued)*

separating from lower density leaf particles. However, particles in this size range represented only a small fraction of the total seston.

Median particle size was significantly correlated with time since disturbance (Table 21.3). Particle size was generally larger on more recently disturbed watersheds (Figure 21.2). Median particle size ranged from 78 to 183 μm (Table 21.7), which is substantially larger than what has been found for many other streams (e.g., Naiman and Sedell 1979), but similar to other studies of Coweeta streams (Gurtz et al. 1980; Wallace et al. 1982a). Though the general trend with time since disturbance was statistically significant, WS 34, a mid-elevation reference watershed, had larger particles than most other streams (Table 21.7). Also, WS 10, a disturbed, low-elevation watershed, had relatively small particles.

Particle density varied with particle size (Kazmierczak et al., in preparation); larger particles were less dense than smaller particles (ANOVA, $p = 0.0001$; multiple t test

Table 21.7. Median Seston Particle Size

Stream	Median Particle Size (μm)
WS 40	183 (14)
WS 17	179 (21)
WS 34	166 (15)
WS 7	153 (15)
WS 41	137 (7)
WS 37	133 (10)
WS 18	133 (11)
WS 13	133 (9)
WS 6	131 (15)
WS 21	131 (16)
WS 14	122 (7)
WS 2	119 (6)
WS 27	106 (11)
WS 19	104 (7)
WS 10	92 (13)
WS 22	89 (8)
WS 36	78 (10)

Values are means over all dates. $n = 12$ for each mean. Standard errors of the means are in parentheses. Vertical bars indicate values which are not significantly different based on multiple t tests with a protected α of 0.05.

Table 21.8. Mean Density (g/cm³) of Seston Particles

Stream	Particle Size Class			
	ML	S	F	VF
WS 6	1.43 (0.02)	1.49 (0.03)	1.58 (0.11)	1.77 (0.04)
WS 7	1.32 (0.02)	1.45 (0.03)	1.45 (0.01)	1.57 (0.01)
WS 10	1.41 (0.06)	1.44 (0.03)	1.48 (0.04)	1.57 (0.04)
WS 13	1.38 (0.05)	1.57 (0.02)	1.52 (0.02)	1.67 (0.06)
WS 17	1.22 (0.03)	1.26 (0.03)	1.35 (0.04)	1.57 (0.08)
WS 19	1.30 (0.02)	1.39 (0.02)	1.41 (0.04)	1.55 (0.02)
WS 22	1.28 (0.03)	1.43 (0.04)	1.53 (0.07)	1.60 (0.05)
WS 37	1.22 (0.02)	1.37 (0.01)	1.49 (0.04)	1.61 (0.05)
WS 40	1.33 (0.03)	1.23 (0.03)	1.36 (0.02)	1.56 (0.02)
WS 41	1.25 (0.03)	1.31 (0.02)	1.45 (0.01)	1.67 (0.06)
WS 2	1.34 (0.02)	1.34 (0.02)	1.58 (0.04)	1.66 (0.07)
WS 14	1.22 (0.02)	1.39 (0.03)	1.46 (0.01)	1.52 (0.02)
WS 18	1.28 (0.02)	1.36 (0.03)	1.36 (0.04)	1.49 (0.08)
WS 21	1.26 (0.02)	1.33 (0.02)	1.34 (0.01)	1.47 (0.02)
WS 27	1.19 (0.01)	1.30 (0.01)	1.43 (0.01)	1.53 (0.03)
WS 34	1.24 (0.01)	1.27 (0.02)	1.43 (0.02)	1.56 (0.03)
WS 36	1.37 (0.01)	1.45 (0.02)	1.50 (0.02)	1.49 (0.02)

$n = 12$ for each mean. Standard errors of the means are in parentheses.

comparison of means with a protected alpha of 0.05, Table 21.8). For three out of four particle sizes, density decreased with time since disturbance (Table 21.4, Figure 21.3). This trend was partially explained by the increase in percent ash that resulted from disturbance (Table 21.3). Density and percent ash were significantly correlated for the smaller two size classes, though not for the two larger size classes (ML, $p = 0.37$; S, $p = 0.06$; F, $p = 0.05$; VF, $p = 0.004$). However, percent ash accounted for at most 16% of the variance in density, suggesting that other, unmeasured factors may influence particle density.

Fall velocities of particles also varied with particle size; larger particles had faster fall velocities than smaller particles (ANOVA, $p = 0.0001$; multiple t test comparison of means with a protected α of 0.05, Table 21.9). Fall velocities of all four size classes were significantly negatively correlated with time since disturbance (Table 21.5, Figure 21.4). This is consistent with the trend for percent ash, and there was a significant correlation between fall velocity and percent ash for all size classes except the largest (ML, $p = 0.17$; S, $p = 0.0003$; F, $p = 0.001$; VF, $p = 0.0005$).

Our analyses of the effects of date of sample collection, elevation, and time since disturbance on various seston parameters showed many statistically significant relationships (Tables 21.3 through 21.5). However, these three variables only explained (as indicated by coefficient of determination) 21 to 62% of the variance in seston measurements. While some of the variance was certainly measurement error, several unmeasured watershed and stream characteristics could also explain some of the variance. These include such factors as the nature of the streambed material, e.g., some streams have extensive sections of bedrock, and other streams cut through old debris

Table 21.9. Fall Velocity (cm/s) of Seston Particles

Stream	ML	S	F	VF
		Particle Size Class		
WS 6	0.886 (0.104)	0.450 (0.040)	0.199 (0.011)	0.117 (0.008)
WS 7	0.741 (0.138)	0.382 (0.057)	0.146 (0.007)	0.070 (0.006)
WS 10	0.635 (0.078)	0.392 (0.024)	0.168 (0.011)	0.062 (0.003)
WS 13	0.748 (0.080)	0.377 (0.023)	0.185 (0.012)	0.080 (0.004)
WS 17	0.490 (0.030)	0.265 (0.013)	0.137 (0.008)	0.071 (0.004)
WS 19	0.343 (0.047)	0.200 (0.018)	0.110 (0.009)	0.067 (0.004)
WS 22	0.518 (0.029)	0.364 (0.037)	0.187 (0.018)	0.060 (0.003)
WS 37	0.485 (0.062)	0.263 (0.020)	0.135 (0.011)	0.054 (0.004)
WS 40	0.637 (0.077)	0.300 (0.018)	0.153 (0.013)	0.059 (0.005)
WS 41	0.635 (0.076)	0.298 (0.021)	0.134 (0.006)	0.083 (0.004)
WS 2	0.528 (0.042)	0.341 (0.018)	0.190 (0.013)	0.094 (0.005)
WS 14	0.411 (0.039)	0.272 (0.014)	0.119 (0.007)	0.056 (0.006)
WS 18	0.430 (0.060)	0.215 (0.009)	0.129 (0.006)	0.076 (0.003)
WS 21	0.434 (0.042)	0.244 (0.018)	0.112 (0.005)	0.074 (0.006)
WS 27	0.443 (0.052)	0.272 (0.022)	0.098 (0.007)	0.052 (0.003)
WS 34	0.428 (0.035)	0.226 (0.012)	0.127 (0.009)	0.050 (0.002)
WS 36	0.596 (0.045)	0.292 (0.013)	0.128 (0.006)	0.063 (0.003)

$n = 30$ for each mean. Standard errors of the means are in parentheses.

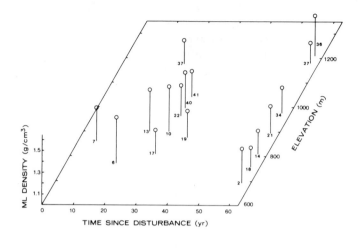

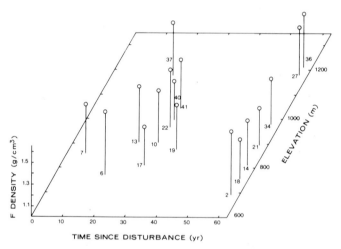

Figure 21.3. Densities of seston in the four larger size classes as functions of time since disturbance and stream elevation. Numbers with each line are watershed numbers. Each point is the
mean of 15 samples.

avalanches. Pre-1900 history might also be important, e.g., there was once a crop field
on WS 2, WS 7 was heavily damaged by a wind storm in 1835 (Hertzler 1936), and
remains of an old logging road can still be seen in sections of the stream on WS 14. We
have also not evaluated effects of watershed and stream morphology.

General Discussion

Results of this study show that watershed disturbance increases seston concentrations
in streams. Concentrations of both organic and inorganic materials are increased;

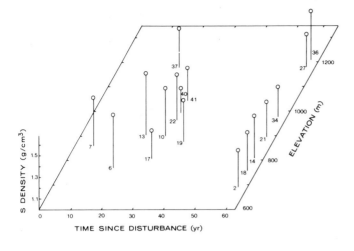

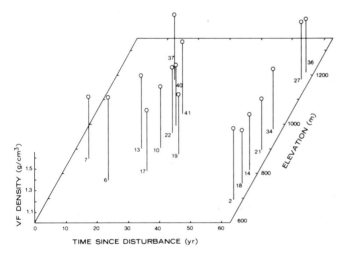

Figure 21.3. *(Continued)*

however, since the inorganic fraction increases more, the percent ash of seston carried by streams draining disturbed watersheds is higher. In addition, the particles are generally larger and have higher density and fall velocity. Depending on the nature of the disturbance, effects on seston concentration and composition may be seen for 30 to 40 years following disturbance.

For commercially logged watersheds, increases in particulate inorganic material might be explained as the transport of sediment that entered the stream during logging. This appeared to be the case for WS 7 for the first few years after logging (Gurtz et al. 1980). Lieberman and Hoover (1948b) and Tebo (1955) attributed the high turbidity of the stream draining WS 10 primarily to material eroded from logging roads and skid trails. However, it is less easy to explain why concentrations of particulate organic

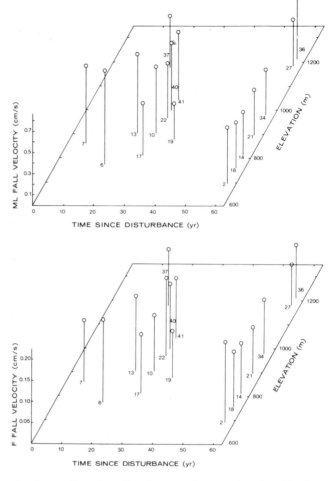

Figure 21.4. Fall velocities of seston vs. time since disturbance and stream elevation. Numbers with each line are watershed numbers. Each point is the mean of 30 samples.

materials are elevated. This might partly be the result of erosion of soil organic materials, but that is probably not a major source of organic seston for more than 2 or 3 years following logging. Also, in-stream biological production of seston should be down, as studies have shown a reduction of invertebrate seston producers (shredders) on disturbed watersheds (e.g., Woodall and Wallace 1972; Molles 1982; Silsbee and Larson 1983; Gurtz and Wallace 1984) and a reduction of allochthonous inputs. Algae may contribute a small amount of organic seston. We suggest that the continued elevated concentrations of both inorganic and organic seston beyond the first few years after logging is primarily due to downcutting of the stream channel and erosion of parti- cles stored in the stream bed. This downcutting can be largely attributed to the decrease in the amount of woody material in the stream (Table 21.1).

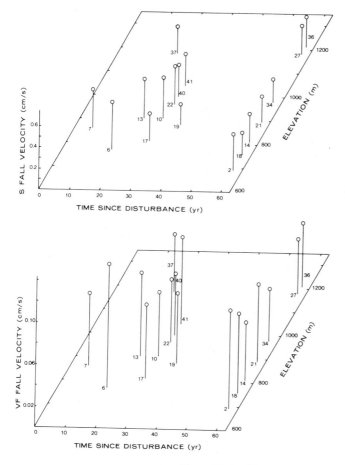

Figure 21.4. *(Continued)*

Seventeen years after disturbance, WS 6 carries very high concentrations of seston (Table 21.6). This is correlated with the absence of woody debris in this stream (Table 21.1). For over 40 years there has been little input of wood to this channel; most old wood has decayed, and there is little to stabilize the stream bed. A similar situation exists on WS 13, except that some slash was left in the stream during the original clear-cutting (1939) and a few of the older logs remain in the channel. Tree regrowth has been rapid and some death of smaller trees is already occurring; in a few years we may begin to see a substantial decline in seston transported by this stream. In other streams, such as WS 7, we may begin to see increasing seston concentrations a few years from now as old debris dams decay and break down.

Based on studies cited above, we assume that much of the seston in reference streams results from biological processing of allochthonous inputs. Our results suggest that more of the seston collected from disturbed watersheds is produced by physical forces. Since our samples were taken during baseflow periods, our results probably minimize effects of disturbance. Samples taken during storms would probably show much greater differences. For example, samples taken from WS 7 and WS 14 during a storm in 1981 showed a greater than sixfold difference in peak seston concentrations — much greater than the threefold difference in prestorm seston concentrations (Webster and Golladay, unpublished data).

Seston carried by small streams eventually reaches larger streams. As shown by studies of WS 10 (Lieberman and Hoover 1948a; Tebo 1955), input from just one small turbid stream can significantly affect the turbidity and invertebrate community of a larger receiving stream. Seston is also an important food resource for many filter feeding invertebrates. The less organic-rich seston of disturbed streams is probably lower quality food than is the seston from the undisturbed streams. Also, because of its higher density and fall velocity, seston from disturbed streams would tend to drop from suspension faster and accumulate in pools of lower gradient streams.

In summary, disturbances to forest ecosystems are reflected in the transport of particulate materials draining these watersheds. In the first few years after disturbance, inputs of sediment directly from the disturbed watershed are probably the major impact. In subsequent years, indirect effects caused by the decline of woody debris dams within the stream are of greater importance. The resulting increased erodibility of the stream channel may cause a stream disturbance that lasts much longer than any observable disturbance of the adjacent terrestrial ecosystem.

7. Man and Management of Forested Watersheds

22. Streamflow Changes Associated with Forest Cutting, Species Conversions, and Natural Disturbances

W.T. Swank, L.W. Swift, Jr., and J.E. Douglass

An original research objective in the establishment of Coweeta was to measure and evaluate the effects of man's use of the forest on the quantity and timing of streamflow. Over the past 50 years at least 40 publications and numerous presentations have addressed this topic. Fifteen individual watershed-scale experiments have been conducted in the basin, involving various intensities of forest cutting and harvest and conversions of hardwood forest to white pine or grass. A description of the treatments is summarized in Chapter 1. The purpose of this chapter is to provide a synthesis of findings on (1) responses in annual and monthly streamflow quantities following cutting, species conversions, and natural disturbance; (2) changes in storm hydrograph characteristics that accompany clearcutting; and (3) the application of results to water resources planning on forested watersheds.

Changes in Annual and Monthly Streamflow

Hibbert's (1966) survey of catchment experiments throughout the world indicated that deforestation increased and afforestation decreased streamflow, but the magnitude of response was highly variable and unpredictable. In that analysis he also pointed out the nearly threefold increase in first year responses after clearcutting between north- vs. south-facing watersheds at Coweeta. In an attempt to improve the relationship between first year water yield increases as a function of basal area cut, Douglass and Swank (1972) expanded the data base to 22 forest cutting experiments in the Appalachian Highlands Physiographic Division (Figure 22.1).

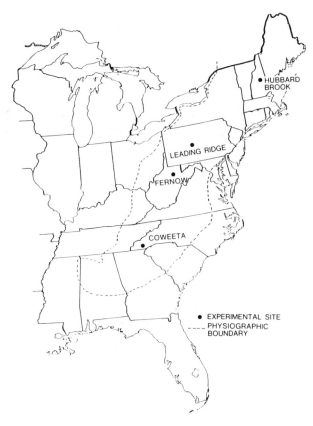

Figure 22.1. Location of the Appalachian Highlands Physiographic Division and experimental sites.

Although the relationship between increases and percent basal area reductions was improved, there was still a wide scatter of points about the regression; in fact, for clear-cutting (100% reduction in basal area), Coweeta responses bracketed all of the experimental results and ranged from 13 cm to 41 cm. An improved version of the first-year increase model was derived by Douglass and Swank (1975) by adding an energy variable (insolation index), which represents energy theoretically received by watersheds of different slopes, aspects, and latitudes (Figure 22.2). The rationale for adding the insolation index as a variable was based on Swift's (1960) findings that solar radiation theoretically available for evapotranspiration is greater on south- than on north-facing slopes in the dormant season, with little difference in the growing season. Additional studies of total and net radiation patterns supported the theory (Swift 1972), and it was hypothesized that streamflow response to cutting was inversely proportional to solar energy input. The curvilinear relationship provides a reasonable representation of the actual response to partial cuttings. The final equation derived for estimating first-year yield increases for hardwoods is:

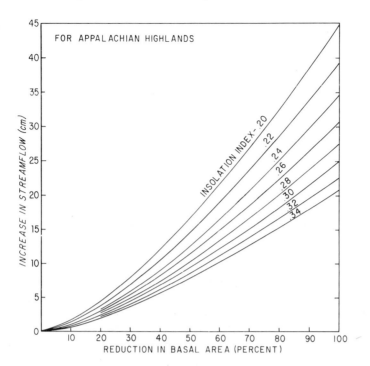

Figure 22.2. First year increases in streamflow after cutting mature, mixed hardwood forests based on the precent reduction in stand basal area and watershed insolation index.

$$\Delta Q = 0.00569 \ (BA/PI)^{1.4462} \qquad (1)$$

where ΔQ is the first year increase in centimeters for hardwoods, BA is the percent basal area cut, and PI is the annual potential insolation in langleys $\times$ 10^{-6} calculated by methods given by Swift (1976). Total variation explained by the two variables was 89%.

As the forest regrows and the evaporating surface area (hence Et) increases, streamflow increases decline as a function of the logarithm of time (see examples, this chapter). This relation provides a model for predicting the yield for any year after harvest if the duration of measurable increases is known. An estimate of mean duration was derived by Douglass and Swank (1972) as 0.62/year for each cm of first-year flow increase. The declining log equation for flow increase in any year (ΔQy) is

$$\Delta Qy = \Delta Q - b \log y \qquad (2)$$

where y is the number of years after harvest and the coefficient b is derived by solving Eq. 2 at the point when $\Delta Qy = 0.0$:

$$b = \Delta Q/\log \ (0.62 \ \Delta Q) \qquad (3)$$

The limitations on the use of these equations have been given by Douglass (1983). An application of the equations to the most recent clearcut at Coweeta is given in

Table 22.1. Comparison of Annual Predicted vs. Observed
Increase in Water Yield Following Clearcutting on Coweeta
WS 7

Year After Clearcutting	Predicted Increase by Model (cm)	Observed Increase[a] (cm)
1	25	26
2	17	20
3	12	17
4	8	12
5	6	4
6	3	4
Total	71	83

[a] Observed increase based on annual calibration regression.

Table 22.1 for Watershed 7 (WS 7), a 59 ha south-facing catchment. In the first year after cutting, streamflow increased about 25 cm and the model predicted 25 cm. In the ensuing 2 years, observed increases were substantially above values predicted from the equations, and these years coincided with the wettest and second driest years in the past 50 years. Fourth and fifth year predictions were in closer agreement with observed values, and the total change predicted for the entire 6-year period was 71 cm compared to 85 cm observed. We would expect the most reasonable performance of the equations over several years rather than for individual years. Large deviations for individual years could be expected for exceptionally wet or dry years and, of course, both observed and predicted values have an error term. Actual use of these equations in forest management planning in the East has been documented (Douglass 1983) and later in this chapter we will provide another example of potential application.

The effects of clearcutting on mean monthly flows are illustrated by results on WS 17 during the 7 years when regrowth was cut annually (Figure 22.3). The timing of monthly increases are representative of other clearcutting experiments in the basin. Clearcutting has little effect on flow during later winter and early spring when soil moisture beneath undisturbed forests is fully recharged. With the onset of the growing season, increased flow from reduced Et begins to appear and increases in magnitude as the growing season advances. In the lowest flow months of September, October, and November, average monthly streamflow is about 100% greater from the clearcut than uncut forest. This is the period when water demands are greatest and flow from undisturbed forests is lowest. Large increases continue into December and then begin tapering off as storage differences between cut and uncut forests diminish. On catchments with less soil water storage, recharge occurs more rapidly and increases are not delayed as late into the winter.

Streamflow Responses to Regrowth

Long-term effects of clearcutting on water yield are important in both water resource planning and evaluation of nutrient export from forest ecosystems. Four experiments

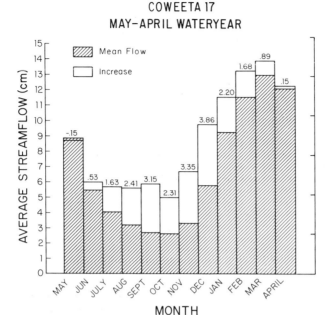

Figure 22.3. Mean monthly flow on Coweeta WS 17 during preclearcutting years and mean monthly increases during a 7-year period of annual recutting.

on three experimental catchments totaling 304 ha of land, almost 200 years of streamflow records, an equivalent number of precipitation records, and detailed vegetation measurements, provide a unique documentation of flow recovery and major variables that affect responses. These experiments were previously summarized by Swift and Swank (1981) and most details will be omitted here. The longest record available is for WS 13, which has been clearcut twice, first in 1940 and again in 1963 without removal of forest products and with natural regrowth allowed. Changes in flow as estimated from differences between measured flow and flow predicted from the control catchment are shown in Figure 22.4 for both cuttings along with trend lines fitted to these changes. Thirteen years after the first cutting, Kovner (1956) proposed the following trend model: flow increase $= a + b$ (log of years since cutting) and the trend line was still valid when the stand was 23 years old. The initial response of the second cutting was nearly identical to that of the first cut, but the shape of trend curves of the two differ substantially in the early years. The estimated termination of yield increases based on data in Figure 22.4 is about 18 years earlier for the second cutting.

Swank and Helvey (1970) partially attributed the more rapid decline in water yield of the second cutting to a more rapid recovery of stand density and perhaps leaf area (Table 22.2), and, hence, evapotranspiration after the second cutting. Seven years after cutting, stem density was 2000 stems ha^{-1} greater after the second treatment. The higher stem density was attributed to greater sprouting potential of the even-aged stand with numerous small stems that were present prior to the second cutting. Leaf biomass

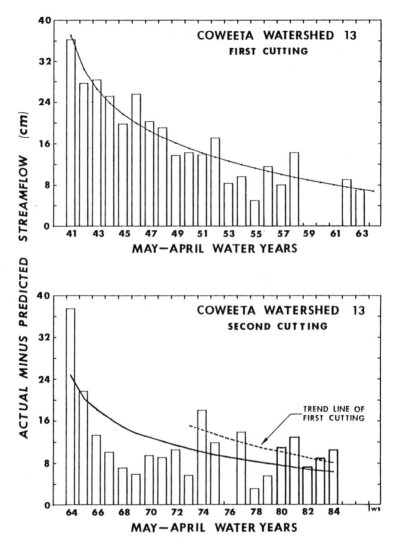

Figure 22.4. Streamflow increases following two clearcutting experiments on the mixed hard-wood covered Coweeta WS 13. Streamflow was not measured in water years 1959, 1960, 1961, and 1976.

and surface area also recovered rapidly during the first 7 years after the second cutting (Table 22.2). At age 7, leaf biomass was nearly equal to leaf production at age 14 after the first cutting, and leaf area index (LAI) was similar to values expected for mature forests. In the next 2 years, leaf surface area and biomass declined sharply due to the mortality from competition in the dense stand, and this was associated with higher streamflow increases during the same period. Fifteen years after the second cutting, LAI was 6.0 and streamflow was near pretreatment levels. These long-term experiments illustrate the strong relationship between the temporal variations of the quantity

Table 22.2. Characteristics of Forest Vegetation at Different Stand Ages for Coweeta WS 13

Water Year and Phase of Treatment	Stand Age (Years)	Stand Density (Stems ha^{-1})	Leaf Biomass (g m^{-2})	Leaf Area Index (m^2 m^{-2})
1934	Mature, uneven-aged	2596	N/A	N/A
1940, First clearcut	Cut	0	N/A	N/A
Regrowth years	8	7630	N/A	N/A
Regrowth years	12	5322	N/A	N/A
Regrowth years	14	N/A	316	N/A
1962	22, even-aged	4196	N/A	N/A
1963, Second clearcut	Cut	0	N/A	N/A
Regrowth years	7	9659	268	5.2
Regrowth years	8	N/A	192	3.7
Regrowth years	9	N/A	191	3.8
Regrowth years	15	7338	309	6.0

N/A, not available.

of evaporating surface, evapotranspiration, and water yield. However, Swift and Swank (1981) found that the addition of a rainfall variable improved the fit of the trend line in the second cutting.

Logarithmic trend lines were fitted to two additional cutting and regrowth experiments (Figure 22.5). Streamflow increases on WS 37 temporarily returned to pretreatment levels during the fourth and sixth years after cutting. During the next 5 years, there was a general streamflow increase and then a second return to baseline; a response pattern similar to WS 13. Detailed discussion of the temporal trends of streamflow and regrowth characteristics of WS 13 compared to WS 37 is given by Swift and Swank (1981). On WS 28, the multiple-use demonstration experiment, 51% of the basin was clearcut, 22% was thinned, and 27% was left uncut. The recovery trend curve is not different in either slope or level from that of WS 37, but the level of flow increases for both experiments is different (less) than WS 13. Both watersheds are higher in elevation, with steeper slopes, generally shallower soils, and shorter growing seasons than WS 13. Hence, we postulate that Et is less and changes in flow due to cutting are lower on these watersheds.

Effects of Cutting on Storm Hydrograph Characteristics

Of the cutting experiments conducted at Coweeta, only three have received rather thorough analysis of changes in storm hydrograph characteristics. Differences in the extent and type of watershed disturbance, the period of record included in post-treatment assessment, inherent watershed hydrologic response factors, and the magnitude and frequency of storms following treatment complicate summary evaluations of storm hydrograph responses. Nevertheless, the percent increases in four important storm hydrograph parameters calculated for the mean storm over the period of analysis are summarized in Table 22.3. Increases in quickflow volumes for cutting experiments with minimal disturbance to the forest floor (WS 7 and 37) are similar during the first

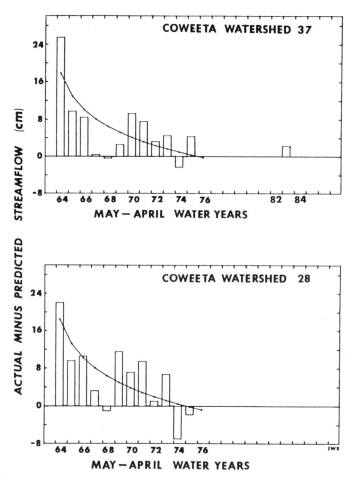

Figure 22.5. Changes in streamflow following cutting the mixed hardwood forest on two Coweeta watersheds. Streamflow measurements were discontinued in 1975.

3 to 4 years after treatment, with a mean increase of about 10%. Commercial clearcutting with cable logging and some road construction (WS 7) produced peak flow rates about double (15% vs. 7%) the mean increases from clearfelling alone (WS 37). Initial flow rates and recession time were also increased 10 to 14% on WS 7. The largest changes occurred on WS 28, where commercial logging involved tractor skidding and a high road density. Quickflow increases averaged over the 9-year post-treatment period increased 17%; mean peak flow rates increased 30% the first 2 years after treatment and declined in subsequent years. It appears that the absence, or low density, of carefully constructed roads, combined with little disturbance to the forest floor during harvest, produced the smallest changes in stormflow responses. More specific and detailed interpretations have been reported by Swank et al. (1982) for WS 7, Douglass and Swank (1976) for WS 28, and Hewlett and Helvey (1970) for WS 37.

Table 22.3. Summary of Storm Hydrograph Responses to Clearcutting

Watershed Number	Size (ha)	Response Factor Mean Stormflow Volume ÷ Mean Precipitation	Treatment	Inclusive Years After Treatment	Mean Storm Increases for Selected Hydrograph Parameters			
					Total Quickflow Volume (%)	Peak Flow Rate (%)	Initial Flow Rate (%)	Recession Time (%)
7	58.7	0.04	Clearcutting, cable logging, product removal and minimal road density	3	10	15	14	10
28	144.1	0.09	77 ha clearcut, 39 ha thinned, 28 ha no cutting; products removed; high road density	9, 2[a]	17	30	N/A	N/A
37	43.7	0.19	Clearcut, no products removed	4	11	7	NS	NS

N/A, not available; NS, not significant.
[a] Inclusive years after treatment used for total quickflow and peak flow rates, respectively.

Natural Disturbance

The partial defoliation on WS 27 by fall cankerworms from 1969 to 1977 was of sufficient magnitude and duration to detect effects on streamflow during part of each year.
Using WS 36 as a control for WS 27, preinfestation calibration regressions were
derived for annual, monthly, and a variety of combined monthly periods. No significant changes in flow were detected in annual flow, individual monthly yields, or most
combined monthly analyses. However, significant (0.05 level) flow reductions for the
November through January period were observed in most years of infestation (Table
22.4). Reductions varied from 5 to 15 cm or 7 to 18% below expected flow levels until
1978, which coincided with the decline of cankerworm populations. These flow
responses are attributed to stimulation of leaf production by defoliation. Instead of
litter production of about 230 g m^{-2} and a leaf area index (LAI) of 3.1 typical for high
elevation Coweeta forests, leaf production during most years of infestation exceeded
425 g m^{-2} with a LAI of at least 6.0. The elevated LAI apparently produced higher
evapotranspiration during the summer months, but streamflow reductions were not
observed at the weir until the winter period. This lag in timing between evaporative
changes within the watershed and measured effects at the weir agree with other
experimental results at Coweeta. Anticipated increases in flow during the spring
months were not detectable, possibly because of the brief period of defoliation at a
time of relatively low evaporative demand. To our knowledge, this is the first
documented evidence where an insect infestation has been shown to increase Et and
reduce streamflow.

Species Conversions

Two major long-term studies of the effects on water yield of converting mixed hardwoods to different vegetative covers have been conducted at Coweeta. One is the
replacement of hardwoods with white pine, and the second a conversion to grass fol-

Table 22.4. Reductions in Flow During the November–
January Period by Year Due to Hardwood Defoliation by
Fall Cankerworm

Year	Change in Flow (November-January)	
	cm	%
1969	−6	11
1970	−9*	14
1971	−7	9
1972	−9*	11
1973	−15*	16
1974	−8*	15
1975	−5*	7
1976	−8*	18
1977	−15*	16
1978	+2	4

*Significant at 0.05 level.

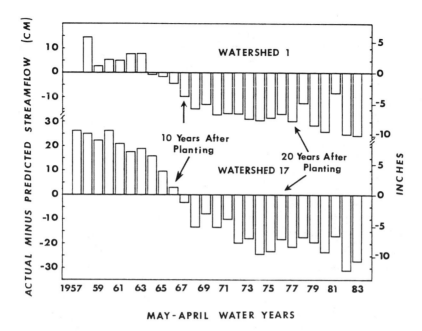

Figure 22.6. Annual changes in flow on two Coweeta watersheds following conversion from mixed hardwood to white pine.

lowed by succession. Both of these conversions have produced dramatic changes in streamflow, which are of importance to forest managers, hydrologists, and ecologists.

The white pine experiments were conducted on both north-facing and south-facing watersheds. For about 6 years after planting of white pine, streamflow increases remained relatively constant and near those expected for clearcut hardwoods (Figure 22.6). Thereafter, as the pine stands developed and hardwood competition was reduced, streamflow increases declined at a rate of 2 to 5 cm per year until about 1972. About 10 years after planting, streamflow was below levels expected from mature hardwoods and, by age 15, water yield reductions were about 20 cm (20%) less than expected for a hardwood cover. In the ensuing 9 years, annual reductions in flow fluctuated between 10 and 20 cm, depending upon annual precipitation. For example, during an exceptionally wet year such as 1980, pine evapotranspiration was much higher than hardwoods, but in the following year (1981), precipitation was quite low and small streamflow reductions were observed. Flow decreases during the more normal rainfall years of 1982 and 1983 exceeded 25 cm on both watersheds.

Reasons for greater evaporative losses from young pine than from mature hardwoods have been given in several papers (Swank and Miner 1968; Swank and Douglass 1974; Swift et al. 1975). Interception and subsequent evaporation of rainfall is greater for pine than hardwoods, particularly during the dormant season. Interception loss varies with LAI, and in the dormant season, LAI for hardwoods is less than 1 compared to 10 for white pine. Thus, less precipitation reaches the soil under closed pine stands and the result is lower streamflow. Simulation of evaporation for pine and hardwoods using the PROSPER model (Swift et al. 1975) shows greater dormant season transpiration

Table 22.5. Simulated Interception and Transpiration Totals for Oak–Hickory and White Pine Forests During the Growing and Dormant Seasons

Year and Vegetation Type	Interception (cm)			Transpiration (cm)		
	May– October	November– April	Total	May– October	November– April	Total
1972–1973						
Oak–hickory	13.83	9.01	22.84	56.05	9.98	66.03
White pine	18.28	13.64	31.92	54.68	21.84	76.52

After Swift et al. Water Resourc. Res. 11:667–673, 1975. Copyright by the American Geophysical Union.

losses from pine (Table 22.5). On an annual basis, simulations indicate that interception and transpiration are equally important from a quantitative viewpoint.

The influence of precipitation quantities on annual responses can be normalized by expressing reductions in flow for any given year as a percent of flow expected if the watershed had not been treated (Figure 22.7). Flow reductions appear to culminate at about age 12 and tend to remain rather constant until ages 25 and 26, when decreases are 35 to 45% below hardwood levels. The original hypothesis in 1968 was that the rapid streamflow reduction resulting from hardwood to pine conversion would tend to level off when LAI culminated and thereafter gradually decline as the total plant surface area (including stem and branch area) slowly increased with stand age. This appeared plausible, because surface area is a structural characteristic closely related to the major evaporative processes of interception and transpiration. The culmination of LAI development shown in the lower part of Figure 22.7 does, in fact, correspond to the leveling off of streamflow reductions and the total surface area index has continued

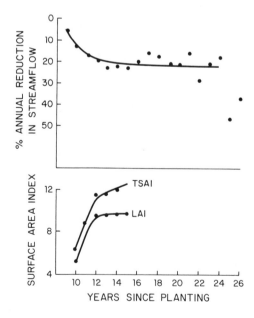

Figure 22.7. Percent reduction in annual flow after the planting of white pine on Coweeta WS 1 in relation to the development of plantation leaf area index (LAI) and total surface area index (TSAI) including foliage, branches, and boles.

Figure 22.8. Mean monthly flows predicted for a mature hardwood forest on Coweeta WS 1 and actual flows measured from a white pine plantation on the watershed during a 4 year period (1980–1983).

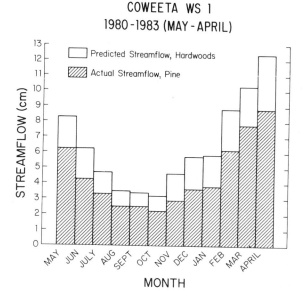

COWEETA WS 1
1980-1983 (MAY - APRIL)

to increase. The surface area graph is incomplete; the population has been sampled for both biomass and surface area in subsequent years, but estimates are unavailable at this time. Some very large and sudden increases in pine evapotranspiration are indicated by reduced flow in both 1982 and 1983; the stand will be resampled again to identify current changes in structural characteristics.

The average monthly flows for pine compared to hardwoods on WS 1 for 1980–83 are shown in Figure 22.8. Significant reductions occur in every month. Large reductions of 2 to 3 cm occur in March, April, and May when flows are highest, and at a time before and during hardwood leafout when evaporative demands can be high. In the low flow months of August, September, and October, reductions average about 1 cm or 40% below levels expected for hardwoods.

The conversion from hardwood to grass, followed by deadening with herbicides and then succession, also significantly altered annual streamflow (Figure 22.9). Merchantable timber was harvested and a seedbed was prepared for planting Kentucky 31 fescue grass. At the time of seeding, lime and fertilizer were applied. In the first year after conversion, grass production was very high, with 7.85 t ha^{-1} of dry matter (3.5 ton acre^{-1}), and there was no significant change in streamflow. In the ensuing years, water yield increased and at the end of the fifth year was about 14 cm above the flow expected for the original hardwood forest. During the same period, grass production declined to 4.04 t ha^{-1} (1.8 ton acre^{-1}). To further test the inverse relation between change in streamflow and grass productivity, the grass was again fertilized in 1965; productivity increased to 7.85 t ha^{-1} (3.5 ton acre^{-1}) and streamflow again dropped to the level expected for hardwoods. A more detailed analysis of the early phases of this experiment is described by Hibbert (1969). These results further demonstrate that both type and amount of vegetation have a major influence on Et and streamflow.

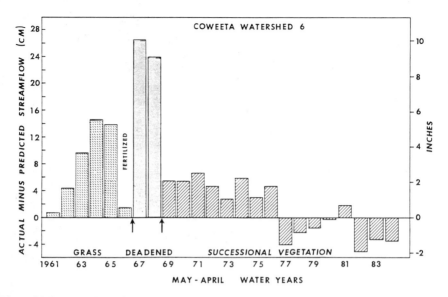

Figure 22.9. Annual changes in streamflow on Coweeta WS 6 during a 6 year period after conversion from hardwood to grass, followed by 2 years of herbicide applications to deaden the grass and the ensuing 16 years of succession.

In 1967 and 1968, the grass was herbicided and flow increased about 25 cm, a response similar to that expected from clearcutting (Figure 22.9). During the first year of succession, the watershed was dominated by a thick, lush cover of herbaceous species, and streamflow increases rapidly declined to only about 6 cm above hardwood levels. This level of increase continued for the next 7 years and then returned to levels expected for hardwoods. Thus, Et from a mixture of young hardwoods dominated by black locust in association with blackberries, herbaceous species, and some grass is equivalent to a mature mixed hardwood forest.

Special Case Application

The hydrologic principles derived from Coweeta studies of water yield and timing have been utilized in a variety of ways by resource managers, administrators, and scientists. A simple example will serve to further illustrate how results can be utilized to evaluate alternative silvicultural prescriptions in the real world (Figure 22.10). In this illustration, we demonstrate the effects of hardwood even-aged management vs. white pine plantation management on long-term water yield. The prescription conditions were taken from current guidelines for the Nantahala National Forest management plan in North Carolina.

We begin by clearcutting a mature mixed hardwood forest with a solar insolation index of 0.27, mean annual precipitation of 180 cm, and a mean annual flow of 90 cm.

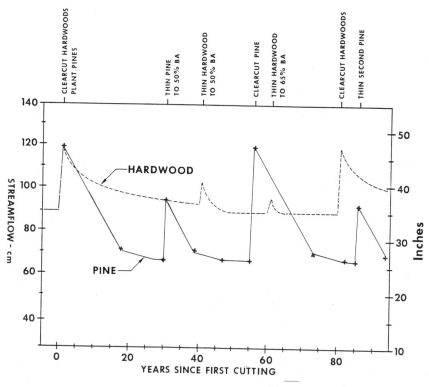

Figure 22.10. Estimates of long-term streamflow responses to hardwood even-aged management and white pine plantation management based on experimental results for Coweeta climatic conditions.

In one case, hardwoods regenerated from sprouts and seeds, and in the other case, white pine were planted. During the rotation, commercial thinning of hardwoods is applied at ages 40 and 60 years with removal of 50% and 35% of the basal area, respectively. The stand is then again clearcut at age 80 years. The prescription for pine is a commercial thinning at age 30 years with 50% of the basal area removed and a final harvest by clearcutting at age 55, and replanting with white pine followed again by thinning at age 30 (Figure 22.10). For the 80 year period depicted, the average annual streamflow under hardwood even-aged management is 95 cm, and for pine the average flow is 81 cm. The cumulative difference in flow between the two prescriptions is 1120 cm (441 inches). If we assume the treated watershed is 10 ha (25 acres), the total difference is 1.12×10^6 m^3 (919 acre-feet) of water in 80 years.

This illustration shows that alternative silvicultural prescriptions can have a substantial and very real effect on water yield. Although a variety of prescription scenarios could be selected, the point is that functional relationships are available to make decisions about silviculture and water yield.

Conclusions

Long-term streamflow records for control and experimental forested watersheds at Coweeta provide a solid foundation for evaluating hydrologic responses to vegetation management. Results from these experiments and those elsewhere in the Appalachian Highlands provide equations for predicting changes in annual water yield following cutting and regrowth of hardwood forests. Only two parameters, proportion of the stand basal area cut and potential insolation of the watershed are needed to solve these equations. Increases are produced in most months, with about a 100% increase during the low flow months when water demands are usually high. The recovery of streamflow to preharvest levels associated with hardwood regrowth shows the interactions between Et, stand dynamics, and watershed physical characteristics. Experiments indicate that commercial clearcutting, with carefully located and designed roads, produce only small and acceptable (about 15%) increases in mean stormflow volumes and peak flow rates. Natural alteration of vegetation such as insect defoliation can also influence water yield by stimulating leaf production and increasing evapotranspiration, thus reducing winter streamflow by 7 to 18%.

Other long-term experiments show the striking dependence of streamflow volume on type of vegetative cover. Within 25 years, hardwood to white pine conversion reduces annual flow by 25 cm and produces significant reductions in every month of the year. Greater Et for pine is due to a higher LAI for pine compared to hardwoods, hence, greater interception and transpiration by pine. Hardwood to grass conversion also alters streamflow depending upon grass productivity. There is no significant change in flow with a vigorous grass cover, but as grass productivity declines streamflow increases. Evapotranspiration from a luxuriant herbaceous cover is slightly lower than in mature hardwoods, but later in succession with a mixture of hardwoods, grass, and herbs, Et is equivalent to hardwoods. In conclusion, forest managers should recognize that silvicultural prescriptions influence evapotranspiration, and hence streamflow, and that hydrologic changes are either a cost of doing business or an added benefit from management decisions.

23. Forest Access Roads: Design, Maintenance, and Soil Loss

L.W. Swift, Jr.

The Regional Guide for the South (United States Department of Agriculture 1984b) recognizes that roads and skid trails are the major sources of sediment from forestry-related activities. The overall environmental impact statement for Region 8 (United States Department of Agriculture 1984a) estimates an existing national forest road network of 56,300 km (31,000 miles) with approximately 200 km (125 miles) of new construction or reconstruction each year. About 70% of this annual construction is classed as local road; i.e., the low-standard, limited-use road that terminates the transportation system. Local roads are often developed for access to timber sales. More than 40 years of road studies and land management demonstrations at Coweeta show both an early recognition that roads were a problem and a continuing effort to describe the magnitude of soil loss and develop technologies to control it. This chapter gives a history of road-related research at Coweeta and summarizes the findings of that research.

Roadbank Stabilization Tests: 1934–1958

About the same time that Coweeta was established, C. R. Hursh began a long series of roadbank stabilization tests and demonstrations. This was a period of greatly expanding road construction, stimulated by Federal and state programs of the Depression era. The use of newly developed earthmoving equipment in the Appalachians exposed large areas of soil in cuts and fills, creating problems not visualized by construction engineers. Hursh recognized the maintenance and environmental costs of these eroding slopes, and, in the spirit of the times, used natural materials and labor-intensive methods to stabilize them. Cut grass or weeds were laid on slopes and held in place with stakes cut from local materials (Hursh 1935, 1939). This mulch broke the eroding force of raindrops, halted the sloughing due to frost action, and encouraged growth of planted grass or naturally seeded vegetation. In forested areas, brush from trees was laid on top of leaf or needle litter collected from the forest. Sometimes, poles were laid horizontally across the slopes and pinned by stakes to hold down weeds or brush. Another technique was to move topsoil to the slope, encouraging growth of local plants (Hursh 1942c). Shrubs and trees provided more permanent stabilization (Hursh 1945). Many of the publications describing these methods contained instructions and photographs to aid work crew supervisors (Hursh 1938, 1939), while others were directed to decision makers (Snyder and Hursh 1938; Hursh 1942a, 1942b, 1949).

Bank stabilization work continued at Coweeta in the 1950s. Test plots of orchard grass, fescue, and ladino clover mixtures were planted at Chestnut Flats. These hay mixtures were then utilized as mulch on Coweeta roads. Bank stabilization by planted grasses was tested when the Dryman Fork roads were built. Rye and love grass were sown in 1957 on cut slopes after contour furrows were cut in relatively steep banks with fire rakes. Love grass provided the densest cover.

Exploitative Logging Demonstrations: 1941–1956

In 1940, Coweeta began a series of watershed treatments to demonstrate three typical but poor land management practices common on homesteads in the southern Appalachian Mountains. The mountain farm and woodland grazing demonstrations were reported elsewhere (Dils 1952, 1953; Johnson 1952; Sluder 1958). The exploitative logging demonstration (WS 10) continued for 15 years (Lieberman and Hoover 1948a, 1948b; Dils 1957). Few requirements were placed on the logging contractor; he only had to confine his operations within the watershed boundary and was not allowed to construct a road or skid trail through the weir site.

Initially, logs were skidded from the area by horse teams. About 5.6 km (3.5 miles) of road and skid trail were constructed between 1946 and 1956 on the 85.8 ha watershed, with most of the roadway lying in or adjacent to streambeds. Skid trails and spur roads were often steep, and little effort was made to divert storm waters off the roads or to vegetate disturbed soil. Logs skidded downslope, often in the natural drainages, further contributed to soil erosion. Based upon transects across roads, about 408 m^3 of soil were lost from each kilometer of road length (860 yards^3/mile) (Lieberman and Hoover 1948b). Because of road proximity to flowing and intermittent streams, most of the eroded soil entered the stream. Turbidities were high, sediment concentrations peaking at 5700 ppm during a storm in 1947. Tebo (1955) reported reductions in stream fauna in Shope Fork below the junction with the muddy WS 10 stream.

By 1958, the eroded roads had become impassable and were closed, cross-ditched, fertilized, and stabilized with grass. The exploitative logging demonstration clearly showed that logging, using the methods typical of the times, severely degraded water quality. Observers concluded that watershed damage had little to do with the poor silviculture of exploitative logging, but was principally due to road design and methods used to remove logs from the woods.

Integrated Forest and Watershed Management Demonstration: 1954–1955

The next logical step was a demonstration that logging roads could be built in the mountains without diminishing water quality. Two treatment watersheds (40 and 41) were selected for this demonstration. Both were marked for sale under standards for individual tree selection silviculture, but with the intent that WS 41 would be managed primarily for intensive timber production (Jones 1955) and WS 40 primarily for water production. After the first cutting, Walker (1957) surveyed the residual stands and reviewed potential future management options. He observed that "the similarity of treatments for the two watersheds, though for contrasting reasons, illustrates the point that good timber management will usually imply good watershed management."

Road construction and logging operations were tightly controlled in order to protect water quality. Skid trails were not permitted; all logs were winched to the road by A-frame skidder or tractor. In this way, most soil disturbance occurred on or adjacent to roads where exposed soil easily could be seeded to grass. Uphill skidding was preferred, because downhill skidding disturbs more soil and creates converging channels which concentrate surface flow during storms. Confining logging equipment to the roadway required the construction of a series of regularly spaced contour roads across the pair of watersheds. Road density was 8 km/100 ha (2 miles/100 acres).

The intent was to develop a road design which protected streams, yet could be laid out with hand level and compass to avoid the expense of a complete engineering design. Contour roads did not parallel the streams, but crossed at right angles with a slight dip or lowered road elevation at the crossing point. All streams, perennial or intermittent, were carried through corrugated metal pipe. The dip at the crossing prevented the stream from flowing down the road if the pipe became blocked. Roads were relatively narrow, 3 m (10 feet) wide, slightly outsloped without an inside ditch, and seeded with grass after logging was completed. Fills were covered by brush, with grass planted on those exceeding 2 m in slope length. Surface drainage was achieved by opentopped culverts or narrow water bars.

These drains required weekly cleaning by shovel during use. Sometimes a water bar had to be reinforced by imbedding a log in the raised berm. This road system successfully met water quality goals, but the weekly maintenance demand and high initial cost discouraged its acceptance by managers and loggers.

Management Tests: 1956–1960

The practicality and economics of this Coweeta road design were tested in two sales on national forest ranger districts. Research personnel served as consultants to the districts for road layout and water sample collection, but were not active in sale administration. A change in personnel reduced district interest in one of these demonstrations, and the logging contractor deviated from the road plan.

On the Tallulah District in North Georgia, both the national forest personnel and the operator remained interested in completing the Stamp Creek sale as planned. Black and Clark (1958) described the road design, logging methods, and site rehabilitation actions. The logger found the road costs acceptable because his savings in lower equipment maintenance and higher work efficiency compensated for the initial construction investment. Turbidity measurements showed water quality was unimpaired, but the success of the demonstration was best shown in a statement by the operator: "All the time we were logging, my men and I drank water out of the stream." Apparently this was an unusual work experience for them.

The Stamp Creek sale demonstrated the feasibility of having district personnel lay out and control a transportation system, and of expecting the logging operator to economically build roads and log an area without destroying water quality. Conversely, the companion sale clearly demonstrated a need for commitment and supervision by all involved to implement changes in roading and logging methods on National Forest lands.

Multiresource Management Demonstrations: 1962–1964

The demonstrations described to this point have dealt with only two resources of the mountain forest, the timber and the water. The Multiple-Use Sustained-Yield Act of 1960 established a legal requirement for that concept of management, often talked about in general terms, but difficult to demonstrate on the ground. Coweeta selected the 144 ha, high-elevation WS 28 for a demonstration of the concept of multiple-use management.

The decision was made to conduct this demonstration as if it were part of a municipal watershed—but not one where access, timber management, and recreation uses were excluded. If all resources in the watershed were to be made available, then access should be provided to most of the basin, not just to some limited area scheduled for the next timber sale. Thus, an important part of the WS 28 demonstration was an early illustration of transportation planning for long-term access into forested mountain land. Hewlett and Douglass (1968) described the transportation plan and design criteria used. Road density was somewhat less than on WS 40 and 41; about 5.2 km/100 ha. Goals were (1) to improve earlier designs so that maintenance requirements such as frequent cleaning of narrow-based water bars could be reduced, and (2) to demonstrate that timber access roads are permanent investments and not temporary expedients.

A solution was the broadbased dip, the design feature that has become a part of nearly every forest road guideline in the eastern United States and has influenced logging road standards internationally. Instead of the partial obstruction to traffic that the water bar presented, the broad-based dip was a gentle roll in the centerline profile of either a contour or climbing road (Figure 23.1). A 3% reverse grade over 6 m provided a relatively permanent block to any water flowing down the road. The dip was outsloped 3% to divert storm waters off the roadbed and onto the forest floor, where transported soil would be trapped by forest litter. Dips are not effective for draining wet soils or cut-bank springs. The broadbased dip is similar in concept to the intercepting dip in the

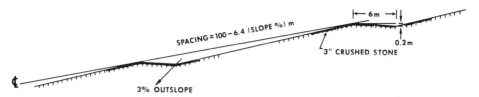

Figure 23.1. Diagram of the broadbased dip design for forest access roads.

California Region supplement to the Forest Truck Trail Handbook (United States Department of Agriculture 1935). Guidelines for the critical placement of the dip have developed with experience. The average spacing between dips is about 60 m. Initial standards for dip spacing proved impractically short on grades steeper than 8%, but even on steep sections dips are recommended at the top of a grade. Water diverted off a roadbed may flow across loose fill, but erosion is reduced if a dip is placed where the fill is short. Dips should not be placed to drain into perennial or intermittent streams where sediment would be carried further by storm waters.

The broadbased dip was linked to two other design features of the Coweeta road—vertical cut banks and no inside ditches. By its very nature, a ditch collects water, and the volume and velocity of storm runoff erodes the ditch and undercuts the bank. Sometimes the ditch carries this sediment to a stream. If a culvert empties onto loose fill, the concentrated water creates even more erosion. For a ditch to remain effective, debris must be scraped out, often disturbing the vegetation protecting the ditch and undercutting the cut slope above the ditch. The road width must allow for the ditchline, thereby exposing more disturbed soil during construction. However, where the roadbed can be drained by outsloping and broadbased dips, the problems of inside ditches can be avoided. Outsloping is most effective when roadbed rutting is controlled. Sometimes a short ditch may be needed to intercept seepage from the cut and drain the roadbed.

If the road has no inside ditch, then the cut bank can slough onto the inner edge of the roadbed without contributing loose soil into the path of storm runoff. Vertical cut banks are less expensive because less right-of-way clearing is required, less soil is moved, and smaller fills created. Cut bank soils slump to the angle of repose carrying roots, seeds, and topsoil to vegetate the exposed surface. At Coweeta, vertical cuts up to 2 m high have stabilized naturally on moist, fertile sites, but 1 m seems to be the limit for drier, less fertile banks. Although clearing a narrow right-of-way lowers construction costs, other factors argue for wider road clearings. For example, "daylighting" accelerates the drying of roads in wet and winter weather (Kochenderfer 1970), and wide right-of-way clearings can be linear wildlife openings (Arney and Pugh 1983).

Although each of the Coweeta road demonstrations was installed to test improved construction methods, they also served to train and educate private citizens and land managers, field technicians, and policymakers from industry and government. As a group, forest industries seemed quickest to adopt the concepts embodied in the Coweeta road design. A significant step was taken by Forest Service Region 8 at the Timber, Water, and Road Work Conference held in October 1968. Staff officers from each national forest and the region reviewed newly issued road design guidelines, saw a presentation on Coweeta WS 28, and visited demonstration roads on several ranger districts. Two goals were achieved: (1) the Conference encouraged wider use of broad-based dips and other design features to raise the standard of timber purchaser-built roads and (2) participants agreed that well-built timber access roads could be accepted into the permanent Forest Service road system without requiring full engineering design services and supervision.

Best Management Practices

The 1972 Amendments (Section 208) to the Federal Water Pollution Control Act require the management of nonpoint sources of pollution from forest activities. The Environmental Protection Agency encouraged each state to identify Best Management Practices (BMP's) as a voluntary means to reduce nonpoint sources of pollution. The principal source was recognized as soil disturbance and erosion. BMP's were chosen as superior to expensive regulation and rigidly defined practices because the individual timber producer could develop methods best for his terrain, equipment, and size of operation and achieve a practical balance between water protection and economic production of wood products.

Most states produced one or more pamphlets promoting and describing BMP's. These documents are illustrated and designed for use by loggers and small timberland owners. Some states further encourage adoption of BMP's through active extension service programs, county forester and consulting forester contacts, and inducements such as tax advantages for managed forest lands. BMP's deal with pollution from pesticides, fertilizers, other chemicals, and increased water temperature as well as soil erosion. However, the greatest effort is directed to encouraging BMP's for erosion control caused by roading, logging, and site preparation. Almost without exception, BMP guidelines for forest access roads include design features based on Coweeta experience. Use of the broadbased dip is not limited to the eastern United States, but also appears in state, industry, and national forest guidelines in the West.

Transportation Planning

One outgrowth of the multiresource management demonstration on Coweeta WS 28 was a realization that long-range planning of a forest transportation system should include intermittent-use access or local roads along with fully engineered forest development roads. Managers have been accustomed to referring to these two broad and sometimes poorly defined classes of roads as temporary and permanent. With the recognition that even the lowest class of road could be a permanent capital investment came the understanding that planning was necessary to assure that each mile of road was constructed on the best possible location. In the 1960s, the Southeastern Forest Experiment Station proposed to cooperate with Forest Service Region 8 in a major demonstration of multiresource management, with the key first step of planning and initiating development of a forest access system. The Upper Nantahala River basin adjacent to Coweeta with its then-maturing second forest was proposed as the demonstration site. The proposal was not acted upon and, thus, 16 years later a portion of the area could be reserved by the 1984 North Carolina Wilderness Law.

In an allied effort, Yandle and Harms (1970) produced a computer model which identified the best alternatives for each successive increment of road construction as active forest management developed over a unit of land. Decisions were driven by timber, wildlife, and recreation management opportunities. The model required a transportation plan and management information on timber stand maturity, location, and value as well as parameters describing other resource values and road construction costs.

Although research on many of these early ideas for transportation planning never passed the proposal stage, the principles which were raised and discussed are now being used operationally by timber, engineering, and other resource staffs working together on both short- and long-range plans for the National Forests.

Bridge and Culvert Size

Capitalizing on the large amount of streamflow data from small watersheds at Coweeta, Douglass (1974) developed relationships between flood frequency and the area and elevation of a watershed. These two factors accounted for 98% of the variation in discharge data. Because a standard method of estimating storm flows, Talbot's formula, gave considerably larger values than the Coweeta equations, new tables were presented to aid in selection of culvert and bridge sizes to handle maximum storm flows with recurrence intervals of 2.33, 5, 10, 20, 30, and 50 years. The value and applicability of this work is further strengthened by the agreement and overlap with flood frequency equations presented by Jackson (1976) and Whetstone (1982) for the mountains, Piedmont, and Coastal Plains of North and South Carolina. Larger structures are indicated for the shallow-soil watersheds of the central Appalachians (Helvey 1981).

Operational Application of Road Design Guidelines: 1976–1984

The elements of road design standards developed at Coweeta have stood the test of time and have been adopted in variation by many government and industry groups. The broadbased dip has found the widest acceptance, while the concept of vertical road cuts has less application. Since 1976, Coweeta's road-related research effort has been devoted to measuring the success of operational applications of current design standards rather than developing new designs (Douglass and Swift 1977).

In cooperation with the engineering staff of Region 8 and the National Forests in North Carolina, Coweeta measured the amount and timing of soil loss from several collector-class roads on the Wayah Ranger District. A collector is a light- to medium-traffic road that connects several local roads to primary or arterial routes and is generally constructed to a higher standard than the local road. Questions asked in this study were:

• In general, how effective are the current design standards for controlling soil loss from roads in steep mountain land?

- Which portion of the total soil loss comes from cut slopes, roadbeds, or fill slopes and how much can each loss be reduced by grass and gravel?
- When during the life of a road does major soil loss occur and what is the influence of season on rate of soil loss?
- How much and what type of surfacing is required for intermittent-use roads?
- How far downslope from the roadway does soil move and are present filter strip standards appropriate?
- If funds or time are limited, what critical part of a road should receive erosion control action first?

These roads were wider, 4.5 to 6.5 m, with deeper cuts and longer fills than the local roads previously studied at Coweeta. They were typical of currently constructed national forest access roads designed for sales using larger trucks. The details of study findings are given elsewhere, so only a summary is presented here. Soil loss rates differed among cut slopes, fill slopes, and roadbeds and were influenced by season and vegetation. Predictably, a graveled roadbed with well-grassed slopes had the lowest soil loss (Swift 1984b). Without any grass cover in early winter, freeze and thaw cycles loosened the cut slopes and large amounts of soil accumulated at the toe of the slope. Without an inside ditch, the debris stabilized and contributed little to sediment leaving the roadway. With the ditch, however, road maintenance and storm runoff moved the loose soil off the site, undercut the newly formed debris pile, and increased the potential for further soil loss. In four winter months, 150 to 360 t/ha were lost from the ungrassed cut slope. After grassing, soil loss was negligible.

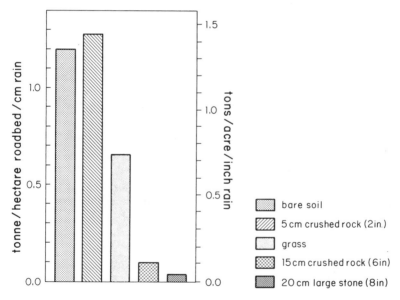

Figure 23.2. Soil loss rate for roadbeds with five surfacing treatments. Roads all constructed of sandy loam saprolite.

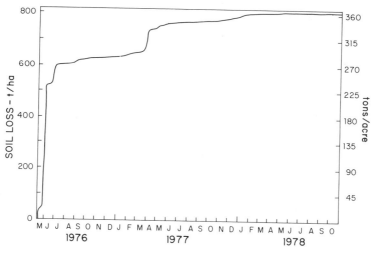

Figure 23.3. Cumulative soil loss from a forest road at a stream crossing in the first 2.5 years after start of construction.

Fill slopes, although uncompacted and unvegetated, eroded only where storm runoff from culverts or dips flowed over loose soil. In early spring, however, when high soil moisture levels are typical, some fills did slump onto the forest floor or against an obstruction downslope such as a brush barrier. Size of fill, steepness of the terrain, and texture of soil influenced slump occurrence and how far soil would move. Slumps were much fewer and smaller in volume on well grassed fill slopes.

Less soil was lost on a unit area basis from an ungraveled roadbed (< 8% grade) than from either cut or fill slopes. After graveling, only small soil loss occurred from storm water flowing in ruts or along the lightly graveled shoulder of the roadbed.

Gravel surfacing is the largest single cost item for forest roads; consequently, lower standard, intermittent-use roads often receive only thin coatings of gravel, spot treatments, or no gravel at all. Soils with high coarse fragment content, such as occur in some roads in the central Appalachians (Kochenderfer et al. 1984) can develop a natural gravel surfacing after an initial loss of finer soil particles. Test sections on a collector-class road at Coweeta (Swift 1984b) showed that soil loss from a lightly graveled roadbed was equivalent to loss from an ungraveled one (Figure 23.2). In contrast, soil loss from a grassed roadbed was half that of the bare soil road, both carrying the same traffic load. Soil loss from fully graveled roadbeds (15 to 20 cm thick) was only 3 to 8% of that from the bare soil roadbed of otherwise similar construction.

Typically, newly constructed roads lose the most soil, primarily during the short period before grass becomes well established and the roadbed is graveled or compacted. Three-quarters of the soil eroded during 2.5 years of observation was carried into the stream immediately below a road crossing in the first two months. Another 15% was measured a year later during the 3 months when the road was used for yarding and hauling logs (Figure 23.3). Thus, 90% of the soil loss from this road section occurred during only 15% of the 2.5-year period.

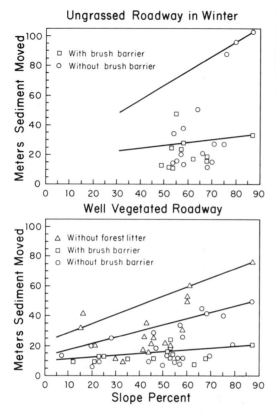

Figure 23.4. Lengths of sediment deposits downslope from roads are greatest where roadway is unprotected by vegetation, and least where brush barriers and forest litter trap sediment flows.

The outsloped road with broadbased dips protects water quality, because sediment-laden storm water is dispersed onto the forest floor rather than into a stream. The method is reasonable if the forest floor is protected by root mat and litter and has high infiltration rates typical of Appalachian Mountain soils. To protect water quality, sediment should be deposited on the forest floor before it reaches a stream channel.

Existing guidelines specify the width of undisturbed forest floor to be reserved for filter strip. Improved road construction methods allow reductions in filter strip widths from past guidelines if certain practices are followed (Swift 1986). A survey of 3.4 km of newly constructed forest road in the vicinity of Coweeta identified 76 major sediment deposits (longer than 7 m) (Figure 23.4). The longest three deposits extended over 85 m downslope, but these drained from portions of road left unfinished and ungrassed throughout the winter. Current management guidelines call for filter strip of 84 m on a 60% slope in moderately erosive soils. Where the road was finished and cuts and fills grassed before winter, deposits were all less than 45 m long, even on slopes over 60%. Brush barriers at the toe of fills held all but the longest deposit to less than 30 m for all slopes and to under 20 m if fills were vegetated. The lack of forest litter or brush barrier in a burned area allowed sediment to move up to 60 m on a 60% slope. These results emphasize that mitigating practices will reduce movement of sediment downslope, thus allowing greater flexibility when selecting road locations.

Wherever a road is built across a perennial or intermittent stream, loose soil falls into and around the channel. There is no filter strip, so unless vegetation or erosion-resistant materials cover the road fill, storms unavoidably wash soil directly into the stream system. Brush barriers extend only to a stream's edge. Thus, road crossings over defined channels are the most critical points on a road because fills are larger, the road drains directly into the stream system, and opportunities for mitigating practices are limited.

In 1976, three roads were built on WS 7 at a density of 5 km/100 ha. During the first year, all sediment collected in the weir originated from the roads; most of it from eight stream crossings during the first 2 months after construction began. For example, sediment measurements immediately below one crossing showed a cumulative total of 600 metric tons of soil entering the stream from each ha of roadway (Figure 23.3). A storm on May 27–29, 1976, produced the maximum streamflow event for the 50-year history of WS 7. Most of the sediment came from fill erosion before it was covered with grass. About 80% of the soil washed into the stream remained in the channel and had not reached the weir 720 m downstream after 2.5 years. However, portions of these deposits are still being transported out of the stream system 8 years later. The sections of roadway observed to contribute storm water and sediment directly into the stream are only 1% of the entire watershed area.

Based on these findings, practical water quality protection can be achieved by (1) designing roads with near vertical cut banks, no inside ditches, and broadbased dips; (2) completing construction and revegetation of cut and fill slopes before winter; (3) installing brush barriers at the toe of fills if the fills are located within 150 feet of a defined stream channel; and (4) fully graveling roadbeds that drain into stream channels.

Road Maintenance

Coweeta research has emphasized design and construction of forest roads, but, in the course of these studies the influence of road maintenance upon soil erosion also has been noted.

A goal of the intermittent-use road design incorporating the broad-based dip was to reduce costs. The design reduces maintenance costs during the period of heavy use, the long-term costs for standby maintenance, and the expenses of reopening a closed road. Experience has shown that grassed roadbeds carrying under 20–30 vehicle trips a month require a very low level of maintenance; primarily annual mowing of roadbed and periodic trimming of encroaching vegetation. The outlet edges of broadbased dips need to be cleaned of trapped sediment to eliminate mudholes and prevent the bypass of storm waters. The frequency of cleaning depends upon the traffic load. WS 28 roads required servicing every 5 to 10 years, but the roads on WS 7 carried more traffic and dip cleanout was required after only 2 years.

Maintenance by motor grader is difficult for this type of road. Scraping tends to fill in the dips, and often the blade cannot be maneuvered to clean the dip outlet. Cut banks are destabilized when the blade undercuts the toe of the slope. Small bulldozers or front-end loaders appear to be more suitable for periodic maintenance of intermittent-use forest roads.

Summary

The design and construction of, and soil loss from forest roads have been continuing areas of research and demonstration by the Southeastern Forest Experiment Station since Coweeta Hydrologic Laboratory was established. The low-cost, low-maintenance intermittent-use road pioneered by Coweeta is widely accepted and adapted to local conditions by government and industry land managers, and strongly recommended by state agencies with the aim of reducing sediment, the principal nonpoint source of pollution from forestry activities.

Several principles can be drawn from the Coweeta studies. An inexpensive design and field layout procedure can produce a serviceable and environmentally acceptable road. The most effective road system results from a transportation plan developed to serve an entire basin rather than the sum of individual road projects constructed to serve short-term needs. Soil exposed by construction should be revegetated quickly. Where possible, storm waters should be removed from the road at frequent intervals and in small amounts by outsloping and dips, rather than by consolidation into ditchlines and culverts. Contour roads and gentle grades require less maintenance and produce less sediment. Gravel surfacing is best, but a grassed roadbed is good where traffic is light and can be controlled to exclude use in wet weather. If only a small quantity of gravel is available, it should be applied on climbing grades, poor trafficability soils, in dips, and near stream crossings. The stream crossing is the most critical part of the entire road, and every effort should be made to protect and vegetate fill slopes and divert storm waters on the road away from the stream. Filter strips and brush barriers prevent sediment from reaching streams. Unnecessary maintenance must be avoided.

Guidelines for forest road design are available which minimize the impact of construction and use on water quality. The task is to apply these in land management.

24. Effects of Pesticide Applications on Forested Watersheds

D.G. Neary

Although Coweeta has not emphasized research on the impact of pesticides on forest watersheds, some fascinating results on the physical, chemical, and biological effects of herbicides and pesticides have been generated in the past 20 years. These findings come from studies using herbicides or insecticides to achieve some other objective, or from a few recent studies aimed specifically at the fate of herbicides in forest ecosystems.

Insecticides have been applied to regulate outbreaks of terrestrial insects (Grzenda et al. 1964) and to study instream detritus processing dynamics of aquatic insects (Wallace et al. 1982). Some of the original research on the effects of forest treatments on water yield used herbicides to simulate cutting (Hewlett and Hibbert 1961) or induce vegetation succession (Hibbert 1966). The water quality impacts of pesticide applications to Southern Appalachian forests were investigated in some of these studies (Grzenda et al. 1964; Douglass et al. 1969). More recently, a series of projects have conducted research to determine the impact of site preparation herbicides on water quality (Neary et al. 1983; Neary et al. 1985). With increased use of herbicides and insecticides in modern forest management, and the rising sensitivity of the public to the fate of pesticides in the environment, these studies have become more important.

The objective of this chapter is to highlight current knowledge of the effects of pesticide application on biological and hydrological processes in Southern Appalachian forest watersheds. This will provide a basis for understanding and delineating future research needs.

Streamflow

Herbicides have been used at Coweeta to eliminate vegetation and to determine the effects on water yield. Watershed 22 (WS 22) was subjected to a simulated strip cut with alternate 10 m strips treated with sodium arsenite to reduce the watershed basal area by 50% (Hewlett and Hibbert 1961). The herbicide was applied by frilled stem injections and by backpack spraying. No records on the rate of application are in existence. The deadened strips were established perpendicular to the stream channel in 1955 and maintained for 5 years by cutting, frilling, and respraying. No woody material was removed from the watershed and no soil disturbance occurred. On WS 6, the original hardwood forest was cut in 1960 and converted to grass to determine the effects of plant species and composition change on water yield (Hibbert 1969). Then, after 6 years in a grassed condition, the watershed was sprayed in 1966 with atrazine (2-chloro-(4-ethylamino)-6-(isopropylamino)-S-triazine), paraquat [1,1-dimethyl-4,4-bipyridinium ion (dichloride salt)], and 2,4-D[(2,4-dichlorophenoxy)acetic acid] to eliminate the grass cover (Douglass et al. 1969).

The 50% basal area reduction on north-facing WS 22 produced a first-year increase in water yield of 198 mm. This is about one-half of that produced by a 100% clearcut on north-facing WS 13 and 17 (Hewlett and Hibbert 1961; Hibbert 1966). However, in terms of the distribution of water yield increases, WS 22 responded similarly to south-facing WS 1. More WS 22 yield increase occurred during the growing season rather than dormant season, as is characteristic of WS 13 and 17. This response difference was possibly due to residual herbicide suppression of understory vegetation and sprouts during the first growing season after herbicide application. Application of herbicides to the grass stand on WS 6 resulted in a large first-year increase (260 mm) in water yield (Chapter 22). However, this increase was lower than the 370 and 408 mm increases measured the first year after cutting on WS 13 and 17, respectively. Water yield was very similar in the second year due to residual weed control. However, by the third year water yield dropped off drastically as a dense regrowth of grass, herbs, vines, sprouts, and tree seedlings increased evapotranspiration.

Water Quality

Until the recent surge in herbicide application for site preparation, the most extensive pesticide applications in forest ecosystems have been for insect control. Pests such as the elm spanworm (*Ennomos subsignarius*), spruce budworm (*Choristoneura fumiferana*), and gypsy moth (*Lymantria dispar*) have been sprayed with various insecticides in large-scale programs. More limited but frequent applications occur on specialized areas such as seed orchards and nurseries. However, information is still very limited on the effects of insecticides on water quality and nontarget organisms.

Insecticides

The first study on the effect of pesticide usage on water quality was done during application of DDT (dichlorodiphenyltrichloroethane) in 1961 and 1962 to control the elm spanworm (Grzenda et al. 1964). In late May 1961, DDT was sprayed over most of

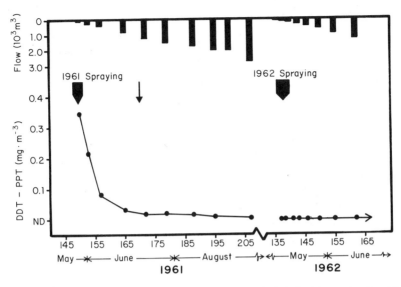

Figure 24.1. DDT residues in streamflow of Coweeta Creek below the laboratory after 1961 and 1962 spraying to control the elm spanworm. (Grzenda et al. 1964.)

Coweeta's Ball Creek and Shope Fork basins by fixed wing aircraft. The planes applied a 12% DDT No. 2 fuel oil solution at a rate of 1.0 kg ha^{-1} technical grade DDT. Residues of DDT in streamflow of Coweeta Creek, sampled at the laboratory boundary, were highest during the application (0.346 mg m^{-3}) (Figure 24.1). The minimum concentration (0.005 mg m^{-3}) was observed 2 months later, when sampling stopped at the beginning of August. The DDT application was repeated in 1962 using a helicopter and a lower rate (0.6 kg ha^{-1} DDT). Also, only 49% of the basin, involving only upper elevations and ridges, was treated. No DDT residues were measured in streamflow sampled over a 7-month period, from May 16 to November 23, 1962. Monitoring did not continue beyond the last date, so there are no data to indicate if the extremely persistent DDT or its primary metabolites DDD and DDE were present in subsequent streamflow.

A study by Wallace et al. (1982b) used methoxychlor [1,1-trichloro-2,2 bis (paramethoxyphenyl) ethane] to exclude macroinvertebrates from a small first-order watershed adjacent to WS 10. The purpose of their study was to assess the influence of macroinvertebrates on detrital processing. The methoxychlor (24% emulsifiable concentrate) was introduced into the stream source areas at a rate of 5 g m^{-3}, based on stream discharge in February 1980. Supplemental applications of 10 g m^{-3} were conducted in May, August, and November. During the period of the study concentrations of methoxychlor in streamflow never exceeded 33 mg m^{-3}. Despite these low concentrations, some drastic impacts on the invertebrate population occurred (Chapter 19).

Herbicides

A primary public concern about the use of herbicides as a silvicultural tool is contamination of surface and groundwaters. There is frequently confusion over the term pesticide. To many people "pesticide" is equated with deadly poison. Also, the lack of adequate information on the environmental fate of herbicides can hinder forestry professionals in explaining the relatively low risks and small environmental impacts of herbicides used in forest management. Herbicide use is rapidly increasing in the South, because chemical site preparation is more economical and less damaging to site productivity than intensive mechanical methods.

Atrazine, Paraquat, 2,4-D

Three herbicides were used on WS 6 in 1966 to eliminate the fescue grass (*Festuca arundinacea*) cover established after clearcutting of the watershed in 1958–1960. Although the herbicide was applied to study the hydrologic effects of removing the grass cover, it also provided an opportunity to gather information on water quality impacts (Douglass et al. 1969).

Atrazine and paraquat were applied in early May 1966 at a rate of 3.9 kg ha^{-1} atrazine and 1.1 kg ha^{-1} paraquat in 760 L of water carrier. A fire tanker was used to spray the herbicides on the grass. The entire watershed except for a 3 m buffer strip on either side of the perennial channel was treated. Atrazine and 2,4-D were sprayed in a second application in early June 1966 to control some of the more persistent grasses and woody plant species. In the second application, atrazine was used at a higher rate (5.0 kg ha^{-1}) and 2,4-D was sprayed at a rate of 3.4 kg ha^{-1} in the same amount of carrier.

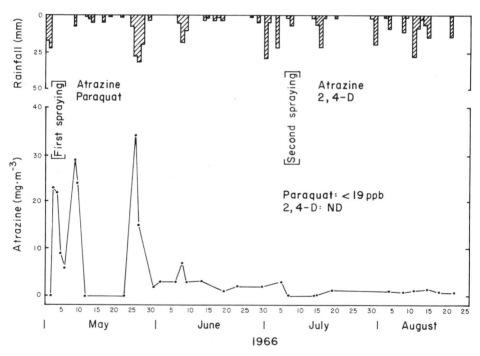

Figure 24.2. Atrazine concentrations in streamflow from WS 6, May–August, 1966. (Douglass et al. 1969.)

Despite the narrow stream buffer strip, only low concentrations of atrazine were detected in stream flow (Figure 24.2). The first two peaks (23 and 29 mg m^{-3}) were the result of spray drift into the ephemeral channel and washoff from the buffer during a small storm on May 9, 1966. From May 12 through 24 no atrazine was detected in streamflow. Rainfall of 60 mm on May 26 to 27 produced the third and highest concentration peaks (34 mg m^{-3}). These residues most likely originated in the buffer strip; thereafter, concentrations dropped off rapidly. The second spraying in July did not cause any increase in atrazine residues in streamflow, despite a higher application rate. This appeared to be a result of a better application with less drift. Atrazine has a low water solubility (33 g m^{-3}), so significant that residue movement in baseflow would not be expected.

Unlike atrazine, paraquat is completely water soluble, but is rapidly and completely inactivated in soil. It is highly bound by complexing with negatively charged sites on clay minerals and organic matter. Thus, only five of 35 samples collected within 45 days of the herbicide application contained residues of paraquat (Douglass et al. 1969). Of these, the peak concentration was 35 mg m^{-3}. After the initial 1½-month period, paraquat was not detected in streamflow. The few residues detected in streamflow were attributed to drift or foliar washoff during storms.

2,4-D was never detected in streamwater after the July 1966 application. Like atrazine, it indicates that spray drift into the stream buffer was not a problem during the second application. Since 2,4-D degrades fairly readily in most soils, there was little opportunity for movement into the streams.

Picloram

A 4 ha portion of WS 19 was converted to white pine (*Pinus strobus*) using herbicide site preparation (Neary et al. 1984). The specific objective of this study was to determine the water quality impact of using picloram (4-amino-3,5,6-trichloropicolinic acid) to deaden a poor quality mixed hardwoods and rhododendron-laurel (*Rhododendron maximum* and *Kalmia latifolia*) stand. Picloram is an extremely effective herbicide on a range of woody and broadleaf plant species, but it is low in toxicity to biota. Therefore, its long persistence in ecosystems, soil activity, and high solubility have raised questions on its potential impact on water quality in forest streams.

Picloram was broadcast manually in May 1978 at a rate of 5.0 kg ha^{-1} in a 10% active ingredient pellet formulation. Porous cup tension lysimeters were installed at depths of 30, 60, and 100 cm to determine movement in Humic Hapludult soils found on the steep slope of WS 19. Two springs 140 m downslope of the picloram treatment boundary were developed for sampling. Streamflow occurred at a temporary flume and the main weir was 200 and 760 m downslope. The drainage areas for these latter two sites are 10 and 28 ha, respectively.

Picloram residues in mineral soil exhibited a typical degradation pattern during the first 7 months after application (Figure 24.3). Residues of picloram in the upper 7 cm of soil were initially higher than expected (11.6 mg kg^{-1}), but exhibited a very short half-life of 4 weeks. Residues in mineral soil declined to near detection limits (0.06 mg kg^{-1}) by the end of sampling (Neary et al. 1985). Deeper in the soil, picloram residues generally followed surface trends, with decreasing concentrations with depth.

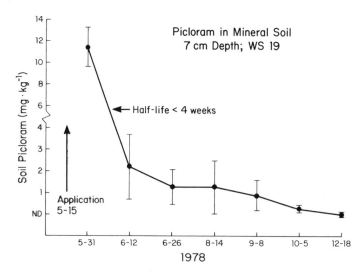

Figure 24.3. Picloram residues in the upper 7 cm of mineral soil, WS 19, May–December 1978 (Neary et al. 1985).

Samples collected below the 0 to 7 cm depth showed small increases (0.1 to 0.3 mg kg^{-1}) in residues between the August and October samplings due to downward migration of picloram. The half-life for picloram, based on these data, was about 4 weeks. This is a very short period compared to 6-month values reported elsewhere.

Trees immediately downslope (60 to 100% slopes) of the picloram treated area did not show any indication of lateral herbicide movement. Despite dry conditions which limited picloram's effectiveness and vegetative uptake (Neary et al. 1979), picloram residues on the mineral soil were not enough to interfere with survival of white pine planted the following March (Neary et al. 1984).

Picloram moving in soil solution was sampled at three depths: 30, 60, and 120 cm (Figure 24.4). The downward moving pulse of residues peaked at 30 cm (174 mg m^{-3}) about 3 months after herbicide application. At 60 cm, the pulse did not pass until 10 months postapplication. The peak concentration at 60 cm was considerably higher (381 mg m^{-3}). At 120 cm, picloram concentrations were considerably reduced (<25 mg m^{-3}) and had two peaks at 6 and 14 months after herbicide application. Movement of picloram in the Humic Hapludult soil was slowed by dry weather in the early summer of 1978, and apparently attenuated by the argillic horizons. Accumulation of organic matter on soil pore surfaces and root uptake may account for the large reduction of picloram residues between 60 and 120 cm. Concentration of picloram at all three sampling depths was at or below detection limits by the end of November 1979.

Retention of picloram residues in the soil system on the application site prevented any adverse impact on water quality. Two springs below the treatment site showed only trace levels of picloram for an 18 day period in December 1978 (Figure 24.5). Intensive sampling of these springs (period D) during the winter of 1978–1979 did not detect any additional picloram. Moreover, intensive samplings at the H-flume and main WS 19 weir showed no picloram residues.

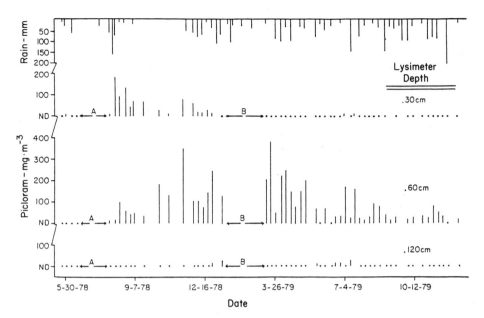

Figure 24.4. Picloram residues in soil solution at 30, 60, and 120 cm in a Humic Hapludult soil, WS 19, 1978 and 1979 (Neary et al. 1985).

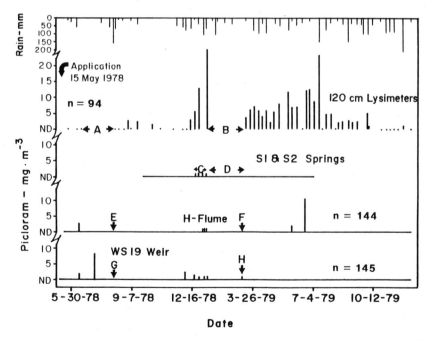

Figure 24.5. Picloram residues in soil solution, springflow, and streamflow, WS 19, 1978 and 1979 (Neary et al. 1985).

Hexazinone

One of the first integrated studies to investigate the environmental fate and effects of forest pesticides prior to U. S. Environmental Protection Agency registration involved the herbicide hexazinone [3-cyclohexyl-6-(dimethylamino-1-methyl-1,3,5-triazine-2,4(1H,3H)-dione]. Although the study was conducted off the Laboratory in the foothills of north Georgia, it has direct implication for site preparation use in the Appalachian mountains and involved a considerable commitment of Coweeta personnel.

Hexazinone was broadcast manually to four small watersheds (1 ha each) in the upper reaches of the Broad River, Chattahoochee National Forest, Habersham County, Georgia. A 60 to 80-year-old stand of shortleaf pine (*Pinus echinata*) and mixed hardwoods (oak–hickory) was treated in late April 1979 with 1.68 kg ha^{-1} hexazinone in 10% active ingredient clay granules. On each watershed, the ephemeral channel was instrumented with 30 cm H-flumes and Coshocton wheel flow proportional samplers to determine flow volumes and subsample storm runoff. Downstream sampling points encompassing nested watersheds of 10 and 100 ha were also instrumented or gaged. Sampling was conducted on a storm event basis to determine the persistence and movement of hexazinone and two metabolites in mineral soil, stormflow, and baseflow. Application of the hexazinone pellets was followed by ample rainfall, which ensured effective action of the herbicide (Neary et al. 1981).

The mean concentration of hexazinone and its two primary metabolites in stormflow followed a typical residue decay curve (Figure 24.6). Hexazinone residues peaked at

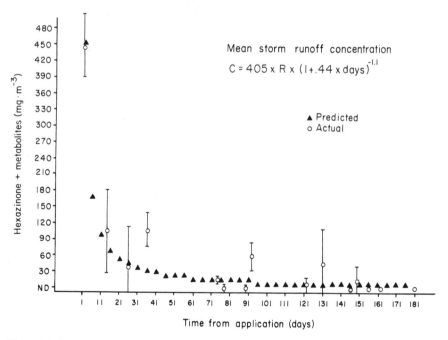

Figure 24.6. Actual and predicted hexazinone and metabolite concentrations in storm flow from small forest watersheds in the Appalachian foothills, 1979 (Neary et al. 1983).

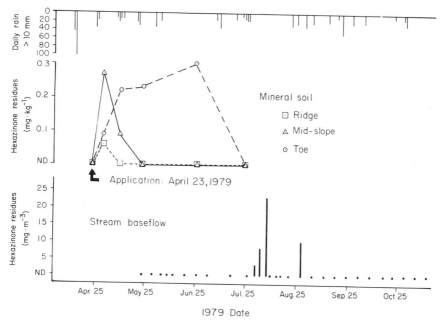

Figure 24.7. Hexazinone and metabolite residues in mineral soil and baseflow (Neary et al. 1983).

442 ± 53 mg m^{-3} in the first storm and fell to one-fourth of the peak by the next storm 2 weeks later (Neary et al. 1983). Concentrations of hexazinone in a series of 26 storms during the year after application followed a modified maximum concentration estimation model. Storms 4, 8, and 10, which carried average residues 20 to 70 mg mg^{-3} higher than predicted, were probably affected by cycling of hexazinone residues. Concentrations of 20 to 120 mg m^{-3} have been observed in rainfall coming through the crowns of hexazinone treated trees. Two storms, 1 and 17, accounted for 71% of the total residue loss (0.53% of the applied herbicide). Storm 1 was characterized by low runoff (0.6 mm) and a high residue concentration (442 mg m^{-3}), while Storm 17 had a high runoff (18.2 mm) and low residue concentration (18 mg m^{-3}).

Hexazinone residues in litter reported by Neary et al. (1983) were 0.18 mg kg^{-1} 3 days after the first rainfall, dropped to nondetectable by 30 days, and then climbed to 3.42 mg kg^{-1} after defoliation was completed. The half-life of hexazinone in mineral soil was on the order of 10 to 30 days. Residues in the upper 10 cm of mineral soil indicated that pulse moved downslope during the 90 days after herbicide application (Figure 24.7). Samples from ridge tops and midslope locations contained peak residues 3 days after rainfall activation of the herbicide. At the slope toe, concentration climbed steadily, peaking 60 days after application. Further evidence to support the existence of the downslope-moving pulse is evident in baseflow residue data.

Baseflow was sampled at a H-flume with a drainage area of 10.4 ha, including 4 to 5 ha of herbicide treated area. Hexazinone and primary metabolite residues were not detected until late July (Figure 24.7). Two pulses with peak residue concentrations of 23 and 10 mg m⁻³ were measured immediately after mineral soil levels dropped off below detection limits. For the remainder of the next 12 months only two trace levels were detected. The small size and short duration of the hexazinone residues in baseflow indicate no potential long-term water quality problems for use of the chemical in forest ecosystems.

Further downstream on a major perennial stream at the point draining 104 ha, hexazinone residues were detected mainly during some of the first eight storm events (Figure 24.8). Residue concentrations never exceeded 44 mg m⁻³. Samples collected from December 1979 through October 1980 did not contain detectable hexazinone residues. As discussed in the following section, these hexazinone levels were not of sufficient magnitude or duration to impact even the most sensitive aquatic species (Fowler 1977) and were well below a suggested water quality standard of 600 mg m⁻³ in drinking water (Leitch and Flinn 1983).

The nonpoint source pollution effects of using herbicides for site preparation were compared with those produced by mechanical site preparation (Neary et al. 1986). Chemical site preparation in the Appalachian foothills, as likely elsewhere, resulted in considerably lower sediment losses. Some anion and cation concentrations in storm-flow were higher than normally measured following clearcutting. Peak nitrate nitrogen levels were about twice those reported for either burning, clearcutting or mechanical site preparation (Figure 24.9). Nitrate nitrogen concentrations were still elevated 1 year later, but returned to a more normal range the following year.

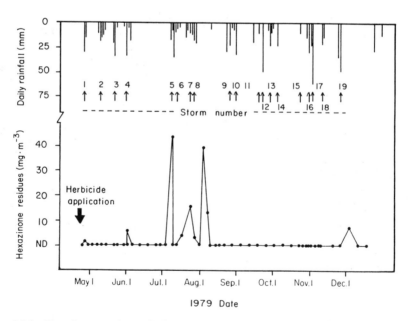

Figure 24.8. Hexazinone and metabolite residues in streamflow from a 104 ha watershed encompassing 5 ha of herbicide treated area (Neary et al. 1983).

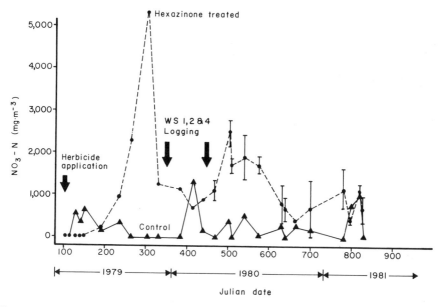

Figure 24.9. Nitrate nitrogen in stormflow from hexazinone treated watersheds (Neary et al. 1986).

Biological Impacts

The topic of biological impacts of forest pesticide applications has not been adequately addressed. Most of the research emphasis has been on residue degradation, water quality, and possible toxicological impacts. Herbicide efficacy on target species has been examined as an integral part of each study, but there is still much to learn about responses of other plants, invertebrates, and microorganisms. Studies by Wallace (this volume) have addressed some of the biological impacts of insecticide applications.

Herbicides

Both the picloram and hexazinone studies involved separate investigation of the effects of each herbicide on woody plants (Neary et al. 1979, 1981, 1983). Both herbicides proved to be effective on the range of tree species in the oak–hickory forest ecosystem. Hexazinone is a unique herbicide in that most pines are tolerant of herbicide rates which kill hardwoods. Some species such as red maple (*Acer rubrum*) and black gum (*Nyssa sylvatica*) show slight resistance to both herbicides. The effects of drought in reducing the activity of both herbicides were noted. Although the herbicides produced shifts in vegetation structure from forest to grass-broadleaf herbaceous plant system, they were not investigated in any detail.

A study of the impact of picloram on sensitive stream algae was installed as part of the WS 19 picloram study. However, no conclusive results were reached since residue levels were too low and sporadic to induce any treatment.

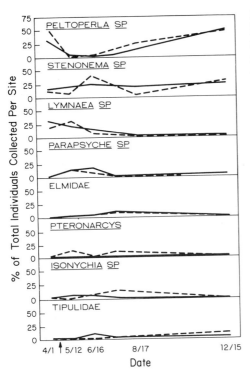

Figure 24.10. Percent of total number of individuals for eight major taxa of invertebrates in stream sections receiving flow from hexazinone treated and control watersheds, 1979 (Mayack et al. 1982).

During the hexazinone forest ecosystem study, terrestrial and aquatic invertebrates were monitored to determine if any changes in species abundance or diversity occurred (Mayack et al. 1982). Although hexazinone or metabolite concentrations were frequently measured in most terrestrial macroinvertebrate samples, results were highly variable. *Acarina* composed the largest fraction of the microarthropod community and the abundance did not vary appreciably between control and herbicide treated watersheds. Considerable variability was noted due to watershed microhabitat differences. Also, the canopy removal by the hexazinone defoliation of the hardwoods produced a natural Berlese funnel heat effect on litter arthropods. Aquatic invertebrates and macrophytes contained hexazinone residues only occasionally and at very low levels. No herbicide treatment effect was measured on members of a variety of aquatic species (Figure 24.10). Calculation of the Shannon-Weaver index for invertebrates showed that a stream portion receiving hexazinone residues consistently had higher index values than an untreated portion. Treatment effects, if any, were evidently masked by natural stream habitat variability.

Future Research

Many gaps still exist in our current knowledge of the fate of pesticides in forest ecosystems. Extensive research on forest pesticides has been conducted in the Pacific Northwest (Norris 1981), but data from the South is more recent and limited. Only one chemical, hexazinone, has a good data base behind it. As new forest pesticides are registered and older ones are reregistered, this situation will improve somewhat. New regulations set by the U. S. Environmental Protection Agency require forest ecosystem fate studies to support toxicological and efficacy data. However, many chemicals are being used in forest management which still have very limited impacts on movement, fate, and nontarget organism. Most of those pesticides have important roles in assuring adequate and economical production of wood and fiber products.

Future research should focus on potential impacts on surface and groundwater quality. Much work is needed to assess subtle impacts on species diversity, microbiological changes, and nutrient and energy flux dynamics. The most efficient means of accomplishing these objectives are through an interdisciplinary team working on integrated watershed level studies. The research approach should involve (1) identification of problem chemicals; (2) focus of effort on field-level studies; and (3) verification of existing models or development of new models to describe pesticide behavior or ecosystem response to pesticides. This approach would ensure extrapolation of the results to the widest area possible, and directly input into the risk analyses programs being started for forest use pesticides.

25. Stream Chemistry Responses to Disturbance

W.T. Swank

The vegetation on 12 watersheds at Coweeta has been altered by experimentation during the past 50 years. Disturbances include light selection cutting and logging, clearcutting without roads and no products removed, clearcutting with various methods of commercial logging, agricultural cropping, conversion of mixed hardwoods to white pine, and conversion of hardwoods to grass accompanied by applications of lime, fertilizer, and subsequent herbicide application. Also, hardwoods on two of the control watersheds have been partially defoliated by insects during the spring for varying periods of time. Two stream-gaging sites are on fourth-order streams and these larger drainages contain a combination of undisturbed and treated watersheds. A brief summary of each vegetation type represented during the period of stream chemistry record is given in Table 25.1. (see Chapter 1 for details of treatments).

In this chapter the objectives are (1) to characterize the chemistry of streams draining the variety of disturbed watersheds at Coweeta in terms of mean annual inorganic nutrient concentrations; (2) to compare long-term net nutrient budgets of selected disturbances with their controls; (3) to describe changes in stream chemistry in response to commercial clearcutting and succession; and (4) to examine stream chemistry responses to natural disturbances.

Mean Annual Solute Concentrations

In most cases the initiation of stream chemistry studies postdate watershed treatments. Results therefore represent conditions at varying time periods since disturbance. Mean annual nutrient concentrations of streams have been divided into two groups; those for

Table 25.1. Summary of Treatments for Coweeta Watersheds, Vegetation Types, and Years of Stream Chemistry Records

Watershed No.	Area (ha)	Vegetation Type Corresponding to Period of Stream Chemistry
1	16.2	White pine at ages 15 and 16 years and at 24 and 25 years
3	9.3	Yellow poplar, white pine and coppice mixture at age 20 and 30 years
6	8.9	Grass-to-forest succession at age 2 through 15 years
7	58.7	Mature hardwoods for 4 years and coppice regrowth at ages 1 through 7 years
13	16.2	Coppice at ages 7 and 8 years; 16 and 17 years
8, 9, 16	759.6, 723.6 and 381.6 respectively	Mixture of mature hardwoods and variety of altered vegetation. WS 8 for 1972 through present; WS 9 and 16 1972–73 and 1980–81.
17	13.4	White pine at ages 13 through 25 years
19	28.3	Mature mixed hardwoods, 24 years and 34 years after understory cut
22	24.3	50 percent mature hardwoods; 50 percent coppice at age 18 and 28 years
27	38.8	Mixed hardwoods, period from 1972 through 1983
28	144.1	Coppice regrowth, poplar cove and mature hardwoods, 9 & 10 years after harvest
36	48.6	Mixed hardwoods, period from 1972–1983
37	43.7	Coppice at ages 9 and 10 years, 19 years
40	20.2	Mostly mature hardwoods, 18 and 28 years after selection cutting
41	28.7	Mostly mature hardwoods, 18 and 28 years after selection cutting

which associated discharge records are available and streams where flow measurements were discontinued in the past. The first group represents flow volume weighted means and the latter are simple arithmetic averages.

The long-term mean flows and flow weighted chemical concentrations of treated watersheds indicate several interesting patterns (Table 25.2). On watersheds where the hardwood vegetation was most recently clearcut (7, 13, 28, 37), NO_3-N concentrations 7 to 20 years after treatment range from 0.02 to 0.18 mg L^{-1}, which substantially exceed concentrations for control watersheds. Catchments converted from hardwoods to white pine (1, 17) also exhibit elevated NO_3-N concentrations more than 25 years after treatment. On WS 40, the light selection cut, NO_3-N levels are equivalent to control streams. The highest NO_3-N concentrations (0.67 mg L^{-1}) observed in the Coweeta Basin occur on WS 6, the most severely disturbed watershed. With regard to other solutes, concentrations of Cl, Ca, and Mg are elevated on WS 6 in comparison to the other streams, partly in response to previous lime and fertilizer applications. Concentrations of NH_4-N and PO_4 are very low and similar to control streams. Levels of nitrate in the fourth-order stream of WS 8 reflect the mixture of water from control and treated

Table 25.2. Average Annual Flow Weighted Concentrations[a] (mg/L) of Dissolved Inorganic Constituents and Flow (cm) for Streams Draining Treated Watersheds at Coweeta

Watershed Number	Vegetation Type	Flow	NO_3-N	NH_4-N	PO_4	Cl	K	Na	Ca	Mg	SO_4	SiO_2
1	White pine	67	0.02	0.003	0.008	0.68	0.52	1.06	0.65	0.36	0.41	5.42
6	Grass-to-forest succession	100	0.67	0.005	0.007	1.04	0.55	1.00	0.99	0.60	0.44	6.59
7	Coppice regrowth	118	0.04	0.004	0.007	0.65	0.52	0.95	0.90	0.38	0.52	7.67
8	Combination of control and treated	132	0.01	0.003	0.005	0.52	0.38	0.78	0.81	0.32	0.65	6.65
13	Coppice regrowth	122	0.04	0.003	0.004	0.49	0.38	0.64	0.45	0.26	0.39	5.73
17	White pine	80	0.13	0.004	0.006	0.53	0.39	0.79	0.51	0.23	0.48	6.72
28	Combination coppice regrowth, even-aged poplar cove, and mature hardwoods	196	0.13	0.004	0.005	0.51	0.43	0.85	0.96	0.44	0.47	NA
37	Coppice regrowth	201	0.18	0.004	0.004	0.47	0.38	0.64	0.73	0.31	1.14	4.64
40	Uneven-aged mixed hardwoods	147	0.005	0.004	0.008	0.61	0.54	1.12	1.04	0.40	0.45	NA

[a] Period of record (June–May water year): For WS 6, 7, 8 and 17 all ions for the period 1973–1983 except for SO_4, 1974–1983 and SiO_2, 1975–1983. For WS 1 and 13 all ions for the periods of 1973–74 and 1982–83 except no SO_4 data in 1973 and SiO_2 data in 1973–74. For WS 28 all ions for the period 1973–74 except no SO_4 in 1973. For WS 37 all ions in 1973–74 and 1983 except no SO_4 data for 1973 and SiO_2 data in 1973–74. WS 40, data available only for 1973.

watersheds. The high SO_4 concentration on WS 37 is in agreement with high concentrations observed for controls, also located at high elevations. Other differences across ions and watersheds are difficult to interpret due to influences of bedrock mineralogy (see Chapter 6).

The second group of treated watersheds represent ungaged streams for the period of chemistry record which typically includes two years of record; i.e., 1972–73 and 1980–81 (Table 25.3). Compared to the other group of treatments, these watersheds received less disturbance and experiments were initiated in earlier years. Nitrate-N concentrations for most streams equal or only slightly exceed control levels. The highest Ca concentrations are found on WS 3 and are influenced in part by past lime applications during agricultural management and in part by bedrock composition. Chloride concentrations are rather uniform across watersheds, while cations show variation that may be related to bedrock mineralogy. The relatively high SO_4 concentrations for all streams compared to the previous group of treated watersheds occur because only one year of record was available (1980–81) and fortuitously coincided with the driest year on record, when SO_4 concentrations in all Coweeta streams were highest.

Taken together, none of these disturbances have produced elemental concentrations sufficient to have an adverse impact on water quality for municipalities or downstream fisheries resources. Even the drastic alterations on WS 6 produced rather modest increases in nutrient concentrations compared to disturbance responses in other regions of the United States. Elevated concentrations of NO_3-N appear to be a sensitive indicator of forest disturbance, and increased concentrations from clearcutting appear to last for at least 20 years following disturbance.

Input–Output Budgets

A second level of watershed response can be derived from nutrient budget (input-output) data (Table 25.4), in which the net water budget is an important consideration in interpreting nutrient fluxes. Due to much greater evapotranspiration from pine than hardwoods, WS 1 and 17 show the lowest discharge and largest net water accumulations. Compared with paired hardwood covered controls, other disturbed watersheds have flows which are elevated above undisturbed conditions that vary with treatment and time of recovery; these have already been quantified (Chapter 22).

Accumulations of NO_3-N are also indicated for some treated watersheds, but the net differences are lower than for controls. WS 6 shows the most obvious contrast in comparison to other ecosystems with a net N loss of 3.9 kg ha^{-1} yr^{-1}. Nitrate inputs and outputs are in balance on clearcut WS 37. Net budgets for Ca and K show the smallest net losses and conversely, largest within system accumulations for the white pine ecosystems. Highest elevation treated watersheds also produced the highest net losses of Ca and K, which are mainly attributed to greater annual discharge. All of the disturbed watersheds had large net annual SO_4 accumulations with values comparable to control watersheds.

Table 25.3. Average Annual Concentrations[a] (mg/L) of Dissolved Inorganic Constituents for Streams Draining Treated Watersheds at Coweeta

Watershed Number	Vegetation Type	NO_3-N	NH_4-N	PO_4	Cl	K	Na	Ca	Mg	SO_4	SiO_2	pH
3	White pine and yellow poplar plantations with coppice regrowth	0.013	0.004	0.005	0.55	0.42	1.05	6.28	0.48	0.74	9.30	7.00
10	Uneven-aged mixed hardwoods	0.006	0.004	0.004	0.55	0.49	0.94	1.44	0.41	0.71	9.55	6.91
19	Uneven-aged mixed hardwoods	0.002	0.003	0.004	0.50	0.32	0.70	0.40	0.23	0.80	5.85	6.69
22	50% coppice regrowth and 50% uneven-aged mixed hardwoods	0.010	0.004	0.004	0.49	0.38	0.74	0.55	0.23	0.85	6.13	6.66
41	Uneven-aged mixed hardwoods	0.010	0.005	0.004	0.58	0.49	0.99	0.97	0.40	1.05	8.06	6.79
Combination control and treated												
9		0.034	0.003	0.004	0.53	0.39	0.83	0.72	0.35	0.66	6.64	6.62
16		0.028	0.004	0.003	0.52	0.38	0.83	0.71	0.32	0.74	6.75	6.75

[a] Averages based on 2 years of records, 1972–73 and 1980–81 for all ions on all watersheds with the following exceptions: SiO_2 for all watersheds, 1980–81; SO_4 for WS 10, 19, 22 and 41, 1980–81.

Table 25.4. Average Annual Nutrient Budgets (kg/ha) for Treated Watersheds at Coweeta Hydrologic Laboratory

Watershed Number	Water (cm)			NO_3-N			NH_4-N			PO_4-P		
	Input	Output	Net Difference	Input	Output	Net Difference	Input	Output	Net Difference	Input	Output	Net Difference
1	189	67	122	2.83	0.16	+2.67	1.98	0.02	+1.96	0.25	0.05	+0.20
6	195	100	95	2.82	6.68	−3.86	1.89	0.05	+1.84	0.26	0.07	+0.19
7	188	118	70	2.69	0.42	+2.27	1.77	0.04	+1.73	0.25	0.08	+0.17
8	212	132	80	3.04	0.19	+2.85	2.02	0.04	+1.98	0.28	0.06	+0.22
13	218	122	96	3.19	0.50	+2.69	2.17	0.03	+2.14	0.30	0.05	+0.25
17	211	80	131	3.03	1.00	+2.03	2.01	0.03	+1.98	0.29	0.05	+0.24
28	277	196	81	4.10	2.50	+1.60	3.06	0.08	+2.98	0.43	0.09	+0.34
37	244	201	43	3.66	3.66	0.00	2.44	0.09	+2.35	0.35	0.09	+0.26
40	244	147	97	3.58	0.08	+3.50	2.69	0.05	+2.64	0.37	0.11	+0.26

Watershed Number	Cl			K			SO_4			Na		
	Input	Output	Net Difference	Input	Output	Net Difference	Input	Output	Net Difference	Input	Output	Net Difference
1	5.70	4.54	+1.16	1.66	3.44	−1.78	30.32	2.71	+27.61	3.59	7.08	−3.49
6	5.42	10.38	−4.96	1.86	5.53	−3.67	30.58	4.33	+26.25	3.37	10.06	−6.69
7	5.10	7.64	−2.54	1.78	6.11	−4.33	29.20	6.07	+23.13	3.19	11.18	−7.99
8	5.81	6.87	−1.06	2.03	4.98	−2.95	33.10	8.44	+24.66	3.62	10.32	−6.70
13	6.76	5.98	+0.78	1.84	4.58	−2.74	32.25	4.52	+27.73	4.42	7.86	−3.44
17	5.85	4.24	+1.61	2.02	3.11	−1.09	32.83	3.58	+29.25	3.64	6.33	−2.69
28	10.43	9.98	+0.45	2.87	8.41	−5.54	48.01	8.94	+39.07	7.34	16.67	−9.33
37	7.98	9.40	−1.42	2.20	7.62	−5.42	38.78	21.72	+17.06	5.19	12.77	−7.58
40	8.84	8.93	−0.09	2.41	7.90	−5.49	—	—	—	6.21	16.36	−10.15

Watershed Number	Ca			Mg			SiO_2		
	Input	Output	Net Difference	Input	Output	Net Difference	Input	Output	Net Difference
1	3.43	4.34	−0.91	0.77	2.41	−1.64	0.38	36.03	−35.65
6	3.80	9.93	−6.31	0.78	5.99	−5.21	0.58	62.54	−61.96
7	3.66	10.60	−6.94	0.76	4.50	−3.74	0.56	87.57	−87.01
8	4.13	10.63	−6.50	0.87	4.25	−3.38	0.62	83.98	−83.36
13	3.91	5.45	−1.54	0.87	3.22	−2.35	0.46	58.25	−57.79
17	4.10	4.04	+0.06	0.87	1.86	−0.99	0.63	49.79	−49.16
28	5.45	18.92	−13.47	1.17	8.66	−7.49	—	—	—
37	4.56	14.59	−10.03	1.03	6.18	−5.15	0.57	86.47	−85.90
40	4.74	15.23	−10.49	1.00	5.86	−4.86	—	—	—

Comparison of Control and Treated Watershed Net Budgets

A more quantitative analysis of nutrient responses to treatment can be made by comparing net budgets with the long-term net budgets of adjacent, undisturbed control watersheds (Table 25.5). Comparisons are based on identical periods of record for each treated watershed and its control. However, it should be pointed out that such analyses are somewhat tenuous, because pretreatment calibrations are unavailable. On the other hand, inferences can be made about treatment effects, because this comparative method minimizes differences in geology and weathering rates between watersheds.

Compared to controls, the young white pine plantations show small losses of NO_3-N, no differences in NH_4-N, PO_4, and SO_4, but accumulations of 1.2 to 4.4 kg ha^{-1} yr^{-1} for other ions (Table 25.5). Accumulations of some nutrients such as Ca, Mg, and K on the pine catchments are 50 to 70% greater than net budgets for control watersheds. The net gain of nutrients in the pine ecosystems is partly due to greater evapotranspiration and reduced flow from pine compared to hardwoods (Chapter 22). But, the average flow reduction of 25% observed for pine accounts for less than half of the estimated cation accumulations. Biological processes are also apparently important, and the high rates of pine biomass and nutrient accretions (Swank and Schreuder 1973) are also major reasons for nutrient conservation by pine ecosystems.

Nitrate-N net budgets for young coppice stands shows a net loss of 2.2 kg ha^{-1} yr^{-1} compared to mature hardwoods, and little difference in NH_4-N and PO_4. Other ions, except Mg, show small accumulations, and it appears that biogeochemical cycles are rapidly reestablished after clearcutting. The grass-to-forest succession watershed (WS 6) showed large losses of NO_3-N, Cl, Ca, Mg; these losses were a response to fertilization and liming as previously noted.

Table 25.5. Net Loss or Gain of Ions for Treated Watersheds with Different Vegetative Cover Based on a Comparison with the Net Budgets of Adjacent Undisturbed Hardwood-Covered Watersheds.

Vegetation Type, Age and Watershed Number	NO_3-N	NH_4-N	PO_4	Cl	K	Na	Ca	Mg	SO_4
Eastern white pine, Age 16 through 26 years (WS 1 and 17)	−0.7	−0.1	0.0	+1.8	+1.7	+4.4	+2.3	+1.2	+0.7
Coppice, age 11 through 21 years (WS 13 and 37)	−2.2	−0.3	0.0	+1.0	+0.9	+2.8	+0.2	−0.6	+0.8
Grass-to-forest succession, age 6 through 16 years (WS 6)	−6.8	−0.1	0.0	−4.4	−0.7	−0.1	−3.3	−2.6	−1.0

Values in kg ha^{-1} yr^{-1}.

Responses to Clearcutting, Logging, and Early Succession

The most precise documentation of effects of clearcutting and logging on stream chemistry conducted at Coweeta is for WS 7, where pretreatment calibrations were established. Water sample collection in the second order stream draining the 59 ha watershed was started in 1975 using a flow proportional sampler and also weekly grab samples; similar methods were used on the adjacent control watersheds (WS 2). Discharge on both watersheds have been measured continuously since 1935, and provide a firm basis for predicting changes in flow volumes on WS 7 due to treatment.

Management of the watershed was separated into three major operations: (1) road construction and stabilization; (2) tree felling and logging; and (3) site preparation. Three logging roads totaling 2.95 km were constructed on the watershed between mid-April and mid-June 1976. Immediately after construction, road cuts and fills were stabilized by seeding grass and applying commercial fertilizer (10-10-10) and lime. Seed, fertilizer, and lime were again applied to cuts and fills in July 1977, and to the running surface of the road in June 1978. Logging began in January 1977 and was completed in June. The majority of the logging was conducted from the roads with a mobile cable system and the forest floor generally remained intact. The site was prepared by felling the stems remaining after logging, and this operation was completed in October 1977.

Pre- and post-treatment concentrations of several selected nutrients in stream water of WS 7 and the adjacent control (WS 2) are shown in Figure 25.1. Throughout the five year calibration period, mean monthly K concentrations on WS 7 were typically 10 to 15% lower than on WS 2. Elevated concentrations were observed during several months early in the summer of 1976 in response to road fertilization. Following cutting and logging in 1977, K concentrations on WS 7 usually exceeded those on the control by 10 to 40% and remained elevated during most months through early 1981. Based on regression analysis, most of the monthly increases on WS 7 were statistically significant (0.05 level). By late 1981, the fourth year after cutting, K concentrations were near expected pretreatment levels during all seasons except the summer, when concentrations were still moderately elevated. Baseline monthly Ca concentrations were typically 30 to 40 mg L^{-1} higher on WS 7 compared to WS 2, but the temporal increases in Ca concentrations associated with the treatment were similar to K temporal trends (Figure 25.1). The influence of road fertilization was observed early in 1976. Beginning in the summer of 1977, Ca concentrations increased significantly (0.05 level) and remained elevated through late 1981. The largest changes in Ca concentrations on Ws 7 (30 to 50 mg L^{-1}) occurred in the third full growing season (1981) after cutting. Examination of Na, Mg, and Cl concentration trends showed a similar lag between the time of treatment and maximum response. Baseline concentrations of NO_3-N in streams draining undisturbed watersheds at Coweeta are typically near analytical detection limits of 2 μg L^{-1} (Figure 25.1). In contrast to K and Ca, no measurable increase in NO_3-N was observed at the weir when roads were fertilized in 1976, due to within stream processes which probably depleted NO_3 before it reached the weir (Swank and Caskey, 1982). Increases in NO_3-N on WS 7 began in early fall, about 9 months after the initiation of cutting. Concentration changes remained low (50 to 75 μg L^{-1}) into the following summer and then peaked (100 to 150 μg L^{-1}) during the

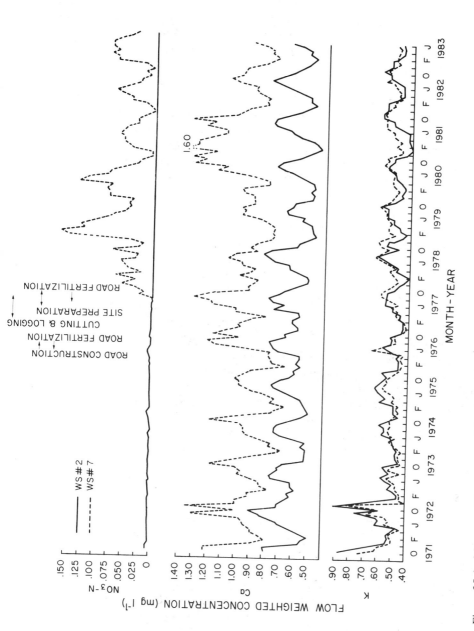

Figure 25.1. Mean monthly concentrations (flow weighted) of K, Ca, and NO_3-N in streamwater of WS 7 during calibration, road construction, clearcutting and logging, site preparation, and post-harvest recovery periods. Data are based on samples collected weekly.

second winter after treatment (Figure 25.1). Thereafter, NO_3-N increases declined toward baseline values, but were still elevated the fifth year after cutting.

Taken collectively, stream solutes showed small but measurable changes in concentrations. These changes indicate disruption of nutrient recycling processes, but losses must be quantified to evaluate the magnitude of management impacts. Concentration data were combined with flow volumes to calculate total flux. Annual changes in solute fluxes were estimated from pretreatment calibration regressions of monthly fluxes between WS 7 and its control, WS 2, for each ion. Correlations between the two watersheds were very good with r^2 values ≥ 0.92 for most ions. Exports of NO_3-N, NH_4-N, and PO_4 were very low on both watersheds and mean monthly values during the calibration period were nearly identical for the two catchments. Thus, treatment effects for these ions were simply derived by difference between pre- and post-treatment periods. Annual increases in streamflow and nutrient export during the first 5 years after logging are shown in Table 25.6. Increased export was greatest during the third year after treatment, with annual values for NO_3-N, K, Na, Ca, Mg, SO_4, and Cl of about 1.3, 2.4, 2.7, 3.2, 1.4, 1.2, and 2.1 kg ha^{-1}, respectively. Cutting had little effect on NH_4-N and PO_4 exports, with only small increases observed in most years. Increases in nutrient export were substantially diminished by the fifth year and appeared to be approaching prelogging levels. A major factor in increased nutrient export is the increased flow response that results from reduced evapotranspiration following cutting (Swank et al. 1982). In the first two years after cutting, annual flow increased an average of 25.5 cm and then declined to 13 cm in the third year. By the fifth year, flow was only 8 cm above the value expected if the watershed had not been cut. Although significant increases in solute export occurred in the first two years, maximum flow export occurred in the third year, when flow increases showed substantial decline. Thus, the timing of nutrient losses is not entirely related (on an annual basis) to the timing of hydrologic responses and may reflect the influence of decomposition and other nutrient recycling processes.

These relatively small changes and trends in nutrient losses clearly illustrate the concepts of ecosystem resistance and resilience. Previous theoretical analyses suggested that mixed hardwood forests at Coweeta exhibit both high resistance and high resilience as related to changes in the N cycle associated with forest harvesting activities (Swank

Table 25.6. Annual Changes in Streamflow and Solutes Following Clearcutting and Logging on Coweeta WS 7

Time Since Treatment (May–April Water Year)	Flow (cm)	Increase or Decrease in Streamflow and Solute Export[a]								
		NO_3-N	NH_4-N	PO_4	K	Na	Ca	Mg	SO_4	Cl
						(kg ha^{-1})				
First 4 months	0.5	0.01	<0.01	0.04	0.43	0.42	0.24	0.26	1.17	0.68
First full year	27.3	0.26	0.03	0.12	1.98	1.37	2.60	0.96	0.81	1.13
Year 2	23.8	1.12	<0.01	0.03	1.95	2.22	2.51	1.15	−0.24	1.62
3	13.1	1.27	0.05	0.06	2.40	2.68	3.16	1.42	1.16	2.08
4	8.6	0.25	0.15	0.06	0.80	1.07	1.63	0.46	0.93	0.59
5	8.0	0.28	0.01	0.02	0.52	0.13	1.19	0.18	0.11	0.10

[a] Increase or decrease based on monthly calibration regressions.

and Waide 1980; Chapter 28). Ecosystem resistance is related to the presence of large storage pools of organic matter and elements, which turn over slowly (Webster et al. 1975) as illustrated by previously reported Coweeta data and relatively small increases in nutrient losses shown in this study. Rapid recovery or return of nutrient losses to baseline levels are indicative of high ecosystem resilience. Reasons for rapid recovery of ecosystem biochemical cycles are related to high rates of net primary production (NPP) and incorporation and storage of nutrients in successional vegetation (Chapter 12). By the third year after cutting on WS 7, aboveground NPP on mesic sites was about 60% the NPP of the original mature hardwood forest. Early successional species exhibit high nutrient concentrations and in the first year after cutting, nutrient pools in NPP for N, P, K, Mg, and Ca were already 29 to 44% of the NPP of the mature forest (Boring et al. 1981). The potential roles of woody litter, soil organic matter, and nitrogen transformations in ecosystem nutrient retention for this clearcutting experiment are discussed in Chapters 12 and 16.

Grass-to-Forest Succession

The long-term experiment on WS 6 provides a unique opportunity to further examine the influence of successional vegetation on nutrient recycling processes as reflected by trends in stream chemistry. As noted previously, WS 6 has substantially elevated concentrations of some solutes due to the nature of past disturbance. The period of weekly stream chemistry determinations span vegetation succession from initial domination by herbaceous species (1969 to 1972) to present day domination by woody species.

Mean monthly concentrations (flow weighted) of several cations for the periods 1969 to 1972 and 1979 to 1982 for WS 18, the nearby control stream, and WS 6 are shown in Figures 25.2 and 25.3 respectively. The seasonal trends of K, Ca, and Mg concentrations for the hardwood forest (WS 18) are replicated for most years, both within and across time periods. Maximum concentrations typically occur during August and October, followed by declining concentrations in November and minimum values in the winter months. Concentrations begin to increase in early spring and continue through the summer until a maximum is reached again in the fall season. This seasonal pattern of stream cation concentrations is typical for other control hardwood watersheds at Coweeta as shown earlier in this chapter, and also for disturbed watersheds with different woody growth such as hardwood coppice and white pine (Johnson and Swank 1973).

Patterns of K, Ca, and Mg concentrations for WS 6 stream water (Figure 25.3) also usually showed distinct seasonal trends during the herbaceous vegetation period (1969 to 1972). However, the timing of minimum and maximum concentrations were shifted. That is, minimum concentrations frequently occurred in August or September with a rapid increase in the fall months, maximum values in early winter, and declining concentrations in the spring. In contrast, 10 years later when the cover was dominated by woody vegetation, concentration patterns appear altered (Figure 25.3) and may be returning toward trends observed for streams draining forested watersheds. For example, Ca, K, and Mg concentrations showed some evidence of decline during the winter months; minimum concentrations frequently occurred during May or June, while

W.T. Swank

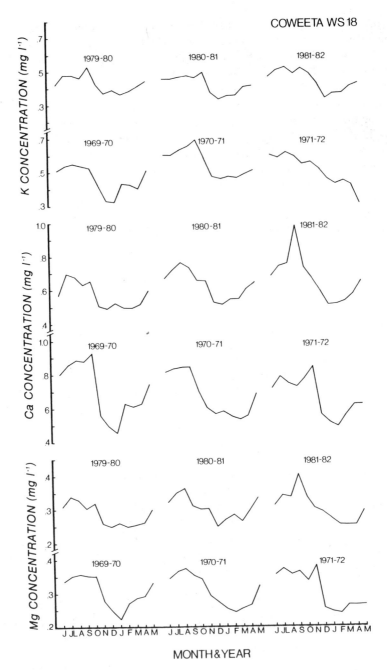

Figure 25.2. Mean monthly concentrations (flow weighted) of Mg, Ca, and K in stream water of control WS 18 during two separate 3-year periods.

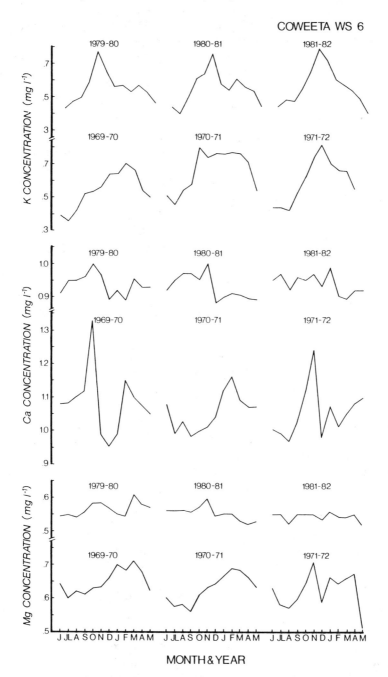

Figure 25.3. Mean monthly concentrations (flow weighted) of Mg, Ca, and K in stream water of WS 6 during 1969–72, when the vegetation was primarily herbaceous species compared with monthly values in the period 1979–82, when the vegetation was dominated by woody species.

subsequent concentration increases occurred earlier in the fall months. Comparison of cation values for the early and later successional periods also indicates that the amplitudes of some concentrations are damped and overall concentration levels reduced during succession. Monthly concentrations of Ca ranged from about 0.95 to 1.25 mg L^{-1} in 1969 to 1972 and 0.90 to 1.00 mg L^{-1} in 1979 to 1982; Mg ranges for the two periods were 0.55 to 0.70 and 0.55 to 0.60 mg L^{-1}, respectively, while K showed little change.

Monthly flow distributions and annual flows on WS 6 during these successional periods were close to those expected for a mature hardwood forest (Chapter 22). Thus, changes in nutrient concentrations are thought to demonstrate the importance of biotic processes in regulating temporal trends of cation concentrations. In the early years of succession (1969 to 1972), vegetation was dominated by a dense cover of herbaceous species such as horseweed (*Erigeron canadensis* L.) and cottonweed (*Erechtites hieracifolia* (L.) Raf.), but by 1980 the vegetation was dominated by woody species with a density of more than 1400 stems ha^{-1}. The magnitude of nutrient uptake, storage, and decomposition are vastly different between herbaceous and woody plants and are postulated to influence both the amount and pattern of nutrient losses. The quantitative contribution of various recycling and other biological processes to nutrient dynamics is unknown and difficult to ascertain, but it is hypothesized that seasonal patterns of stream nutrient concentrations will become reestablished as recycling processes characteristic of woody vegetation continue to develop.

Effects of Natural Disturbances on Stream Chemistry

Natural disturbances are an inherent feature of baseline ecosystems which can affect nutrient supply and mobility. Insect populations and the impact of their associated activities on biogeochemical cycles have been described (Chapter 21) and demonstrated in several Coweeta ecosystems.

In one study, an outbreak population of the fall cankerworm (Lepidoptera: Geometridae), a spring defoliator of hardwood forests, was first observed adjacent to and on the Coweeta Basin in 1969. Watershed 27, a 38.8 ha control catchment, was the primary site of infestation. Defoliation began in 1970 at the higher elevations (1400 m) on the catchment and progressed toward lower elevations in ensuing years. During defoliation, mean monthly concentrations of NO_3-N in stream water of WS 27 frequently rose above 40 mg L^{-1} (Figure 25.4). High concentrations were observed in late winter and also during and immediately following the time period of cankerworm feeding (late April through early June). At the peak of infestation in 1974, 33% of the total leaf mass was consumed and NO_3-N concentrations were elevated throughout the year. Subsequently, levels of defoliation were less severe, and in 1978 the population returned to endemic or nonoutbreak levels. The decline in the cankerworm population was accompanied by a return of NO_3-N concentrations toward baseline levels (Figure 25.4).

Increased stream export of NO_3-N concomitant with defoliation was also observed on WS 36 (Figure 25.4), another high elevation catchment at Coweeta. However, cankerworm infestation was not present on the catchment when stream chemistry analyses were initiated; thus, during 1972 and 1973, mean monthly NO_3-N concentrations were representative of other undisturbed forest ecosystems at Coweeta (generally

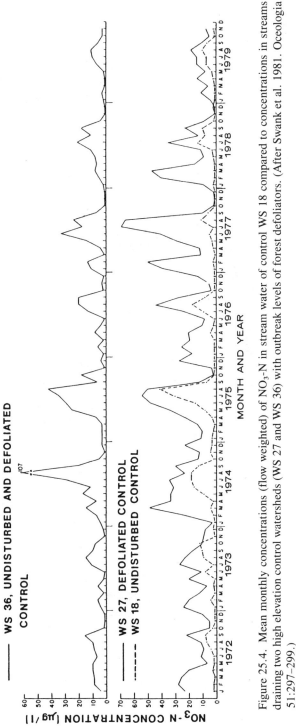

Figure 25.4. Mean monthly concentrations (flow weighted) of NO_3-N in stream water of control WS 18 compared to concentrations in streams draining two high elevation control watersheds (WS 27 and WS 36) with outbreak levels of forest defoliators. (After Swank et al. 1981. Oceologia 51:297–299.)

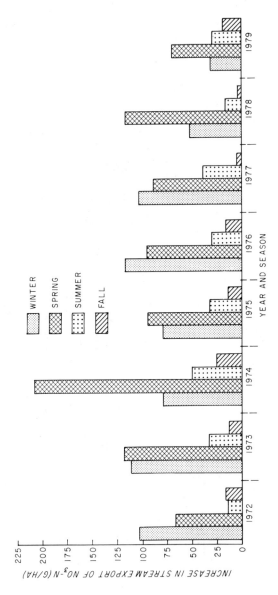

Figure 25.5. Seasonal increases in NO$_3$-N export from the partially defoliated hardwood forest on WS 27 during an 8 year period.

below 10 µg L⁻¹). In the late spring of 1974, concentrations began to rise and egg mass surveys on trees in 1975 confirmed that infestation had occurred on the catchment, although defoliation was not as severe as on WS 27. Again, by 1979, concentrations had returned to baseline levels (Figure 25.4). Elevated NO_3-N concentrations for a "control" stream in association with extensive cankerworm defoliation was also detected for another catchment about 13 km from the Coweeta Basin (Swank et al. 1981). The impact of defoliation on seasonal exports of NO_3-N from WS 27 are shown in Figure 25.5. Annual increased export of NO_3-N for the period shown ranged from about 200 g ha⁻¹ in 1978 to 450 g ha⁻¹ in 1974, and more than 80% of the increased export occurred during the winter and spring months. This pattern was due to the combined factors of elevated concentrations and high streamflow during these seasons compared to the remainder of the year. Discussion of why defoliation leads to increased NO_3-N export have been given elsewhere (Swank et al. 1981).

Further evidence for insect regulation of stream nutrient responses at the ecosystem level is found on another catchment at Coweeta. Characteristics and the treatment history for WS 6 were previously discussed. During the period of grass and herbaceous-to-forest succession, mean annual NO_3-N concentrations in stream water gradually declined from about 0.75 mg L⁻¹ in 1972 to 0.50 mg L⁻¹ in 1978 (Figure 25.6). Then in 1979, NO_3-N concentration showed an abrupt increase to 0.75 mg L⁻¹ concurrent with a heavy infestation of the locust stem borer (*Megacyllene robiniae*). Since black locust was the dominant woody species, the infestation was distributed over most of

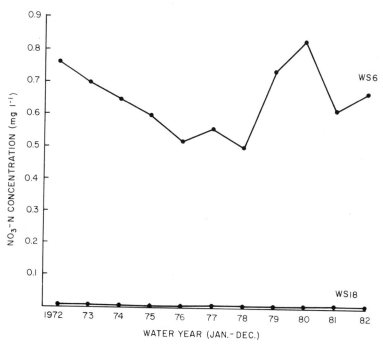

Figure 25.6. Mean annual concentrations (flow weighted) of NO_3-N in streams draining control WS 18 and WS 6 during the period of grass-to-forest succession.

the catchment. In the subsequent year (1980), NO_3-N concentrations increased to 0.90 mg L^{-1} and has remained above preinfestation levels in ensuing years. By 1982, 21% of the black locust trees were dead, 18% were severely injured, and many of the remaining stems showed some evidence of canopy decline (L.R. Boring, unpublished data). A number of hypotheses are currently under investigation to examine the functional relationship between insect stress and increased NO_3 loss from the ecosystem; these studies are focused on the effects of insect populations on nutrient uptake, quantity and quality of litterfall, nitrification, and nitrogen fixation.

Both of the preceding experiments demonstrate functional, ecosystem-level regulation of nutrient recycling by terrestrial insects under outbreak conditions. The catchment responses represent the integrated functional properties of an entire forest ecosystem and reflect changes in biogeochemical processes within the ecosystem. In both cases, changes in concentrations of other dissolved constituents such as SO_4, K, Ca, and other cations are difficult to detect because baseline values are usually higher than NO_3, and differences in bedrock mineralogy between insect infested and noninfested catchments influence ions sufficiently to mask small changes in ionic composition.

Summary

Long-term measurements of dissolved inorganic constituents of streams draining disturbed watersheds have been made at Coweeta. Disturbances include commercial selection cutting, commercial and noncommercial clearcutting, conversion of mixed hardwoods to white pine and grass covers, agricultural cropping, and natural disturbances comprised of insect outbreaks. Initiation of stream chemistry studies postdate most watershed treatments and represent conditions at varying periods of time since disturbance. Taken together, the responses in stream chemistry to forest disturbances can be summarized by the following points:

1. Over the period of observation beginning in 1972, none of the disturbances produced nutrient concentrations that would have an adverse impact on water quality for municipalities or downstream fisheries.
2. Compared to other forested regions of the United States, increases in nutrient concentrations of streams are small, even for the most drastic vegetation disturbances.
3. Nitrate-N is a sensitive indicator of forest disturbance and although concentrations are quite low (<0.2 mg L^{-1}), elevated levels in streams draining clearcuts appear to persist for at least 20 years after cutting.
4. Comparisons of annual nutrient input and output budgets for control versus disturbed watersheds illustrate the importance of evapotranspiration (Et) processes in regulating biogeochemical cycles. Hardwood to white pine conversion increased Et, reduced annual stream discharge, and consequently reduced the export of some dissolved nutrients. Conversely, clearcutting reduced Et, increased water flux, and increased nutrient export.

5. Budget data, combined with process research, also demonstrate the importance of biological processes in nutrient retention and loss from forest ecosystems. Decomposition, net primary production, and uptake and storage of nutrients in successional vegetation are important factors that regulate the magnitude and timing of nutrient exports.

6. Outbreak infestations of two different insects provide evidence for insect regulation of nutrient recycling and losses at an ecosystem level as revealed by changes in stream chemistry.

26. Acid Precipitation Effects on Forest Processes

B.L. Haines and W.T. Swank

Acid rain can displace essential elements from plant leaves (Wood and Bormann 1975), and soil (Wiklander 1973/74), and it can inhibit element uptake by plants (Arnon et al. 1942; Black 1968). These processes of leaching and inhibition of uptake have the potential to disrupt the cycling of mineral elements upon which forest production is partly dependent. Quantification of the potential magnitude and consequence of this disruption of mineral cycling is critical to the development of alternative management policies. A balance is needed between the cost of reducing emission of acid forming oxides of sulfur and nitrogen into the atmosphere from the combustion of fossil fuels, and the value of forest production affected by acid rain.

Acid rain occurs in industrialized countries, including much of the eastern USA (Likens and Butler 1981; Brezonik et al. 1980) as well as in areas remote from industrial development (Galloway et al. 1982; Haines et al. 1983). The precipitation at Coweeta is acidic.

This paper summarizes and updates the review of Haines and Waide (1980) that described results of experiments designed to determine potential effects of acid rain on forest processes at Coweeta. Specifically, we describe the potential effects of acid rain on canopy processes, litter leaching, soil leaching, and element uptake by roots. Plant growth and the integration of above and below ground processes are explored in relation to the interacting stresses of acid rain, oxidants, and other environmental variables. For reviews of acid rain studies on a global scale see Drablos and Tollan (1980), Hutchinson and Havas (1980), Teasley (1984), D'Itri (1982), Morrison (1984), Evans (1984), and Linthurst and Altschuller (1984).

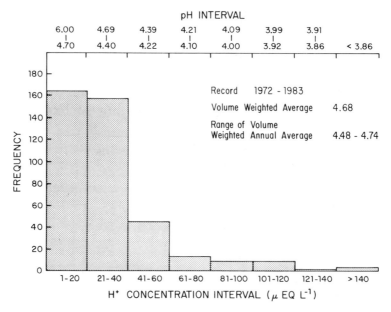

Figure 26.1. Frequency of H$^+$ concentrations for all periodic bulk precipitation collections from 1972–1983.

Acidity in Rain

The frequency distribution of H$^+$ concentrations in bulk precipitation collections is shown in Figure 26.1. Over a 12-year period only 18 observations had pH values lower than 4.0. The highest pH value for an individual collection during the entire period was 6.4 and the lowest was 3.1. The long term volume weighted pH of bulk precipitation is 4.6, with a range in annual values of 4.5 to 4.7.

As precipitation enters and moves through a forest ecosystem, water chemistry undergoes substantial changes (Swank and Swank 1984). The average annual H$^+$ concentration in major hydrologic compartments of a mixed hardwood forest at Coweeta shows H$^+$ depletion as water passes through successive ecosystem components (Table 26.1). The forest canopy and uppermost 25 cm of soil are major sites of H$^+$ depletion. By the time precipitation reaches the stream, H$^+$ concentrations are reduced from 17 μeq L^{-1} to 0.2 μeq L^{-1}, which is equivalent to two pH units (4.8 to 6.7). These in situ data illustrate the potential buffering effects of the ecosystem to acid rain.

Forest Canopy

Potential responses of plant canopies were investigated in separate experiments utilizing various combinations of early secondary successional herb species, mature hardwood forest species, and plantation pines occurring at Coweeta (Table 26.2).

Table 26.1. Average Annual H Ion Concentrations in Hydrologic Compartments for a Mixed Hardwood Forest at Coweeta (WS 7, Pretreatment)

Compartment	H+ Concentration (μeq L−1)
Precipitation	17.1
Throughfall	5.1
Litter water	4.6
Soil water (25 cm)	1.0
Stream water	0.2

From Swank and Swank (1984).

Rates of foliar leaching were quantified for species subjected to simulated acid rains of pH 5.5, 4.5, 3.5, and 2.5. Simulated acid rains consisted of deionized water plus a salt and an acid component. The salt component contained the following amounts of elements in mg/L: Ca 0.23, Na 0.17, K 0.08, Mg 0.05, NH_4-N 0.02, and PO_4-P 0.007. These are similar to the average bulk rainfall element concentrations found at Coweeta. The acid component was made with reagent grade acids to produce the mass ratios of SO_4:NO_3:Cl of 10:7:1 reported by Cogbill and Likens (1974) for New York state acid rain. No attempt was made to simulate organic nitrogen, organic phosphorus,

Table 26.2. List of Plant Species and Experiments Where Plant Processes Were Evaluated Following Application of Simulated Acid Rains

	Leaching[a]	Damage Threshold[b]	Gas Exchange[c]	Growth[c]	Leaf Wettability[d]
Erechtites hieracifolia (L) Raf.	X	X			X
Erigeron canadensis L.	X				
Robinia pseudo- acacia L.	X	X	X	X	X
Quercus prinus L.	X	X˙			X
Carya illinoensis (Wang). K. Koch	X	X			
Liriodendron tulipifera L.	X	X	X	X	X
Acer rubrum L.	X	X			
Cornus florida L.	X	X			
Pinus strobus L.	X	X			
Liquidambar styraciflua L.			X	X	X
Platanus occidentalis L.			X	X	X

[a] Haines, Chapman, and Monk (1985).
[b] Haines, Stefani, and Hendrix (1980).
[c] Neufeld, Jernstedt, and Haines (1985).
[d] Haines, Jernstedt, and Neufeld (1985).

or organic acids found in natural rainwater. Significant differences were found among species, with the herbs *Erigeron* and *Erechtites* showing the highest rates of foliar leaching. No significant differences were found among pH treatments (Haines et al. 1985).

How much below pH 3.5 would leaves need to be acidified to cause damage? Leaves of eight species (Table 26.1) were subjected to droplets of simulated acid rain having pH values of 2.5, 2.0, 1.5, 1.0, and 0.5. Damage, quantified as diameters of necrotic spots at sites of droplet application, did not occur above pH 2.0 (Haines et al. 1980).

The effect of simulated acid rain on gas exchange was examined in four tree species (Table 26.2).Groups of plants received simulated acid rain at pH 5.6, 4.0, 3.0, or 2.0 for 20 minutes per day every third day for a total of 16 exposures. Decreases in net photosynthesis occurred only at pH 2.0; *Liriodendron* showed the smallest reduction, and *Platanus* the most. Photosynthetic rates in *Platanus* decreased after exposure to pH 2.0 mainly due to changes in mesophyll conductance (CO_2 fixation processes) (Neufeld et al. 1985).

Differences in leaf damage susceptibility were related to leaf wettability measured by either water holding capacity or droplet contact angle (Haines et al. 1985). *Platanus*, the most susceptible tree examined had a relatively low contact angle and high water holding capacity, while *Liriodendron*, the least susceptible species, had a high contact angle and low water holding capacity. *Liriodendron* leaves were glabrous, with cells of the adaxial epidermis supporting granular surface wax deposit on the cuticle. Leaves of *Platanus* were covered with branched trichomes on both surfaces, and epicuticular waxes were absent from the adaxial surface. Thus, the wetter a leaf could become the greater the damage by pH 2.0 acid rain.

Litter

Leaf litter was removed from 31 cm diameter circles on the forest floor of mature hardwood forest WS 2 and subjected to simulated acid rain of 5.5, 4.5, 3.5, or 2.5. Leaching of NH_4-N, K, Ca, Mg, and PO_4-P increased with increasing rainfall acidity. However, solution NO_3 concentrations were depleted by passing through litter with greatest depletion occurring at the lowest pH (Haines 1981). Nitrate may have been incorporated into microbial biomass.

If the volume weighted annual average rainfall pH were to decrease by an order of magnitude from the present 4.7 to 3.7, the annual rates of element leaching from litter would probably increase. Elements leached during the growing season might be immobilized by microbial populations, adsorbed to exchange sites in soil, taken up by plant roots, or leached out of the system. Elements leached from litter during the winter when tree roots and microbial processes are less active might be lost via deep leaching to streams or ground water.

Soils

Soils obtained from the white pine (*Pinus strobus*) plantation WS 1 and hardwood WS 7 were leached with artificial solution having a salt component and a simulated acid rain component. The salt component was made to mimic the composition of water collected from beneath the litter with zero tension lysimeters (Best and Monk 1975; Best unpublished observations, Haines et al. 1982). When the study was designed, the magnitude of acid rain induced leaching of salts from leaves and from leaf litter were not known. Therefore, salt strengths of $10\times$ and $100\times$ were also formulated so as to bracket a wide range of salt leaching rates. These salt solutions were then acidified to pH values of 5.5, 4.5, and 3.5 with mixtures of $SO_4:NO_3:Cl$ of a 10:7:1 ratio similar to that reported for New York state acid rainfall (Cogbill and Likens 1974). With ionic strengths of $1\times$, $10\times$, and $100\times$ at 3 pH levels, there were nine treatment combinations to which composite soils from each watershed were subjected. Composite soil samples representing a hypothetical core 10 cm^3 in volume were leached with 216 ml of simulated soil solutions for 20 hr on a shaker, centrifuged, and the supernate decanted for analysis. This was repeated 10 times to simulate the potential leaching over a 10 year period (Haines and Waide 1981).

Generally, the patterns of soil element enrichment and loss were the same in both pine and hardwood soil systems. Most elements showed a cumulative net loss at all pH levels with a $1\times$ salt solution. At $10\times$, Ca, Mg, and P showed adsorption to soils instead of loss. At $100\times$, K, Na, Ca, Mg, and P generally showed adsorption to the soil while Al, Mn, and Fe generally showed a net loss from soil into solutions. In this experiment both pH and ionic strength were changed by factors of 10. A change in salt strength by a factor of 10 had a greater effect than a tenfold change in the acid content. Data from this experiment need further analyses and interpretation, particularly with regard to buffering capacity as defined by base saturation, etc.

Roots

Rates of Ca uptake by excised *Liriodendron* roots were determined for artificial soil solutions that had been acidified to pH values of 5.0, 4.0, and 3.0. The excised roots were incubated in a gradient of synthetic soil solutions ranging from 0.2 to 10 times the 2.4 mg Ca/L found in soil solution draining from the litter layer of a hardwood forest at Coweeta. By the use of ^{45}Ca tracer, the relation of the Ca uptake rate to Ca concentration was determined (Figure 26.2). The rates of element uptake by roots at pH 5.0 and at pH 4.0 at a concentration of 5 mg Ca/L (near the Michaelis-Menten half saturation constants for both pH 4.0 and 5.0 treatments), were 0.15 and 0.12 mg Ca/g dry weight root/30 min. In contrast, roots exposed to pH 3.0 solutions at 5 mg Ca/L had uptake rates on the order of 0.01 mg Ca/g dry weight root/30 min or about 10 times less than roots incubated at pH values of 5.0 and 4.0. If soil solutions were acidified to pH 3.0, Ca uptake and possibly the uptake of other elements would be decreased. These results are from a single experiment on a single species and substantially more experimentation is needed before firm conclusions are reached. The effects of pH altered aluminum speciation on element uptake by roots also need to be evaluated.

Growth

The effect of simulated acid rain on growth was examined in the same plants for which gas exchange was examined (Table 26.2). Height growth decreased only at pH 2.0 in *Platanus* and *Robinia*. Height of *Platanus* at pH 2.0 was 73% that subjected to pH 3.0

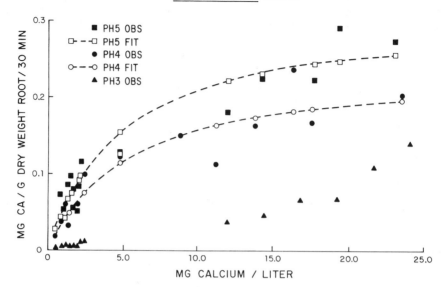

Figure 26.2. Rate of absorption of Ca by *Liriodendron tulipifera* (tulip poplar tree) roots in mg of Ca/gram dry weight of root/30 min as function of concentration of Ca in solution and of solution pH. Dashed lines are least squares best fit to Michaelis model by program of Cleland (1967). Inhibition of Ca uptake at pH 3 precluded fitting data to model.

while height of *Robinia* at pH 2.0 was 50% that subjected to pH 3.0. At pH 2.0, total biomass was significantly decreased in *Platanus* to 68% and in *Liquidambar* to 67% of plants subjected to pH 3.0 simulated rain. Stem and root weights decreased the most (Neufeld et al. 1985) possibly because of decreased carbon uptake due to leaf damage.

Potential Effects of Chronic Acid Deposition

The approximate threshold values of acid rain effects at Coweeta (Table 26.3) are based on short-term experiments. These results suggest that solutions of pH 3.5 will promote element leaching from leaf litter and from soils, but will not affect plant canopy processes. However, these experiments did not address the potential effects of long-term or chronic exposure of the system to rain with pH values of 3.5 and higher. Potentially acid rain sensitive processes which we have not investigated are pollen germination, root–microbe–soil interactions, leaf weathering, and sulfur and aluminum transformations.

Discussion

Plant biomass accumulation and reproductive investment integrate both above and below ground plant–environmental interactions. Essential interactions include (1) the net acquisition of solar radiation at rates avoiding increased respiration or damage to

Table 26.3. Approximate Threshold pH Levels for Effects of Simulated Acid Rain on Terrestrial Process at Coweeta

Compartment	Mineral Element Leaching[a,b]	Threshold for Damage[c]	Mineral Element Uptake[b]	Photosynthesis[d]	Growth[d]
Canopy–Leaves	none at 5.5–2.5	2.0–1.5	–	2.0	–
Litter	2.5	–	–	–	–
Soil	3.5	–	–	–	–
Roots	–	–	3.0	–	–
Whole plant	–	–	–	–	2.0

[a] Haines, Chapman and Monk (1985).
[b] Haines (1981).
[c] Haines, Stefani, and Hendrix (1980).
[d] Neufeld, Jernstedt, and Haines (1985).

leaves by overheating; (2) the net acquisition of CO_2 through stomata to the mesophyll of leaves with minimum loss of water by the reverse pathway; (3) the net acquisition of essential mineral elements while avoiding accumulation of some elements at toxic concentrations; and (4) the net acquisition of water from soil in excess of water loss rates via transpiration. These four acquisition processes may be disrupted by other processes which may be arbitrarily classified as physically and biologically mediated. Some physically mediated processes are variations in rainfall, fire, air temperature, the vapor pressure deficit of the air, and soil water potential. Some biologically mediated processes include plant competition for solar energy, CO_2, essential elements, and water; nutrient availability as controlled by litter decomposition, biological nitrogen fixation, microbial nutrient uptake, and nutrient uptake by plants; loss of plant parts to parasites, pathogens, and to herbivores; and damage from acid rain, ozone, sulfur dioxide, and nitrogen oxides resulting from human modification of biogeochemical cycling.

Acid rain is but one of many potentially interacting factors which can influence the net acquisition of energy, CO_2, minerals, and water. Determining the quantitative contribution of a single mediating factor to increased or decreased plant biomass accumulation or reproductive capacity is extremely difficult for in situ forest ecosystems. Indeed, is quantification of acid rain effects upon tree growth at Coweeta a realistic objective? Experimental work summarized in Table 26.3 suggests that the threshold for damage to some ecosystem processes is between pH 2.0 to 3.0, while acid rain at Coweeta ranges from 3.1 to 6.4 with a volume weighted average of 4.6. If rainfall acidity were to increase from the lowest recorded 3.1 by a factor of four to 2.5, measurable change would probably occur. We also recognize the difficulties in extrapolating laboratory findings to in situ conditions; one must regard short-term, controlled studies as a guideline to expected responses. We cannot unequivocally discount the long-term effects of a moderately acid precipitation regimen on forest productivity at this site. A study is currently in progress to examine trends in tree growth at Coweeta in relation to natural and anthropogenic factors. Such research is appropriate, since long-term stream chemistry trends for control watersheds indicate that hardwood ecosystems at Coweeta may be in the initial phases of response to atmospheric input of airborne chemicals (Chapter 4).

Potential negative effects of gaseous air pollutants are more immediate. The SO_4 and NO_3 in acid rain are generally thought to be derived from SO_x and NO_x liberated to the atmosphere in the burning of fossil fuels. Although the volume weighted rainfall pH at Coweeta of 4.68 is more than 100 times higher than the threshold for leaf damage, we postulate that the ambient concentrations of SO_x, NO_x, and/or O_3 at Coweeta may be high enough to have direct effects on net plant metabolism.

Ambient levels of atmospheric pollutants have been demonstrated, via use of open topped chambers, to decrease yields of crops in England and the USA and to decrease growth of successional plants in the Blue Ridge Mountains of N.W. Virginia by 30 to 50% (see review by Bormann 1982). In open topped chamber experiments, plants received the same solar radiation, rain (or acid rain), insects, drought stress, etc., but some chambers received air which was filtered to remove pollutants. Decreased growth of plants in ambient air compared with growth of plants in filtered air is interpreted as the effect of air pollutants on plant growth.

Extensive browning of current year foliage, particularly on dominant and codominant white pine, was observed on Coweeta WS 17 during the relatively wet summer of 1984 without evidence of insect damage or disease. The threshold for acid rain damage to white pine at Coweeta is about pH 1.5 (Haines et al. 1980) and during the summer period, pH of bulk precipitation was not abnormally low. Examinations of foliage showed evidence of photooxidant damage (C. Berry, personal communication) and at least three major oxidant events were observed between June and September. A long-term data base on biogeochemical cycling and productivity in the plantation provides an outstanding opportunity to document some of the effects of a major oxidant event on ecosystem processes.

A greater resistance of plants to present day levels of acid rain than to levels of photooxidants is not surprising. Plants have evolved with rain in their environment for millenia. They have developed protective cuticles and waxes which minimize both water loss and mineral element leaching. Elevated concentrations of gases such as O_3, SO_2, and NO_x are relatively new to plants in evolutionary time, but these gases diffuse through stomata into the plant leaf by the same pathway as CO_2. Resistance to these toxic gases may have evolved only in floras in regions of volcanic activity.

Our present knowledge about the relative importance of acid rain vs. oxidant impacts on forest productivity at Coweeta is incomplete. Current research is addressing critical questions on this subject. These efforts will continue to be a major component of the long-term research program at Coweeta.

27. Trace Metals in the Atmosphere, Forest Floor, Soil, and Vegetation

H.L. Ragsdale and C.W. Berish

Trace Metals in Natural Environments

Substantial increases in ambient levels of toxic trace metals, such as lead (Pb), are directly attributable to coal combustion, metal smelting, waste disposal, and the twentieth century use of leaded alkyl derivatives in gasoline. For example, about 70% of the annual global Pb emission comes from gasoline combustion (NAS 1983). Most anthropogenically derived trace metals are deposited around point sources (Nriagu 1978; NAS 1983; Andresen et al. 1980). Lead differs from many trace metals in that it is widely distributed through the atmosphere with subsequent long-range transport and deposition of Pb particulates. For example, Schlesinger and Reiners (1974) reported that Pb deposition on the peaks of the White Mountains in New England approached urban deposition levels (Nriagu 1978). Similarly, Jaworowski et al. (1975) and Weiss et al. (1975) reported up to twofold increases in cadmium (Cd) deposition in arctic snow fields over the last 300 years.

Many toxic trace metals, such as vanadium (V), Pb, and Cd, have anthropogenic sources, while other metals, such as boron (B) and magnesium (Mg), are derived from mineral weathering and other natural processes such as volcanic eruption. The industrial revolution with its release of many trace metals has resulted, for example, in 10- to 100-fold increases in Pb burdens for much of the world's living flora and fauna (Elias et al. 1975; Nriagu 1978). Ever-increasing fossil fuel consumption and worldwide industrialization signal a continued increase in surficial metal concentrations as a result of atmospheric deposition.

Impacts on Natural Systems

Concern over toxic metals in the ambient environment, especially in human food chains, is based primarily on the many detrimental impacts metals have on human tissues. For example, the highly toxic metal Cd is the causative agent in itai-itai disease, identified in Japan (Yamagata and Shigamatsu 1970). Lead in urban environments contributes to many health problems, including anemia, kidney damage, mental retardation, and poisoning of children (Needleman 1980).

Beyond detrimental impacts on human tissues, toxic trace metals have been implicated in the "Waldsterben", the forest decline syndrome which affects many German forests (Ulrich et al. 1980). Forest decline has been especially severe in the Black Forest of Germany, and many spruce–fir stands have been decimated (Tomlinson 1983). Public concern for the health of forests in the eastern United States is growing (Vogelman 1982).

In the spruce–fir forests of the northeastern United States there has been a decline in stand basal area and vigor, the primary symptoms of forest decline in Germany (Johnson and Siccama 1983; Johnson et al. 1981). Existing hypotheses attribute forest decline to a variety of causes, including single pollutants; the possible interaction of pollutants such as Pb and ozone; indirect effects which could occur from deposition of excess nitrogen on the forest floor; and the natural stresses of drought and early-late freeze periods which may have predisposed some forests to injury from anthropogenic stress.

The documentation of existing elemental baselines at some relatively pristine locations will be critically important in future evaluations of trace metal impacts on forest dynamics, and possibly in the resolution of the causes of forest decline. Presently in the United States, Long-Term Ecological Research (LTER) sites, such as the Coweeta Hydrologic Laboratory, are excellent locations for such baseline monitoring. In this chapter we present current and historical trace metal data for some major ecosystem compartments for hardwood reference stands at Coweeta.

Study Area

The forest floor, soil, and hickory (*Carya glabra* and *tomentosa*) trees were sampled in Spring 1983, within two Coweeta control watersheds (WS 2 and WS 18). The ambient atmosphere was sampled in a large opening in WS 6, a successional watershed adjacent to WS 18. In addition to the two low-elevation control watersheds, litter (O1) and humus (O2) grab samples were collected from the summit of Albert Mountain (1592 m) at the Coweeta site. Characteristics of WS 2 and WS 18 are given in other chapters in this volume. On both watersheds a square, one hectare reference plot, field marked to 5 m intervals, was previously established based on a design used at the H. J. Andrews Forest. The reference plots were used for sampling, because they facilitate randomization and future relocation of plot sampling sites. Even though the overall vegetation patterns on the two watersheds are very similar, some differences were observed. For example, WS 18 has a higher total tree basal area, diversity, density, and a greater evergreen shrub layer than does WS 2.

Sampling Procedures

In each of the two, one ha reference plots, 16 sites were randomly chosen. Separate litter and humus samples were collected volumetrically and stored in sealed plastic bags. Mineral soil was sampled at the same random locations used for the forest floor collection. After removing the litter and humus, composite soil samples ($N = 3$) were taken with a 5 cm diameter plastic pipe at two depths, 0 to 5 cm (including the fermentation layer) and 25 to 30 cm. Woody root and debris mass were not estimated within the soil depths sampled.

Thirteen hickories (*Carya*) were randomly selected from each 1 ha plot. Hickory was selected because Baes and Ragsdale (1981) found that the Pb concentration patterns in these ring-porous trees are highly reflective of changes in environmental Pb, and because these two species compose approximately 14% of the total tree basal area on the 1 ha reference plots. Sampling standardization included removing two cores at breast height from each randomly selected hickory tree in the dominant or codominant crown classes, and excluding trees smaller than 15 or greater than 32 cm dbh. Field methods stressed elimination of possible metal contaminants (Berish and Ragsdale 1985).

Ambient atmospheric metal concentrations were sampled in a large forest opening in WS 6, adjacent to WS 18. Four vacuum pumps were operated continuously at 2.55 cubic meters/hr flow from March 1983 to February 1984. Spectrographic (Gelman brand) filter paper with low trace element content and low sulfur dioxide retention was used to collect the particulate matter. Filter handling procedures in the laboratory and between the laboratory and the field were designed to eliminate contamination of the filters.

Analytical Procedures

Litter and humus samples were dried to a constant mass at 105 °C, weighed, and ground finely in a stainless steel Wiley mill. Weighed litter, humus, and wood sections were ashed at 400 °C. Extract solutions were analyzed by atomic absorption as described below. Each soil sample was air dried, mixed, oven dried at 70 °C, and passed through a plastic 2 mm sieve. Soil carbon was determined with a Leco carbon autoanalyzer using standard techniques. Total cobalt (Co), copper (Cu), manganese (Mn), zinc (Zn), and Cd were determined by atomic absorption from extracts of soil digested by concentrated nitric-perchloric-hydrochloric acids (5:3:2 ratio) (Jackson 1958; Beaton et al. 1968). Total Pb was determined by atomic absorption on extracts of soils digested in concentrated nitric acid (Sealy et al. 1972; Burrell 1974). Air filters were dried at 60°C, weighed, ashed at 400°C, reweighed, and extracted in 3 M HNO$_3$ according to standard procedures (EPA 1977).

Chemical element analyses were performed on a Perkin-Elmer 306 Atomic Absorption Spectrophotometer. Samples destined for Cd, Co, Cu, and Pb analysis were injected by an automatic pipetter into a HGA-2100 graphite furnace, while Mn and Zn were determined by flame analyses under optimized conditions. Triplicate analyses were performed on each sample. Standards of all elements were compared to National

Bureau Standards reference materials (orchard leaves-SRM1571, tomato leaves-SRM1573) before analyses (NBS 1981). All standard reference concentrations were within 10% of the NBS values.

Standing metal burdens were determined by multiplication of each sample metal concentration by the sample mass. Metal burdens were determined for litter and humus, for soil from the 0 to 5 cm horizon rich in organic material, for a deeper mineral layer between 25 to 30 cm, and for hickory wood from each of the two permanent 1 ha reference plots. Trace metal soil burdens were calculated by multiplying bulk density (0 to 10 cm = 1.24, 20 to 30 cm = 1.30, McGinty 1976) by soil volume and metal concentrations. Atmospheric metal concentrations (mean of four collectors) were calculated as monthly average values; the total metal content of each filter was divided by the volume of filtered air before averaging. Annual average values were calculated as volume-weighted averages. Corrections for filter loading with particles were not necessary due to very low particle mass in the Coweeta atmosphere.

Metal Inputs

Lead and other trace metals of anthropogenic origin are accumulating, possibly at declining rates, in many forest ecosystems (Heinrichs and Mayer 1980; Siccama et al. 1980; Johnson et al. 1982). Metals accumulate because of high meteorologic deposition (Schlesinger and Reiners 1974; Groet 1976; Smith and Siccama 1981; Swanson and Johnson 1980) and because they complex rapidly with soil organic matter (Schnitzer and Skinner 1967). Strong adsorption properties often result in long retention times; in fact, Benninger et al. (1975) estimated a mean retention time of approximately 4000 years for Pb in forested ecosystems at Hubbard Brook, NH.

Wet- and Dry-Fall

The wet-only trace metal input of Cd, Mn, Pb, and Zn was intermittently measured between 1981 and 1982 at the Coweeta Climatological Base Station (Lindberg and Harris 1981; Lindberg and Turner 1983). The wet-input of Pb at Coweeta was 50 g $ha^{-1}yr^{-1}$, while Cd wet-input, 0.77 g ha^{-1} yr^{-1}, was about fiftyfold lower (Table 27.1). Zinc had the highest wet-input, 89 g $ha^{-1}yr^{-1}$, and Mn deposition, 18 g $ha^{-1}yr^{-1}$, was lower than that measured for Pb. At Coweeta, Pb and Cd metal deposition were either lower than or comparable to metal deposition measured at the Walker Branch Watershed, Oak Ridge, Tennessee (Table 27.1), and thus fairly characteristic of a nonindustrialized eastern North American location. Deposition of anthropogenically derived metals in precipitation at Coweeta was, however, substantially higher than found in Turrialba, Costa Rica; a pristine, tropical site assumed to have no industrially derived deposition (Hendry et al. 1984). For example, there was no detection of V, Pb, and Cd (all anthropogenically derived elements) in any of 12 Turrialba samples which were collected over one year and representative of 30% of yearly rainfall. In contrast, the deposition of naturally derived elements in Costa Rican rainfall, such as iron (Fe), B and Mn, was similar to those measured at Coweeta.

Table 27.1. Estimated Annual Input of Selected Metals to Low Elevations of the Coweeta Hydrologic Laboratory

Site[a]	Measured Input	Input (g ha^{-1} yr^{-1})[b]						
		Cd	Pb	Zn	Mn	V	Fe	B
CO	wet	0.77	50	89	17	—	—	—
CO	dry	0.21	18	31	36	—	—	—
CO	wet and dry	0.98	68	120	53	—	—	—
WB	wet	1.0	47	62	17	—	—	—
WB	dry	0.3	26	48	46	—	—	—
WB	wet and dry	1.3	73	110	63	—	—	—
TCR	bulk	bd	bd	16.3	42.5	bd	12.0	7.3

[a]CO, Coweeta Hydrologic Laboratory, Otto, North Carolina after Lindberg and Turner, 1983. Sampled in 1981–82; WB, Walker Branch, Oak Ridge, Tennessee, after Lindberg and Turner, 1983. Sampled in 1981–82; TCR, Turrialba, Costa Rica, after Hendry et al. 1984. Sampled in 1979–80.
[b]bd, below detection; V $<$ 0.5 µg, Pb $<$ 0.8 µg/g and Cd $<$ 0.3 µg/g.

Atmospheric Concentrations

The average monthly particulate concentration varied from 4.4 to 14.7 µg/m^3 from March 1983 through January 1984 (Table 27.2) and exhibited strong seasonal effects. The highest particulate concentrations, about 14 µg/m^3, occurred from June through August. A second period of moderately high particulate concentration, 7 to 9 µg/m^3, occurred in September and October. The lowest particulate concentrations, about 5 µg/m^3, occurred in the winter and early spring. The organic content of the particulate matter ranged from 4 to 14% of the total particulate mass. Organic content was high during the months of the annual leaf cycle of deciduous trees, May through October, and peaked in the summer, June through August, at approximately 13%.

Table 27.2. Monthly Atmospheric Concentrations of Selected Constituents at the Low Elevation WS 6 Sampling Station

Month	Particulates (µg m^{-3})	Organic Matter %	Elemental Concentration (ng m^{-3})	
			Pb	Cd
83 Mar	5.37	5.16	6.97	0.072
Apr	—	4.18	5.50	0.050
May	5.67	10.53	5.48	0.053
Jun	13.41	13.09	10.06	0.131
Jly	14.34	13.82	13.39	0.144
Aug	14.71	13.61	8.97	0.060
Sep	6.77	10.29	7.10	0.057
Oct	8.92	9.70	11.81	0.093
Nov	4.85	7.02	11.40	0.077
Dec	4.37	7.06	9.29	0.067
84 Jan	5.28	6.00	10.67	0.079

During winter and early spring, organic matter declined to 4 to 7% of the total particulate concentration.

Similar to the particulate concentrations, the monthly average concentrations of chemical elements show moderate to strong seasonality, with maxima during the summer to autumn and minima in the late fall to winter months (Figure 27.1). Lead and Cd behaved similarly, with maximum concentrations during June and a secondary peak in October (Table 27.2). These concentration peaks correlate with periods of maximum highway traffic in this region of North Carolina.

Lead and Cd concentrations in urban areas are known to show strong seasonality, with maxima reached in the summer period and minima occurring during the winter (VanHassel et al. 1979). The remote Coweeta Basin at a protected, low elevation site shows an annual pattern of change similar to that observed in urban areas. The only known sources for Pb and Cd in the Coweeta atmosphere are emissions from vehicles, woodburning and coalfired power plants. Long-range transport could explain the presence of these trace elements in the Coweeta atmosphere. The pattern for Zn resembles that for Pb and Cd through July; the post-July Zn pattern differs substantially since its maximum concentrations are observed later in the summer-fall period (August through October). The June through October period of maximum Zn concentrations correlates with higher organic content in particulate matter.

Maximum concentrations of Mg and Ca (not shown) were also correlated with the period of highest organic content. This correlation suggests that the elemental concentration maxima may be caused by organic debris, such as leaf fragments. The supposition is further supported by the immediate decline in these elemental concentrations following autumn leaf drop.

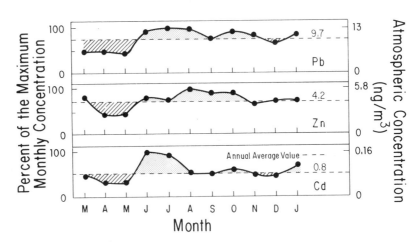

Figure 27.1. Annual change in atmospheric elemental concentrations in the low elevation WS 6 station. Each graph is scaled to 100% of the maximum monthly elemental concentration. Concentration ranges are shown on the right axis. The annual average elemental concentrations are shown as the dashed lines across each graph, with the values written on the dashed line.

Table 27.3. Forest Floor Organic Mass[a] and Elemental Concentrations[b] for Two Control Watersheds

Variable	Sample	WS 2 $\bar{x}$	WS 2 (SE)	WS 18 $\bar{x}$	WS 18 (SE)
Mass	Litter	805.40	(70.58)	921.60	(94.89)
	Humus	339.50	(47.36)	244.30	(49.98)
Soil Carbon*	0–5	4.59	(0.21)	5.48	(0.52)
	25–30	1.10	(0.11)	1.32	(0.20)
Cd	Litter	0.28	(0.01)	0.27	(0.02)
	Humus	0.32	(0.02)	0.32	(0.01)
	0–5	0.05	(<.01)	0.06	(<.01)
	25–30	0.01	(<.01)	0.03	(<.01)
Co	Litter	1.32	(0.15)	1.41	(0.34)
	Humus	1.96	(0.30)	2.31	(0.19)
	0–5	5.09	(0.29)	7.53	(0.61)
	25–30	4.10	(0.41)	5.80	(0.65)
Cu	Litter	5.82	(0.43)	6.41	(0.76)
	Humus	9.00	(0.71)	8.60	(0.92)
	0–5	28.53	(2.23)	34.69	(4.37)
	25–30	35.06	(2.57)	56.41	(7.23)
Mn	Litter*	1130.87	(80.08)	875.18	(77.02)
	Humus*	1694.20	(136.41)	1326.75	(102.21)
	0–5	533.29	(56.39)	495.18	(44.50)
	25–30	521.84	(44.89)	383.06	(41.02)
Ni	Litter	2.78	(0.21)	3.63	(0.91)
	Humus	4.57	(0.76)	5.96	(0.56)
Pb	Litter	13.11	(0.77)	13.15	(0.88)
	Humus	23.86	(1.37)	25.74	(1.56)
	0–5	6.70	(0.44)	9.41	(0.62)
	25–30	5.47	(0.36)	6.83	(0.45)
Zn	Litter	4.69	(0.33)	4.52	(0.28)
	Humus	5.88	(0.41)	5.64	(0.34)
	0–5	69.79	(3.14)	105.42	(5.59)
	25–30	89.52	(7.91)	117.50	(6.87)

[a] (g m^{-2}), n = 16 samples per mean.
[b] (µg g^{-1}), n = 16 samples per mean.
*Comparisons between the watersheds that were significant with alpha equal to, or greater than, 0.05. Significant difference detected between watersheds (t–test, SAS 1979).

The Forest Floor and Soil

Watershed Metal Comparisons

Comparisons between watersheds of forest floor litter and humus showed no statistical differences between watersheds for concentrations of nickel (Ni), Cd, Co, Cu, Pb, and Zn in either of the organic layers (Table 27.3). Manganese concentrations in WS 2 litter and humus layers, however, were approximately 20% higher than those found in the forest floor layers of WS 18. The elements Cd, Co, Cu, Ni, and Zn were present at 0.1 to 10 μg/g; Pb concentrations were between 10 and 30 μg/g; and Mn concentrations ranged from 800 to 1700 μg/g.

Litter and Humus Metal Concentrations

Metal concentrations in humus were significantly ($P = 0.01$) higher than metal concentrations measured in litter. Higher trace metal concentrations in humus relative to litter emphasizes the tendency of trace metals to adsorb onto organic material (Tyler 1972). The only exception to that trend was for Cd, which had similar concentrations in litter and humus. Concentrations in humus exceeded those in litter by 25 to 65% for all other trace metals measured (Table 27.3).

Litter and Humus Metal Burdens

There was no difference between watersheds for metal burdens in litter and humus. Late spring litter mass ranged from about 805 g/m² on WS 2 to 920 g/m² on WS 18; humus varied from 244 to 340 g/m² on WS 18 and WS 2, respectively (Table 27.3). The largest trace element burden in litter and humus was for Pb; total Pb in litter and humus was approximately 18 mg/m² on both watersheds (Table 27.4). Lesser burdens were observed for Cu, Zn, Ni, Co, and Cd, with approximate burdens of 8.5, 5.3, 4.1, 1.6, and 0.2 mg/m², respectively. The Cd burden was less than 2% of the Pb burden. Manganese burdens were approximately 70 times larger than trace metal burdens. Mean forest floor burdens of Cu, Pb, and Zn at Coweeta were about 20 times smaller than comparable measurements for the eastern United States (Andresen et al. 1980).

Soil Metal Concentrations

Analyses of 0 to 5 cm soil samples indicated that soil carbon and concentrations of total Co, Pb, and Zn were higher in WS 18 than in WS 2 (Table 27.3). Similarly, analyses of the deeper (25 to 30 cm) soil samples showed that concentrations of Co, Cu, and Zn were greater in WS 18 than in WS 2 (Table 27.3).

Lead, Cd, and Co concentrations were higher in the 0 to 5 cm soil horizon than in the deeper mineral soil. Enrichment of the upper few centimeters of soil with trace metals indicate that metal deposition was probably aerial (Burton and John 1977). In contrast, Cu, Mn, and Zn concentrations were not surficially enriched; similar metal

Table 27.4. Forest Floor Metal Burdens[a] (g m⁻²)

	Cd	Co	Pb	Cu	Zn	Mn
Litter	0.0002	0.0011	0.011	0.006	0.004	0.87
Humus	0.0001	0.0006	0.007	0.003	0.002	0.48
0–5 cm	0.0100	0.3900	0.500	2.000	5.500	32.00
5–30 cm	0.0100	1.6100	2.000	18.500	33.000	146.00
Total	0.0203	2.0017	2.518	20.509	38.506	179.35

[a]Calculated by multiplying total elemental concentrations by soil bulk density values of McGinty (1976). Concentrations of the 25–30 cm soil horizon were used to conservatively estimate the 5–30 cm elemental burdens.

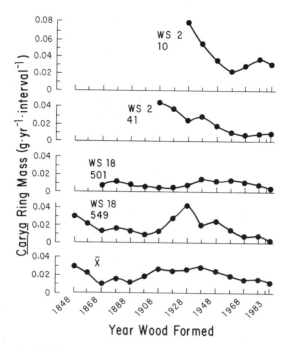

Figure 27.2. Chronological change in *Carya* spp. annual ring mass. The annual mean ring mass is shown on the lower graph. The four major patterns of annual ring mass change are shown in the remaining graphs, as illustrated by trees 10 and 41 on WS 2 and 501 and 549 on WS 18.

concentrations in the A and B horizons indicate that present metal concentrations are primarily the result of bedrock weathering and soil forming processes.

Soil Metal Burdens

Manganese had the highest metal burden at each depth (Table 27.4). Trace metal burdens of Co and Pb were about 1000 times smaller than total Mn burdens. Surficial and deeper mineral soil burdens followed the trend of $Mn \gg Zn > Cu \gg Pb > Co \gg Cd$ for both watersheds. Berish and Ragsdale (1986) compared trace metal pools at Coweeta to various sites and concluded that, to date, Coweeta has slightly elevated trace metal concentrations, and is relatively pristine with respect to trace metal deposition.

Forest Tree Stems

Patterns of Hickory Tree Annual Growth

At Coweeta, as in much of the eastern United States, contemporary forests have developed following agricultural abandonment of land, or after similar major disturbance. Rapid tree growth rates characteristically follow disturbance-release cycles. Gradually increasing *Carya* growth rates at Coweeta (post-1883) were probably associated with the cessation of logging and disturbance on the control watersheds (Figure 27.2). Sustained, rapid hickory growth was facilitated by the demise of the American chestnut

(*Castanea dentata*). For example, Nelson (1955) reported that 11 taxa, including both hickory and oak species, developed rapidly and increased in importance at Coweeta with the loss of chestnut.

Fifty years (1893 to 1943) of rapid and sustained hickory growth rates were followed by steadily declining growth rates, except for some of the younger or crown-released individuals (Figure 27.2). The majority of the dominant or codominant hickory trees at Coweeta are probably in a mature phase of their life cycle, where the amount of wood produced per year becomes relatively uniform. As a bole circumference increases under the conditions of approximately equal annual wood volume production, the successive annual radial increments become smaller.

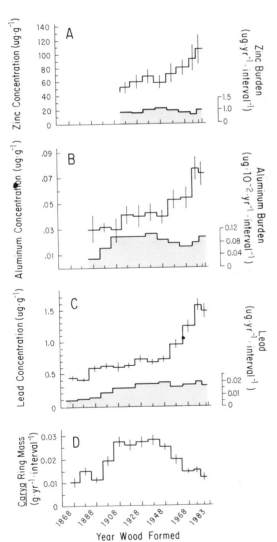

Figure 27.3. Comparison of change in metal concentrations and burdens [Zn (A), Al (B), Pb (C)] for annual woody tissue of *Carya* spp. Graph D is mean annual ring mass with standard error (Berish and Ragsdale 1985).

Table 27.5. Elemental Concentrations (μg g^{-1}) for *Carya* spp. Bole Tissue in Coweeta Watersheds 2 and 18

	Cu	Cd	Zn	Ni	Mn	Mg	Pb	Al
Wood[a]								
Mean	0.98	0.044	67.0	4.99	101.7	815.3	1.01	0.060
SD	0.17	0.007	17.4	0.85	20.3	111.1	0.12	0.008
n	8	8	8	8	8	8	8	8
Bark[b]								
Mean	2.72	0.515	296.2	3.00	382.1	1433.4	0.87	0.107
SEM	0.34	0.053	42.5	0.47	79.7	274.3	0.13	0.013
n	8	8	4	5	5	4	25	12

[a] Mean values for wood are based on average concentrations for 10 year intervals from 1908 to 1982.
[b] Mean values for bark are computed for the number of trees in the sample.

Metal Concentrations in Tree Rings

Lead concentrations in *Carya* boles at Coweeta increased both in historical and recent times (Figure 27.3). For example, Pb from 1840 to 1870 averaged about 0.4 ppm and increased by a factor of 1.5 in the 1880s. Lead concentrations remained at approximately 0.6 ppm until the 1920s, when Pb derivatives were added to U.S. gasoline. Since then, Pb concentrations averaged approximately 0.8 ppm until the 1940s. Since the 1940s, Pb concentrations in woody tissues have more than doubled, with consistent increases up to the present concentrations of over 1.5 ppm. As shown below, the amount of lead in annual wood mass, however, has not increased over the last 50 years.

The 30 year pattern of continually increasing Pb concentrations in recent *Carya* wood at Coweeta contrasts with that found for *Carya* wood in urban Atlanta forests. Over the last decade, Pb concentrations and burdens have declined in *Carya* wood of two urban Atlanta oak-hickory forests (Berish and Ragsdale 1984). The urban Atlanta *Carya* spp. studied by Berish and Ragsdale have generally 2 to 13 times more Pb than is found in *Carya* trees at Coweeta. In Atlanta *Carya* wood, the peak Pb concentrations occurred during the early to mid-1970s. The decline in wood Pb concentrations over the last decade is attributed to the greater utilization of nonleaded gasoline.

Aluminum (Al) concentrations in hickory tree rings at Coweeta are an order of magnitude lower than Pb concentrations, but the temporal deposition patterns of Al and Pb are very similar (Figure 27.3). Aluminum concentrations increased sharply in the post-World War II period, corresponding to the time of marked increases in Pb concentrations.

Zinc concentrations in tree rings were two or more orders of magnitude greater than Al and Pb concentrations (Figure 27.3). The temporal pattern of Zn concentrations in hickory tree rings at Coweeta is generally similar to those discussed for Pb and Al. Zinc differs from Al and Pb in that the post-World War II concentration increase is smaller and statistically undocumented. Many elements (e.g., Mn, Mg, Cu, and Cd) had comparatively high concentrations in the bark, a region of active elemental transport (Table 27.5). Concentrations of the majority of those elements (Mn, Mg, and Cd) remained generally constant since the 1800s (Berish and Ragsdale 1985).

Metal Burdens in Tree Rings

Quantification of trace metals in tree rings as burdens is important since it reveals whether the amount of chemical element deposited in an annual ring has changed over time. For example, the temporal patterns of Pb, Zn, and Al burdens show a threefold increase from the late 1880s to 1910 (Figure 27.3). Since the early 1900s, however, these burdens have remained generally constant. Declining annual ring width is not correlated with increased trace metal burdens in annual wood. The uniform metal burdens indicate that relatively constant quantities of trace metals are being concentrated into the narrower annual growth rings of recent years.

Ring-width decline is not correlated with increasing trace metal burdens, but is correlated with increasing trace metal concentration. These two apparently contradictory results may result from a relatively constant annual uptake of trace metal being deposited within a decreasing annual wood mass. The selective cutting on the Coweeta site prior to 1910 initiated regrowth of the forest, which is reflected in the graph of *Carya* ring mass (Figure 27.3). Ring mass increased by a factor of about 3 in 1910 and declined by a factor of about 3 from 1948 to the present. Present *Carya* ring mass is similar to ring mass produced between 1868 and 1888.

At Coweeta the correlation between annual increment decline and concentration increases is interpreted as a direct result of natural tree growth and development in a forest stand. The concentration-decline correlation, therefore, does not necessarily signify increasing trace element toxicity and is not necessarily an index of increasing atmospheric trace element deposition.

Metals at High and Low Elevations

Lead concentrations in Albert Mountain litter and humus samples were significantly higher than those at the lower elevations of WS 2 and WS 18 (Table 27.6). Lead in the Albert Mountain litter was about 50% greater, while humus Pb concentrations were about 130% larger than those at lower elevations. In contrast, Cd, Co, Cu, Mn, Ni, and Zn concentrations in Albert Mountain litter and humus were lower than for those same metals at lower elevations.

Trace metal burdens at higher elevations within the Coweeta basin may slowly increase because of increasing local urbanization and long-range metal transport and deposition, and the tendency of trace metals to adsorb onto organic matter. For example, Pb in surface litter and humus from the highest peak, Albert Mountain, is 50 to 130% greater than corresponding Pb concentrations from lower elevational sites (Ragsdale et al. 1984). The high Pb quantities result from impaction and deposition of airborne Pb particles.

Trace Element Toxicity at Coweeta

The probability that present levels of trace metals in the Coweeta area are directly causing significant vegetative damage is slight. Andresen et al. (1980) reported average forest floor burdens for Pb (1234 mg/m^2), Cu (170 mg/m^2) and Zn (1013 mg/m^2) for

Table 27.6. Forest Floor Organic Mass (g m^{-2}) and Elemental Concentration (μg g^{-1}) at Low and High Elevations[a]

Variable	Sample	WS 2 and WS 18		Albert Mt.	
		$\bar{x}$	SEM	$\bar{x}$	SEM
Mass	Litter	863.50	(50.10)	−	−
	Humus	291.90	(34.91)	−	−
Cd	Litter	0.27	(0.01)	0.28	(0.03)
	Humus*	0.32	(0.01)	0.39	(0.02)
Co	Litter*	1.36	(0.12)	0.42	(0.10)
	Humus*	2.14	(0.23)	0.59	(0.20)
Cu	Litter*	6.11	(0.43)	3.89	(0.28)
	Humus*	8.80	(0.57)	4.84	(0.73)
Mn	Litter	1003.02	(59.28)	1029.24	(97.45)
	Humus*	1510.48	(90.10)	877.25	(141.42)
Ni	Litter*	3.28	(0.30)	1.39	(0.12)
	Humus*	5.28	(0.60)	2.06	(0.39)
Pb	Litter*	13.13	(0.58)	19.74	(1.12)
	Humus*	24.83	(1.04)	57.24	(4.69)
Zn	Litter*	4.60	(0.21)	3.96	(0.62)
	Humus*	5.76	(0.26)	4.26	(0.33)

[a] Means ($\bar{x}$) and standard errors (SEM) for WS 2 and WS 18 are combined totals ($n = 32$) and were compared with grab samples ($n = 5$) from Albert Mountain.
*Comparisons between the watersheds and Albert Mt. that were significant with α equal to, or greater than, 0.05. (t–test, SAS 1979).

over 50 sites in the northeastern United States. Metal burdens in the forest floor at Coweeta were lower than those in the northeastern U.S. by factors of 68 for Cu and 168 for Zn. Moreover, those same trace metal concentrations in hickory wood at Coweeta were generally far below those found in woody twig tissue of urban deciduous trees by Smith (1973). Smith found that lead in the urban woody tissue averaged about 100 ppm, a factor of 100 greater than found in hickory wood at Coweeta. Cadmium in urban twig tissue averaged around 1.4 ppm, about 35 times higher than at Coweeta. Copper in the urban woody twigs (9 ppm) was nine times higher than that found at Coweeta.

The question of "what is a natural (normal) radial growth rate for mature forest stands" is paramount for the interpretation of the effects of pollutants on tree growth. For example, declining growth rates for other forests have been reported in the northeast (Stone and Skelly 1974, Phillips et al. 1977, Johnson et al. 1981, Tomlinson 1983, Johnson and Siccama 1983), the southeast (Baes and McLaughlin 1984), and some spruce-fir-beech forests in West Germany (Ulrich et al. 1980). At some of those same locations of forest growth decline in Tennessee, for example (Baes and McLaughlin 1984) and in Vermont (Vogelmann 1982), increases in Al wood concentrations have been reported. The generalized trend of increasing concentrations of trace metals such as Pb and Al in tree rings, if interpreted without a consideration of metal burdens in the tree rings, can be misleading, since concentration increases can result from decreasing annual ring width or mass measurements. The most likely interpretation of *Carya* growth at Coweeta is that of a mature, healthy hardwood forest. We do not rule out the possibility, however, that phytotoxic trace metals, such as Pb and Al, may be predisposing trees to further injury or disease, and may be reducing tree growth at a level we presently cannot perceive.

Summary

Trace metals can have detrimental effects in the ambient environment at low levels, while high levels of trace metals can be lethal. The absence of a trace element data base at Coweeta coupled with the realization that trace elements may be exceedingly important in the growth and health of ecological systems were the primary reason for initiating this study.

The soil column (O1 + O2, A, and upper B horizon) of two, low-elevation control watersheds in the Coweeta Basin contains lower concentrations of trace elements (Cu, Zn, Pb, and Cd) than commonly reported for many other North American sites. The litter-humus forest floor burdens of Cu, Pb, and Zn were 20 times smaller than found in forests of the industrialized northeastern United States. The largest trace element burden in the forest floor was for Zn. The low standing stocks of trace metals in low-elevation forest floors at Coweeta indicate that present atmospheric inputs of metals are low.

Lead concentrations in *Carya* boles at low elevations have increased both in historical and recent times. Historical increases in Pb may have been related to regionwide disturbance from metal smelting. Recent increases in trace metal concentrations are the result of smaller annual growth rings. Since the early 1900s, trace metal burdens in hickory wood has remained constant.

Lead concentrations in the high elevation Albert Mountain litter and humus samples were significantly greater than those at the lower elevations of WS 2 and WS 18. The greater lead concentrations at high elevation result from long range transport and deposition of airborne lead particles. Similar trace metal deposition patterns have been found for high-elevation sites in the northeast.

8. Perspectives on Forest Hydrology and Long-Term Ecological Research

28. Forest Ecosystem Stability: Revision of the Resistance–Resilience Model in Relation to Observable Macroscopic Properties of Ecosystems

J.B. Waide

In contrast to previous chapters in this volume, this chapter is conceptual and theoretical. The resistance-resilience model of ecosystem relative stability or response to disturbance (Webster et al. 1975) has provided part of the conceptual foundation for ecosystem research at Coweeta (Monk et al. 1977). It has also provided a point of departure for analyzing responses of stream ecosystems to experimental disturbances (Webster et al. 1983; Chapter 21, this volume), and long-term forest responses to intensive management (Waide and Swank 1976; Swank and Waide 1980). However, both methodological and conceptual criticisms of this model have been published. Also, recent advances in ecosystem science alter the theoretical basis of the model. In response, I present here a revised theoretical interpretation of the resistance-resilience model.

Interest in ecological stability is partly a consequence of earlier natural history approaches to ecosystem analysis and management. The complexity-stability hypothesis (Pimm 1984) provided more recent stimulus for research on this topic. But, in spite of extensive recent research, theories of ecosystem stability remain confused and unsatisfactory. Pimm (1984) suggested several reasons, particularly that investigators have employed inconsistent measures or meanings of basic concepts under study. However, Pimm's review excluded from consideration those macroscopic variables that should be the primary focus of ecosystem stability analyses.

Several major factors contribute to current confusions over ecosystem stability. First, many authors have analyzed and compared ecosystems isolated from the environmental constraints to which they are closely coupled. Such studies – typified by

numerous temperate-tropical comparisons which motivated early interest in stability—
have generated inappropriate conclusions concerning ecosystem dynamics and properties. One of the major points I wish to make is that, in the context of the physicochemical environment to which they are coupled, natural ecosystems are (conditionally)
stable systems. Hence, the topic of ecosystem stability has received improper or undue
emphasis.

Egerton's (1973) insightful review of the "balance-of-nature concept" and its
influence on ecological theory implicitly identified a second confusion in ecosystem
stability analyses. As a direct consequence of this "oldest ecological theory," ecologists
have tended to view the natural world in relation to the human spatiotemporal perspective and life span (see also Allen and Starr 1982; O'Neill et al. 1986). Thus, planktonic
ecosystems which turn over rapidly are typically considered unstable, whereas forest
ecosystems which exhibit less rapid dynamics are viewed as stable. If detailed studies
of natural ecosystems over the past several decades have taught us one major lesson, it
is that ecosystems are scale-dependent systems (see also Chapter 30), and that perceptions of change referenced to the human life span provide an inappropriate basis for
judging ecosystem dynamics and stability. Ecosystem analyses referenced to human
perceptions lead to the improper equating of stability with constancy, and to the failure
to appreciate the significance of ecosystem dynamics in space and time.

Third, theoretical research on the complexity-stability hypothesis has complicated
the development of macroscopic theories of ecosystem stability. This research has
fostered the view that ecosystems can be randomly assembled from constituent populations and resultant stability properties quantified. Such approaches ignore the fundamental structuring imposed on the dynamics and properties of biological populations
by ecosystem constraints of matter and energy processing. Also, legitimate concerns in
applied ecology, such as the behavior of ecological systems simplified by man or the
conservation of genetic diversity, have become intertwined with the complexity-
stability hypothesis (Goodman 1975). This has created problems for analyzing
ecosystem persistence in spatially and temporally variable environments, and for
examining ecosystem responses to anthropogenic disturbance.

Fourth, ecologists have failed to elucidate general relationships between ecosystem
properties and characteristics of constituent species populations. The main points are
that ecosystem properties and dynamics cannot be predicted from those of constituent
species populations, and that ecologists have devised no generally acceptable means of
decomposing ecosystems into subsystems with functional properties which commute
with measures of intact system dynamics (Schindler et al. 1980; O'Neill and Waide
1981; O'Neill et al. 1986). Thus, ecosystem stability is not to be judged solely in relation to species persistence in space or time or to constancy of species composition.

Finally, as in many areas of ecosystem science, discussions of stability are clouded by
the lingering shadow of arguments over older, deterministic theories of succession and
climax. Thus, many purported criticisms of ecosystem stability theory have actually
been criticisms of classical but inappropriate models of succession and climax, as well
as of the underlying balance-of-nature concept.

The major intent of this chapter is to clarify discussions of ecosystem stability. Based
on the theoretical conceptualization of ecosystems as hierarchical biogeochemical
systems, I relate observable macroscopic properties of natural ecosystems to their

stability. Moreover, I reevaluate the relative stability model of Webster et al. (1975) in relation to this theoretical perspective, and suggest that resistance and resilience represent scaling variables which reflect the space-time coupling of entire biotic assemblages (i.e., ecosystems) to specific physicochemical environments. The basic thesis underlying my arguments is that macroscopic properties of ecosystems reflect the space-time coupling of constituent biotic populations to environmental variables which regulate the exchange, transformation, and storage of energy and essential elements by ecosystems. These macroscopic properties reflect both the persistence of natural ecosystems in spatially and temporally variable environments, and the extent and rate of ecosystem response to exogenous disturbance. Arguments presented here represent the refinement and extension of discussions in Waide and Jager (1981).

Brief Review of Mathematical Stability Concepts

This paper focuses on the stability of ecosystems rather than differential equations. However, as a point of departure for what follows, this section provides brief review of mathematical stability concepts. These ideas are covered in ecologically relevant detail elsewhere.

Mathematical stability theory focuses on the behavior of a system of interest as described by a set of dynamic equations defined on an appropriate state space. For such a system, a singular point is defined as a point in state space at which rates of change of all state variables become zero. The concept of stability is concerned with the configuration of the state space as defined by the dynamic equations: are there one or several singular points within this space? And, how do solutions to the dynamic equations behave in the neighborhood of each singular point? If the equations admit only one singular point, concern is with the global stability of the system; if they admit two or more singular points, the focus is on the local stability or behavior of the system in a small region of state space around each.

Intuitively, the concept of stability relates to whether a system, with nominal or undisturbed dynamics described by the dynamic equations, remains essentially unaltered following some imposed disturbance, or whether the system has been changed substantially by that disturbance. Disturbance relates abstractly to some temporary change in the functional form of the equations or to a temporary displacement in the state or in values of system parameters. Mathematical stability theory provides formal definition of what it means for a system to be essentially unaltered. In particular, a system is considered weakly stable if its postdisturbance dynamics are confined (geometrically) to a small neighborhood of state space around the singular point, and asymptotically stable if it uniquely converges or returns to the singular point once it is displaced away from it. Of greatest relevance to discussions of ecosystem stability is the concept of conditional stability: a conditionally stable system is one which converges to a particular singular point, not from anywhere in state space, but only from a bounded region of space around the singular point. The singular point to which the system converges is termed an attractor, and the region of state space within which convergence occurs is termed the basin or domain of attraction. Note that these definitions can be generalized, such that convergence and stability may be judged in reference to a fairly complex and perhaps oscillatory nominal trajectory (e.g., limit cycle behavior; trajectory refers to the time course of system behavior, from an initial to a final state over a defined time interval).

Based on these mathematical definitions, I develop the argument in this chapter that ecosystems may be viewed abstractly as conditionally stable attractors in environmentally defined basins of attraction. Holling (1973, and subsequent elaborations) reached the same conclusion in a population-theoretic context, and employed the concept of conditional stability in his definition of ecosystem resilience (which differs meaningfully from the Webster et al. 1975 use of this term). Central to these ideas are the abstract concepts of state space and basin of attraction. The ecological literature contains concrete examples of abstract state space diagrams. Whittaker's (1975, Figure 4.10) diagram of the boundaries of terrestrial biomes in relation to mean annual temperature and precipitation may be viewed as an abstraction of ecosystem organization, at extremely broad scales of space and time, depicted as basins of attraction within an environmentally defined state space. So too may Holdridge's (1967) life-zone classification of vegetation. Similarly, Whittaker's (1956) diagram of the distribution of vegetation in the Great Smoky Mountains in relation to topography and elevation, and a comparable diagram for vegetation distribution within the Coweeta Basin (Figure 10.1), are examples of ecosystem organization abstracted as basins of attraction within environmentally defined state spaces at finer scales of space and time. Each of these diagrams depicts ecosystem organization as a discrete spatial mosaic of basins of attraction, abstracted at a definable scale of space and time from basically continuous patterns in the biosphere.

Applicability of Stability Definitions in Ecosystem Analysis

The extensive literature on ecosystem stability is characterized by vague and inconsistent uses of basic terms (Pimm 1984) and by ecologically inappropriate applications of the mathematical definitions just reviewed. As a consequence, some ecologists have concluded that classical stability concepts have limited ecological utility. In contrast, a more useful view is that these mathematical ideas must be applied in an ecologically meaningful fashion if continuing research is to produce truly general ecological insights rather than numerical results of mathematical interest only. In this spirit, I briefly elaborate several restrictions on the application of mathematical stability definitions in ecosystem analysis. These restrictions represent constraints on the ecosystem attributes to be considered, and on the evidence to be examined in judging the stability of natural ecosystems. Such restrictions highlight the essential fact that stability characteristics of ecosystems differ meaningfully from those of populations, food webs, and communities.

First, ecosystem stability is not equivalent to constancy of function over time. The tendency to equate stability with constancy results from the influence of the balance-of-nature concept on ecological thinking. Constancy may imply dynamic stability (i.e., ability to recover following disturbance), but it may also imply lack of disturbance, the occurrence of which is essential to the testing of stability. Thus, for example, frequent statements in the literature that forests are more stable than grasslands or plankton are without meaningful ecological content. Such statements reflect man's perception of rates of change in the natural world in relation to his own spatiotemporal perspective and life span. They imply nothing about the functional organization of the three ecosystem types, each of which must be judged conditionally stable at a macroscopic level in relation to a particular physicochemical environment.

Second, ecosystem stability is not necessarily or uniquely related to species persistence or to constancy of species composition over space or time. To the extent that species diversity represents a functional redundancy at the level of the ecosystem (Whittaker and Woodwell 1972; O'Neill et al. 1986), macroscopic ecosystem properties are insensitive to some range of variation in species composition. To this same extent, species turnover at various scales of space and time is an important mechanism maintaining ecosystem integrity in response to environmental variation at those same scales. This restriction does not imply, however, that analyses of ecosystem dynamics are unrelated to species characteristics (see also Chapter 30). To borrow Hutchinson's (1965) metaphor, extant species represent the cast of actors in the contemporary ecological theater; at least some of the biotic components of ecosystems must respond following disturbance or the entire ecosystem will collapse. But, in general, species turnover in space and time is one mechanism preserving ecosystem functional integrity. Therefore, ecosystem stability should not be judged solely upon constancy of species composition. The challenge remains to determine experimentally in what circumstances and at what scales species turnover reflects function-preserving redundancy as opposed to directional change to a functionally different ecosystem.

A direct corollary of this second restriction is that the macroscopic stability of ecosystems is unrelated to tests of the complexity-stability hypothesis. Ecosystem trophic complexity may be related to the functional redundancy of component species, as well as to the spectrum of frequency signals impinging on a given ecosystem in a given physicochemical environment. It may also promote the functional constancy of the ecosystem at certain space-time scales (McNaughton 1985). But, in relation to the macroscopic view developed here, no universal relationship exists between ecosystem stability and trophic complexity.

A third important restriction concerns confusion in the use of the term disturbance. Ecosystems are recognized here as being only conditionally stable within certain bounded regions of environmentally defined state space. In this context, it is important to elucidate specific ecological and environmental factors which set the bounds of scale-dependent ecosystem organization viewed abstractly as basins of attraction. Often these bounds are most clearly revealed by persistent, severe disturbances which completely degrade ecosystem structure, which destroy the scale-dependent coupling between an ecosystem and a particular physicochemical environment, or which represent permanent directional changes in that environment. Such structure-degrading disturbances may be both natural and anthropogenic in origin. However, although ecosystem responses to some types of anthropogenic disturbance may mimic natural disturbance responses (Woodwell 1970), responses to many human impacts may be quite different. Therefore, it becomes tenuous to generalize too far concerning ecosystem responses to both anthropogenic and natural disturbance. This point is particularly relevant to the consideration of anthropogenic insults which cause permanent changes in environments beyond the ability of ecosystems to respond, and which do not parallel impacts which are part of the evolutionary history of natural ecosystems (or to which constituent biotic populations are not "preadapted" in some well-defined sense).

Other confusions concerning the nature of disturbances in the context of stability theory also exist. At the space-time scale at which they occur, statistically regular oscillations in climatic variables do not represent an ecosystem disturbance any more than does the occurrence of wildfire in woodland, savanna, or grassland ecosystems subjected to burning on a statistically regular basis. Such oscillations form part of the signal structure of the physicochemical environment and have become incorporated into the scale-dependent structure of the ecosystem through the evolution of life history characteristics and rate processes attuned to these oscillations (see also O'Neill et al. 1986). Thus, at this specific space-time scale of observation, ecosystem dynamics have become coupled to these abiotic oscillations, which should not be viewed as disturbances at that scale. However, such oscillations do represent disturbances at finer scales of space and time, as do abiotic oscillations which do not exhibit statistically regular properties and to which biotic populations have therefore not had time to adapt. McNaughton's (1985) detailed analysis of the coupling of vertebrate grazing and primary production to space-time patterns of rainfall in tropical savannas nicely illustrates this point.

Another confusion common in the literature reflects a misunderstanding of the basic fact that ecosystem stability is concerned with the configuration of an environmentally defined state space, not with the behavior of any specific trajectory (i.e., time course of ecosystem behavior) within that space. For example, consider the ecosystem present at a specific point on the earth's surface (e.g., oak–hickory forest on a mid-elevation, north-facing slope at Coweeta). Following long-term climatic change associated with glaciation, the ecosystem present at that geographic location would change dramatically (e.g., to spruce-dominated boreal forest). Whereas many authors would interpret this as reflecting ecosystem instability (e.g., Davis 1976), it may reflect just the opposite. If the new ecosystem state at this point now lies (in an abstract geometric sense) within the bounded region of state space defined by the climate newly occurring at that location, then this shift reflects conditional stability. In other words, the configuration of the abstract state space has not changed, but its physical realization on the landscape has been altered due to spatiotemporal dynamics in macroclimate. In reference to this example, Davis (1976) identified life history characteristics which have allowed tree species to invade successfully under closed canopies as being important in the long-term response of eastern forests to glaciation. In the context developed here, such adaptations become important mechanisms which preserve ecosystem functional integrity at the spatial scale of the eastern deciduous forest and at the temporal scale of glaciation events. However, at finer space-time scales (e.g., at the scale of the local forest stand), glaciation clearly represents an exogenous disturbance which triggers a long-term dynamic response.

A final, significant restriction on the application of classical stability concepts in ecosystem analysis is implicit in the points above: stability is a scale-dependent property of ecosystems. Definition of the focal ecosystem of interest is not arbitrary, but requires delineation of the space-time scale over which relevant ecosystem dynamics are to be observed, and elucidation of the processes which regulate observable dynamics at that scale. Thus, dynamic behaviors which are stable at one space-time scale of observation are not at another. In exactly the same way, disturbances must be defined in a scale-dependent fashion (see also Gerritsen and Patten 1985). At a relatively fine space-time scale of observation, low-frequency environmental oscillations disturb ecosystem integrity. At a broader space-time scale of focus, these same physicochemical frequency signals have become incorporated into ecosystem structure through similar life history adaptations of constituent species to common environmental oscillations. At this broader space-time scale the ecosystem is stable in respect to these oscillations, which are not disturbances but rather frequency signals to which ecosystem dynamics are coupled. Allen and Wyleto (1983) discussed this point explicitly in relation to the role of fire in structuring tall-grass prairie ecosystems.

Theoretical Context for Macroscopic Approaches to Ecosystem Stability

The above restrictions on ecological stability analyses reflect the need for a comprehensive synthetic theory of the ecosystem. Because they are essential to arguments developed here, key elements of such an emerging theory are briefly summarized in this section (see Schindler et al. 1980; Waide et al. 1980; O'Neill and Waide 1981; Allen et al. 1984; O'Neill et al. 1986 for further details). This theoretical approach rests upon the macroscopic conceptualization of the ecosystem as a hierarchical biogeochemical system.

Based upon the formalism of irreversible or nonequilibrium thermodynamics (Prigogine 1980), the ecosystem may be viewed as a configuration of matter and energy that persists in far-from-equilibrium states by dissipating biologically elaborated free energy gradients, thereby forming organic structures out of inorganic elements mobilized from the physicochemical environment. Thus, the ecosystem may be viewed as a functional biogeochemical system (Figure 28.1): Energy in solar radiation is converted into chemical bond energy by autotrophic photosynthesis. This chemical bond energy represents a free energy gradient which is dissipated in order to build persistent high-energy organic structures out of low-energy inorganic compounds acquired from the surrounding geochemical matrix. Upon death, organic structures are decomposed, with energy being dissipated as heat and elements returned to the geochemical matrix in a state of low chemical potential.

Hence, it is appropriate to view biogeochemical element cycles both as a necessary consequence of energy dissipation at the level of the ecosystem (Morowitz 1978), and as facilitating a certain level of energy dissipation governed by physicochemical constraints on biological element mobilization and recycling (Webster et al. 1975). The ecosystem may thus be conceptualized macroscopically as an open dissipative structure (Blackburn 1973), and as a persistent organic configuration maintained in far-from-equilibrium states by coupled levels of energy dissipation and biogeochemical cycling. This thermodynamic perspective explicitly provides the theoretical basis for relating ecosystem stability to macroscopic structural and functional properties.

Ecological applications of hierarchy theory (Simon 1962) explicitly provide the basis for scale-dependent analyses of ecosystem dynamics (Webster 1979; Schindler et al. 1980; Waide et al. 1980; O'Neill and Waide 1981; Allen and Starr 1982; Allen et al. 1984; O'Neill et al. 1986). Central to hierarchical approaches is a concern for time

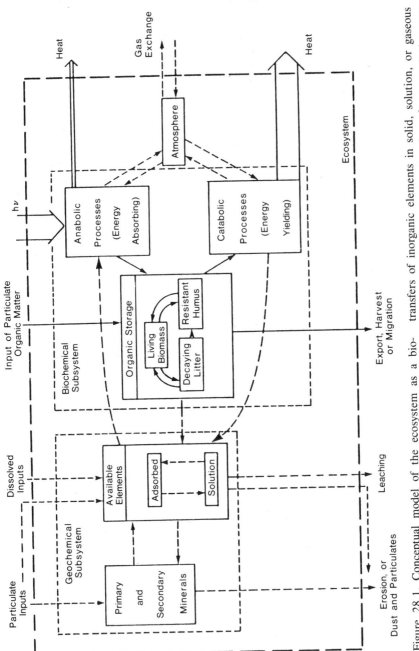

Figure 28.1. Conceptual model of the ecosystem as a bio-geochemical system. Boxes in the diagram represent storage pools of elements within ecosystems, or processes involving the synthesis and degradation of high-energy organic structures out of low-energy inorganic compounds. Double arrows depict the flow of energy through the system, solid single arrows represent transfer of elements bound in organic forms, and dotted arrows symbolize transfers of inorganic elements in solid, solution, or gaseous forms. The biochemical subsystem depends on constant energy input and dissipation, and on continual exchanges of elements with the atmosphere and the geochemical subsystem. The entire ecosystem is thus represented thermodynamically as an open dissipative structure, exchanging both energy and matter with the surrounding biosphere.

constants associated with levels of system organization: definition of system structure is contingent upon the design of sampling strategies or filters which reveal frequencies at which relevant system behaviors occur. Hierarchical systems are structured into a series of vertical levels, with higher levels corresponding to lower frequencies of behavior (i.e., slower time constants) and lower organizational levels to higher frequencies. Further structuring of hierarchical systems is based upon strengths of interaction among components within a given frequency- or rate-defined level: strongly interacting components form subsystems at a given hierarchical level, and are functionally delimited from other groups of strongly interacting components within that same level at points where interaction strengths are reduced in magnitude. In such hierarchical systems, interactions between subsystems at a given level define the dominant time constant at the next higher level, so that subsystems at one level become individual components at the next higher level. Thus, each level is effectively and functionally isolated from other levels, and is both a constraint on lower levels and constrained by higher levels. Each level may thus be viewed as a persistent level of system organization, and as a structure exhibiting dynamics which are essentially independent of detailed, higher frequency behaviors at lower organizational levels.

What is suggested here, and discussed more completely elsewhere (see O'Neill et al. 1986 for a thorough but slightly different treatment of these ideas), is that the structuring of ecosystems as hierarchical or scale-dependent systems is a natural consequence of the evolutionary emergence of these systems as open dissipative structures in spatially and temporally variable environments (the term evolutionary is used in the expanded, hierarchically explicit sense of Gould 1982). Natural physicochemical environments may be characterized by statistically regular oscillations of relevant variables at various scales of space and time (e.g., frequency components in time series of precipitation and temperature, hydrodynamic processes in aquatic environments, statistical recurrence intervals for wildfire or major wind or rain storm). Such oscillations represent frequency signals which become incorporated into the scale-dependent structure of the ecosystem through the evolution of common life history adaptations to similar frequency signals by constituent species populations. Through this process of evolutionary incorporation (Allen and Iltis 1980), ecosystem dynamics become coupled to environmental oscillations at ever wider scales of space and time. As a consequence, high-frequency dynamics reflect ecosystem behaviors at relatively fine scales of space and time, whereas low-frequency dynamics occur over broader space-time scales. Thus, ecological sampling schemes should involve the design of scale-dependent filters or windows which reveal ecosystem organization and dynamics at certain scales of space and time.

The theoretical conceptualization of ecosystems as hierarchical biogeochemical systems provides the rationale for scale-dependent, macroscopic analyses of stability. This approach also provides a basis for resolving conflicting approaches to stability by ecologists working at population and ecosystem levels of organization. That is, the ecosystem must be recognized as a dual hierarchy (O'Neill et al. 1986), both of functional processes and of species populations. The functional or process view of the ecosystem, the central concern in this analysis, focuses attention on macroscopic properties of ecosystems and their coupling to physicochemical environments at definable scales of space and time. In contrast, the species-centered view of the ecosystem is concerned with the dynamics, persistence, and interactions of constituent biotic populations. In relation to the physicochemical environment defined at a specific space-time scale, the processes of energy dissipation and biogeochemical cycling represent functional constraints on the behavior and dynamics of species. But, it is the species themselves which "carry out" these functional processes, and which are responsible for the scale-dependent coupling of the total biotic structure (i.e., the ecosystem) to the physicochemical environment.

This view of the ecosystem as a dual hierarchy resolves otherwise disparate views of stability. Ecosystem ecologists view ecosystems as conditionally stable systems in environmentally defined basins of attraction; population ecologists focus on the persistence of species assemblages and on multiple stable states of ecosystem organization (Sutherland 1974; May 1977; Holling 1981). Such approaches are not inconsistent. Viewed in terms of a hierarchy of processes, macroscopic measures of ecosystem structure and function are exceedingly conservative; i.e., insensitive to considerable variation in species composition. Thus, distinct species assemblages realize comparable levels of structure and function, and a single basin of attraction defined—at a specific scale of space and time—in macroscopic structural and functional terms may encompass several basins defined by the persistence of distinct species assemblages. Much of the purported evidence in support of multiple stable states has been discounted in part because previous studies did not analyze the dynamics of species assemblages over appropriate scales of space and time (Connell and Sousa 1983). However, this does not invalidate the suggested relation between ecosystem- and population-oriented definitions of stability in terms of a dual hierarchy. This approach represents a macroscopic view of functional convergence at the level of the ecosystem. Elucidating relationships between the dynamics and stability of the ecosystem viewed as a hierarchy of functional processes, as opposed to a hierarchy of species assemblages, is a present challenge for ecologists and an area of research currently receiving insufficient attention.

The Ecosystem: A Conditionally Stable System

Previous sections of this chapter provide the theoretical basis for the present macroscopic approach to ecosystem stability. Based upon these ideas, I summarize in this section explicit arguments derived from irreversible thermodynamics, hierarchy theory, and results of experimental ecosystem research to support my contention that ecosystems may be viewed abstractly as conditionally stable attractors within an environmentally defined state space.

Thermodynamic Perspective

From the perspective of nonequilibrium thermodynamics, the biological structures which persist in far-from-equilibrium states are only locally stable (Morowitz 1978; Prigogine 1980). In such dissipative structures, organizational integrity is maintained by a certain level of energy dissipation (in fact, energy dissipation or so-called excess entropy production represents a Liapunov function for dissipative structures which guarantees their local stability; Prigogine 1980). Significantly different levels of energy dissipation (entropy production) result in functionally different organizational states. (These well-defined thermodynamic concepts are extended here to the far-from-equilibrium range of ecosystem dynamics only intuitively, not formally; see Morowitz (1978), Prigogine (1980), and O'Neill et al. (1986) for appropriate restrictions.)

This thermodynamic view of locally stable states is equivalent to the view of the ecosystem as a conditionally stable system. From the thermodynamic perspective, the ecosystem is a persistent structure maintained in nonequilibrium states by coupled levels of energy dissipation and biogeochemical cycling (Figure 28.1). Energy fixation and subsequent dissipation result from metabolic activity by the existing organic structure, and provide for the elaboration of new structure. This metabolic function represents an autocatalytic capacity at the level of the ecosystem, which may result in unbounded and hence unstable growth. It is specifically the coupling of energy dissipation to biogeochemical cycling, and consequent kinetic limitations on biotic mobilization and recycling of essential elements, which places bounds on ecosystem-level autocatalytic growth processes and thus guarantees stability (Webster et al. 1975). These kinetic limitations on element cycling result from the space-time coupling of a specific biotic structure (i.e., ecosystem) to a given physicochemical environment. Factors which break this scale-dependent coupling destroy the balance between the existing organic structure, energy dissipation, and biogeochemical cycling, and cause the ecosystem to collapse or expand to functionally different organization states. Hence, the ecosystem is a conditionally stable attractor within an environmentally defined state space.

Hierarchical Perspective

Implicit in the hierarchical approach to biological systems (Simon 1962) is the concept of stability. This is made explicit in Bronowski's (1972) discussion of stratified stability. Each level in the hierarchy of systems is viewed as a stable structure independent of structural details and high-frequency behaviors associated with lower organizational levels. Each level is thus viewed as a stable level of system organization and evolution, which provides the template for further evolutionary advance to higher organizational levels. Thus, hierarchical organization and stratified stability are inherent in the evolution (in the sense of Gould 1982) of complex biological systems.

The theoretical view of ecosystems as scale-dependent systems is equivalent to this hierarchical concept of stratified stability (see also O'Neill et al. 1986). Because macroscopic characteristics of ecosystems are coupled to environmental oscillations at certain scales of space and time, the dynamics and functional integrity of the ecosystem are governed by processes operating at that scale, independently of processes operating at other scales. Factors which destroy this space-time coupling of ecosystem structure and function to specific frequency ranges of environmental signals result in a functionally different system. Thus, stratified stability is another way of stating that conditional stability is a scale-dependent attribute of ecosystems.

Evidence from Experimental Ecosystem Research

Results of intensive research on natural ecosystems provide explicit experimental evidence for the macroscopic view of ecosystems as conditionally stable systems. Such evidence is of two main types: research on space-time dynamics of ecosystem structure and function in relation to oscillations in relevant environmental variables, and data on patterns, mechanisms, and rates of ecosystem recovery following experimental manipulation.

The first category includes, among others, studies of McNaughton (1985) on coupled dynamics of vertebrate grazing and grass production in relation to space-time patterns of rainfall in African savanna ecosystems; research by Sprugel and Bormann (1981) on dynamics of productivity and biogeochemistry in northeastern wave-regenerated balsam-fir forests; and considerable research on dynamics of coniferous forests in response to space-time patterns in the occurrence of wildfire and defoliation (Holling 1981). In each case, the persistence and functional integrity of the ecosystem is maintained by the coupling of biotic processes to the space-time pattern of oscillation in specific environmental variables. If the space-time pattern of triggering environmental oscillations is altered, as in the case of wildfire suppression, the ecosystem undergoes change to a functionally new organizational state, revealing the conditional stability of the previous state. Such experimental evidence forms the basis for dynamic biological oceanography and limnology (Legendre and Demers 1984), which views the scale-dependent organization of aquatic ecosystems as being coupled to hydrodynamic processes across scales of space and time.

The second category of evidence includes macroscopic studies of ecosystem response to experimental manipulation, such as forest clearcutting experiments at Hubbard Brook (Bormann and Likens 1979) and Coweeta (Swank and Crossley, Chapter 1). At both sites, the state disruption of forest removal has not destroyed the fundamental coupling of the forest ecosystem to the physicochemical environment, and a variety of biological processes have been responsible for rapid recovery as measured by macroscopic structural and functional variables (Swank and Waide 1980).

In the case of Hubbard Brook, results of watershed manipulation experiments, forest simulation models, and studies on surrounding forested stands were formalized as the "shifting mosaic steady state model" (Bormann and Likens 1979). This model accounts both for the persistence of northern hardwood forests in relation to regional environmental oscillations and for forest recovery from experimental disturbance. This model focuses on persistence and recovery in macroscopic structural and functional terms, and views the northern hardwood forest as a mosaic of developmental phases which, at a certain space-time scale, achieves a conditionally stable steady state. This model also identifies a series of phases which together comprise the dynamic recovery process.

In the case of Coweeta, numerous watershed-scale clearcutting experiments have been conducted (Table 1.2). From these experiments, data on a variety of structural and functional measures of ecosystem state are available for assessing rate and extent of ecosystem recovery (e.g., streamflow, stream nutrient concentrations, nutrient export, leaf area and biomass, tree density and basal area). Analyses of postdisturbance dynamics based on this extensive data base reveal common patterns and rates of ecosystem recovery as measured by these structural and functional variables (Table 28.1). These results provide strong evidence for the functional recovery of southern Appalachian forests following clearcutting, for the essentially conservative nature of macroscopic ecosystem properties, and for the high predictability of ecosystem response to experimental disturbance when the fundamental ecosystem-environment

Table 28.1. Estimated Turnover Rates and Time Constants for Various Macroscopic Measures of Forest Ecosystem Recovery from Clearcutting at Coweeta[a]

Ecosystem State Measure	Watershed	Turnover Rate[b] (yr^{-1})	Time Constant[b] (yr)
Streamflow	7	0.17	6
	13 (1st cut)	0.08	12
	13 (2nd cut)	0.11	9
	28	0.17	6
	37	0.14	7
Solute export	7		
NO_3-N		0.20	5
K		0.17	6
Ca		0.17	6
Stream [NO_3-N]	Several[c]	0.14	7
Leaf area index	13	0.12	8
Leaf biomass	13	0.09	11
Stand density	13,28	0.06	16
Basal area	13,28	0.05	20

[a] Data derived from a variety of watershed clearcutting experiments performed at Coweeta. Refer to Table 1.2 for description of the relevant experiments.
[b] Time constant defined as the time required for 0.632 of the total response to occur; the turnover rate is the inverse of the time constant.
[c] Based on stream chemistry from WS 3, 7, 10, 13, 19, 22, 28, 37, 40, and 41.

coupling has not been disrupted. Other chapters in this volume—particularly those by Boring et al. (Chapter 12), Waide et al. (Chapter 16), Webster et al. (Chapter 21), Swank et al. (Chapter 22), and Swank (Chapter 25)—discuss biotic processes responsible for the conditional stability and resilience of forest and stream ecosystems at Coweeta.

Revisions to the Resistance-Resilience Model of Ecosystem Relative Stability

Arguments similar to those presented in previous sections of this chapter provided justification for the original resistance-resilience model of ecosystem relative stability (Webster et al. 1975). However, the theoretical basis for these earlier arguments was not fully developed, and the model itself was based on a somewhat inappropriate mathematical formalism. This has generated conceptual (Botkin 1980) and methodological (Harwell et al. 1977) criticisms of the resistance-resilience model as well as confusions in model interpretation and application. To alleviate these problems, I first reexamine the original model, and then reinterpret the concepts of resistance and resilience in relation to the present theoretical developments.

Original Resistance-Resilience Model of Ecosystem Relative Stability

In defining resistance and resilience, Webster et al. (1975) restricted their attention to the analysis of ecosystem response to disturbance within definable basins of attraction.

These authors were concerned with the comparative or relative stability of ecosystems, and with ecosystem properties associated with different levels of relative stability. Resistance and resilience were defined as two components of relative stability in relation to the time course of change in some macroscopic measure of ecosystem state (e.g., biomass, total nitrogen export) following disturbance; resistance was inversely related to the maximum extent of state displacement following disturbance; resilience was directly related to the rate of recovery to the nominal state. Webster et al. (1975) hypothesized that these definitions identified alternative strategies for ecosystem persistence in spatially and temporally variable environments; resistance to displacement results from the maintenance of large organic structures and element standing crops; resilience following displacement is related to the rate of ecosystem metabolism and element recycling. Thus, the observable structural and functional properties of ecosystems were viewed as a balance realized between factors favoring resistance and resilience, and properties of biogeochemical element cycles were central to analyses of the two components of ecosystem relative stability.

Webster et al. (1975) constructed and parameterized a hypothetical model of ecosystem biogeochemistry for eight idealized ecosystem types. Model responses to comparable disturbances were fit to the equation for a second-order damped oscillator. This allowed the concepts of resistance and resilience to be quantified, the eight ecosystem types to be ranked in terms of relative resistance and resilience, and specific ecosystem properties associated with resistance and resilience to be identified. These analyses confirmed the hypothesized inverse relationship between resistance and resilience: certain environments favor ecosystems maintaining large organic storage which turns over slowly, attributes which contribute to ecosystem persistence by enhancing resistance to displacement. In contrast, ecosystems in other environments maintain low organic storage which turns over rapidly, and persist by responding rapidly following displacement—i.e., by being highly resilient.

Revised Interpretation of Resistance and Resilience

The original resistance-resilience model has proved useful for characterizing ecosystem responses to exogenous disturbance. However, the model as originally proposed is not consistent with the theoretical treatment of the ecosystem as a hierarchical biogeochemical system. Moreover, the original ranking of ecosystems in terms of resistance and resilience obscures more basic principles of ecosystem organization, suggesting the need for a revised interpretation.

This revision is motivated by a point made previously: the biota of every ecosystem have evolved in response to statistically regular environmental oscillations of certain types, magnitudes, and periodicities. That is, the physicochemical environment impinging upon a given ecosystem exhibits a certain frequency or signal structure. In response, the ecosystem coupled to this environment exhibits a certain biotic turnover structure—a complement of sizes, longevities, and metabolic or turnover rates among its constituent biota. In other words, the frequency structure of the environment has become incorporated into the biotic turnover structure of the ecosystem through the life history evolution of constituent biota. This is an ecosystem- rather than a species-level property, and is a specific means of ecosystem-environment coupling.

In this context, consider briefly a temperate deciduous forest and a planktonic ecosystem characteristic of the open ocean as idealized end points along a continuum of ecosystem-environment frequency couplings. In the planktonic ecosystem, minimal biotic structure is present and essential elements are not effectively retained within the planktonic assemblage, which is thus constantly losing elements to lower depths via settling of organisms and particulate matter. Sustained productivity depends on hydrodynamic mixing processes for recycling of essential elements. Thus, the physicochemical environment impinging on this ecosystem is characterized by frequent fluctuations in element availability, largely driven by relatively fine-scale hydrodynamic processes. A biotic structure characterized by small, short-lived autotrophs having high turnover rates represents a clear coupling to this environment. Such an ecosystem-environment coupling, reflected in rapid biotic responses to driving hydrodynamic processes and oscillations in element availability, would be termed highly resilient in the context of Webster et al. (1975).

In contrast, forests are characterized by massive structures which retain large pools of elements in living and decaying organic tissues. These elements are released gradually as organic tissues decompose and are rapidly taken up by root-mycorrhizal associations and free-living microbes, so that essential elements are (compared to the plankton) effectively retained and recirculated within the ecosystem. This large biotic structure, the result of a high rate of net production maintained over long generation times in an environment which oscillates at a much lower frequency than in the planktonic ecosystem, effectively moderates its own environment and buffers itself against oscillations in climatic conditions and in moisture and element availability. Due to the inverse relation between mass and metabolic rate inherent in biological organization, element pools in this ecosystem turn over relatively slowly. Following Webster et al. (1975), this ecosystem would be termed highly resistant.

Although it may initially seem appealing to discuss differences between planktonic and forest ecosystems in the context of the original resistance-resilience model, it is inappropriate to do so in that what is involved is a more fundamental difference between ecosystems and associated physicochemical environments. That is, plankton and forest represent different biotic turnover structures, coupled to environments which differ substantially in the level of productivity and organic mass they will support as well as in the frequencies at which they oscillate (i.e., in frequency structures). It is simply not useful to compare these systems in the context of "extent of displacement" and "rate of recovery" following disturbance. These are two functionally different ecosystems undergoing dynamics at different space-time scales, and it is not reasonable to expect them to respond similarly to disturbance.

To clarify this point and to relate revised concepts of resistance and resilience to the theoretical treatment of the ecosystem as a hierarchical biogeochemical system, I define three macroscopic variables which appear sufficient to characterize fundamental properties of ecosystems and their space-time couplings to specific physicochemical environments. Each variable actually has dual character: a property of the physicochemical environment (Figure 28.2a) as realized by an associated property of the ecosystem (Figure 28.2b).

The first variable may be termed the productive capacity of the environment. Each physicochemical environment, based on characteristic climatic, hydrodynamic, and

J.B. Waide

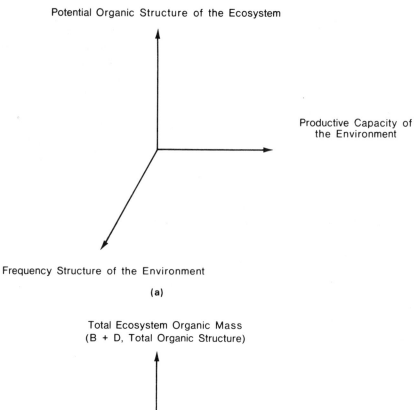

Potential Organic Structure of the Ecosystem

Productive Capacity of
the Environment

Frequency Structure of the Environment

(a)

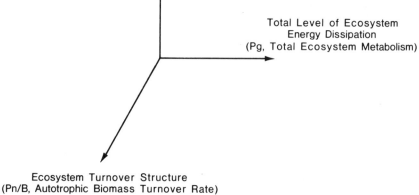

Total Ecosystem Organic Mass
(B + D, Total Organic Structure)

Total Level of Ecosystem
Energy Dissipation
(Pg, Total Ecosystem Metabolism)

Ecosystem Turnover Structure
(Pn/B, Autotrophic Biomass Turnover Rate)

(b)

Figure 28.2. State space diagram for representing ecosystem organization and properties in terms of the three macroscopic variables defined in the text. Each variable has dual character: (a) an abstract property of the environment as realized by (b) an associated property of the ecosystem.

geochemical conditions, will support a certain level of biotic productivity. Because it is impossible to measure this capacity in the absence of an ecological system to realize it, the gross autotrophic productivity of the ecosystem (Pg) provides a useful index. From the thermodynamic perspective elaborated previously, Pg represents the total level of ecosystem metabolism; i.e., the total level of energy fixation/dissipation coupled to a certain level of biogeochemical cycling.

The second macroscopic variable is related to the frequency structure of the environment, as incorporated into and measured by the biotic turnover structure of the ecosystem. Depending on the space–time scale of analysis, various measures of this variable are possible (note that measures of the other two variables must explicitly be referenced to this same scale). For this analysis, the biomass turnover rate Pn/B – the ratio of net autotrophic productivity to autotrophic biomass – is a useful index of this variable, appropriate at the scale of the local ecosystem. That is, this ratio provides a measure of the dominant frequency of oscillation exhibited by a local plankton patch in response to smallscale hydrodynamic controls over biotic growth processes and element availability, or by a local forest stand in response to oscillations (years to decades) in climatic variables or occurrences of wildfire or major windstorm.

The third macroscopic variable is the potential total ecosystem structure which can be supported in a given environment. This is measured by the actual organic matter standing crop maintained by an ecosystem in a given environment, quantified as the sum of autotrophic biomass (B) and total nonliving organic matter (D).

These three macroscopic variables provide the basis for representing ecosystem organization graphically in relation to the theoretical concepts developed herein. Thus, the ecosystem may be represented, abstractly in relation to the three-dimensional state space of Figure 28.2, as a persistent organic structure maintained in far-from-equilibrium states (B + D) by coupled levels of energy dissipation and biogeochemical cycling (Pg) at a certain scale of space and time (Pn/B). Reichle et al. (1975, Figure 2, p. 31) utilized a similar graphical scheme to illustrate ecosystem organization in a different theoretical context.

To apply these concepts, data on the three macroscopic variables have been composited for a variety of terrestrial, wetland, freshwater, and marine ecosystems (Table 28.2). These composites were derived primarily from original summaries in Rodin and Bazilevich (1967), Whittaker and Likens (1973a, 1973b), and Waide and Jager (1981), plus additional supporting sources (Table 28.2). Where direct comparisons are possible, these data appear similar (order-of-magnitude agreement or better) to the more extensive ecosystem inventories of Olson et al. (1983).

The original data summaries which provide the basis for values in Table 28.2 are among the most general and extensive inventories of macroscopic ecosystem properties currently available. These data were composited from a wide variety of original studies. Thus, it is not clear whether the basic data were filtered or averaged over the space-time scale appropriate for the present analysis (e.g., were the forest biomass values averaged across a range of stands of different ages at the appropriate space-time scale?). Conclusions based on this analysis should therefore be accepted with caution. The overall conclusions of this analysis should prove robust, but specific relationships among variables or ecosystems may change as new data become available. It will be

Table 28.2. Summary Data on Macroscopic Structural and Functional Properties of a Variety of Terrestrial, Wetland, Freshwater, and Marine Ecosystem Types[a]

Ecosystem Type	Symbol[b]	Pg	Pn	B	B+D	Pn/B
Tropical rain forest	TRF	9.0	2.7	48.	68.	.056
Tropical seasonal forest	TSF	5.0	2.0	38.	58.	.053
Temperate evergreen forest	TEF	3.25	1.3	35.	62.	.037
Temperate deciduous forest	TDF	2.9	1.2	34.	58.	.035
Boreal forest	BF	2.0	0.8	22.	56.	.035
Swamp forest[c]	SF	2.8	1.1	17.5	35.	.063
Woodland-shrubland	WS	2.0	0.8	5.8	20.	.138
Mangrove swamp	MS	6.0	4.0	6.0	28.	.670
Savanna	Sa	1.5	0.9	4.0	12.	.225
Grassland	Gr	1.2	0.7	1.8	20.	.400
Desert-shrub	DS	0.3	0.2	0.65	12.	.277
Extreme desert	De	0.005	0.003	0.02	0.02	.150
Tundra	Tu	0.26	0.16	1.0	35.	.160
Bog	Bg	0.30	0.18	0.86	34.	.210
Palustrine wetland[d]	PW	2.1	1.3	1.0	1.7	1.30
Salt/tidal marsh	SM	2.5	1.5	3.3	3.5	.454
Lacustrine, littoral	Lt	1.4	1.1	0.46	1.8	2.37
Estuarine, littoral	EL	6.3	3.6	0.75	1.6	4.76
Marine reef/algal bed	RAB	3.8	2.3	1.5	1.6	1.53
Streams, creeks, springs	SC	0.64	0.34	0.097	0.81	3.46
Open estuary	Es	0.58	0.34	0.005	0.12	64.4
Lacustrine, limnetic	Lm	0.33	0.25	0.004	0.073	63.0
Coastal upwelling	Up	2.7	1.6	0.011	0.040	146.
Continental shelf	Sh	0.40	0.24	0.005	0.030	48.0
Open ocean	Oc	0.21	0.12	0.003	0.023	41.7

[a] Relationships among these macroscopic ecosystem properties are graphed in Figure 28.3. The variables Pg, Pn, B, and D, respectively, refer to estimated values of gross and net primary productivity (kg dry matter m^{-2} yr^{-1}), total autotrophic biomass (kg m^{-2}), and total nonliving organic matter (i.e., total detritus, kg m^{-2}). The autotrophic biomass turnover rate, Pn/B, is in units of yr^{-1}. Most of the data shown here were taken or calculated from the earlier summaries in Whittaker and Likens (1973a, 1973b), Rodin and Bazilevich (1967), and Waide and Jager (1981). Supporting data were compiled from Reiners (1973), Wetzel and Rich (1973), Woodwell et al. (1978), Whittaker (1975), and Schlesinger (1977). In general, data on terrestrial ecosystems represent composites of values in the Whittaker–Likens and Rodin–Bazilevich summaries; data on wetland, freshwater, and marine ecosystems were largely composited from the Whittaker–Likens and Waide-Jager summaries. In cases where productivity estimates were available only in terms of Pn, values of Pg were estimated using the following Pn/Pg ratios: for TRF, 0.3; for other ecosystems containing woody vegetation, 0.4; and for all other ecosystems, 0.6. These estimated ratios were taken from sources cited above plus McNaughton and Wolf (1979).
[b] These symbols are used to identify points plotted in Figure 28.3.
[c] Combination of values shown in Waide and Jager (1981) for bog forests and wooded swamps.
[d] Combination of values shown in Waide and Jager (1981) for fens, meadow marshes, and lotic marshes.

some time before emerging theories of scale-dependent ecosystem organization can be rigorously tested against appropriate data sets.

Based upon the values in Table 28.2, relationships among the three macroscopic ecosystem variables are graphed in Figure 28.3. Each "stick and ball" in this figure corresponds to a given ecosystem type, interpreted as a locus of ecosystem organization or as a localized basin of attraction in an environmentally defined state space. Contrasts in organization among ecosystem types are apparent in the full three-dimensional diagram (Figure 28.3a) and are further clarified by projections of points onto each of the three two-dimensional axes (Figures 28.3b–d).

Although the individual points graphed in these diagrams vary along each axis, they also tend to appear as loose clusters separated by nearly order-of-magnitude changes in the value of the appropriate macroscopic variable. This tendency is particularly apparent along the organic structure (B + D) and biomass turnover rate (Pn/B) axes, suggesting nearly order-of-magnitude jumps in ecosystem structure (B + D) due to increases in energy dissipation (Pg) coupled to environments at increased scales of space and time (Pn/B). Each of these loose clusters represents ecosystems grouped according to the physiognomy of dominant autotrophs. Thus, localized at the back right of Figure 28.3a, and characterized by large organic structure and high metabolism but slow turnover, are a series of closed forests (BF, SF, TDF, TEF, TRF, TSF). At successively lower values of ecosystem metabolism (Pg) and organic structure (B + D), but higher rates of biomass turnover (Pn/B), are three additional ecosystem clusters: terrestrial ecosystems dominated by open-canopy woody species, grasses, or other herbaceous species (Bg, DS, Gr, MS, Sa, Tu, WS); shallow or near-shore freshwater, wetland, or marine ecosystems dominated by emergent vascular plants (EL, Lt, PW, RAB, SM); and open-water ecosystems dominated by planktonic autotrophs (Es, Lm, Oc, Sh, Up). Within these clusters other functional relationships are also apparent. For example, within each cluster a consistent positive relationship exists between ecosystem metabolism and biomass turnover rate, suggesting that within a constant physiognomy, accelerated rates of element turnover lead to higher rates of gross productivity and energy dissipation.

Of greatest relevance to discussions here is the strong inverse relationship ($r^2 = 0.94$) between total organic structure and the biomass turnover rate (Figure 28.3d) (excluding the single point for extreme desert, De, which appears as an outlier on all axes). These two variables were shown by Webster et al. (1975) to be strongly correlated with resistance and resilience, respectively. Thus, this diagram represents the postulated inverse relationship between resistance and resilience as originally defined. However, Figure 28.3 and the underlying theoretical arguments show this to involve more general relationships among fundamental, macroscopic properties of ecosystems viewed as hierarchical biogeochemical systems—i.e., Figure 28.3d depicts more than simple differences among ecosystems in terms of resistance to and resilience following disturbance. This figure represents fundamental differences among functionally distinct ecosystems maintaining structure and undergoing dynamics at different scales of space and time. Thus, because mass and metabolic or turnover rate are inversely related at all levels of biological organization (Conrad 1973), resistance and resilience are reinterpreted theoretically as scaling variables which reflect the space-time coupling of the biotic structures we term ecosystems to specific physicochemical environments.

Hence, the terms resistance and resilience should not be applied to analyses of postdisturbance dynamics of functionally distinct ecosystems. Such dynamics are shown here to result from fundamental scale-dependent properties and rate constants inherent in the organization of natural ecosystems. For example, the turnover rate for the recovery of forest basal area following clearcutting at Coweeta (Table 28.1) is effectively equal to the Pn/B ratio shown for temperate deciduous forests (Table 28.2) (a comparable Pn/B value for oak-hickory forests at Coweeta is about 0.06). However, the

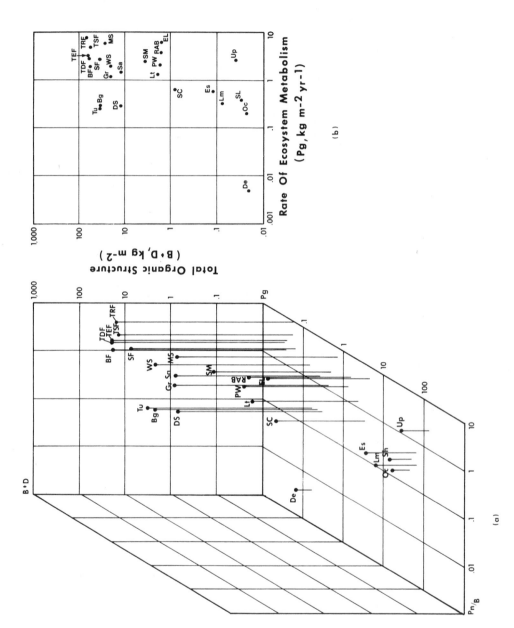

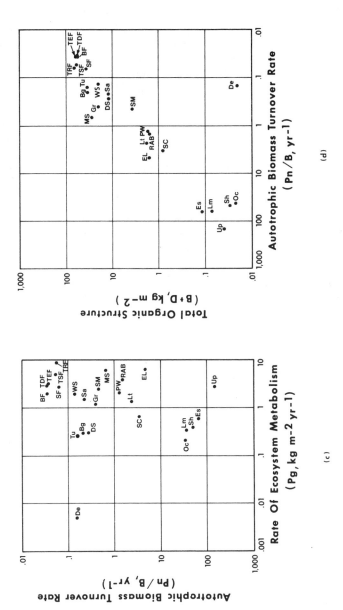

Figure 28.3. Graphical representation of structural and functional properties of natural ecosystems in terms of three macroscopic variables: the total organic structure of the ecosystem (B + D, kg m⁻²), the rate of ecosystem metabolism (Pg, kg m⁻² yr⁻¹), and the autotrophic biomass turnover rate (Pn/B, yr⁻¹). Graphs shown here are based upon data and symbols presented in Table 28.3. In part (a), the full three-dimensional relationship among these variables is graphed. In parts (b)–(d), projections of the individual points onto the three two-dimensional axes are presented.

concepts of resistance and resilience are still quite useful, particularly in a management context, where their application is restricted to postdisturbance dynamics within a given basin of attraction; i.e., when applied to different local realizations of a given ecosystem type which exhibit comparable levels of organic structure and energy dissipation-biogeochemical cycling at a common scale of space and time (e.g., to the recovery of different Appalachian oak forests following management intervention). It is in this revised context that the concepts of resistance and resilience have been applied to the analysis of potential impacts of intensive management on nitrogen cycling in Coweeta forests (Waide and Swank 1976, Swank and Waide 1980). However, even in this context, the definition of resistance requires slight revision. What is important is not the absolute magnitude of the change in state following disturbance, but whether the change is or is not sufficient to push the ecosystem into a new basin of attraction, and hence to assume a new and functionally distinct organizational state. Thus, resistance should be measured in relation to the size of the relevant basin of attraction; e.g., the magnitude of state change or disturbance which can be tolerated and functional integrity preserved, but beyond which recovery is impossible. The definition of resilience remains unchanged: the rate of recovery to the previous macroscopic state, for disturbances which do not push the ecosystem into new basins of attraction. Note that the nominal macroscopic state can be defined abstractly in relation to a singular point, a limit cycle, or other complex trajectory.

Finally, what factors regulate ecosystem resistance and resilience as redefined here? Again, biogeochemical considerations appear critical, and the earlier results of Webster et al. (1975) generally apply. Hence, resistance is enhanced by the presence of large stores of essential elements, partly in abiotic pools but especially in slowly decaying organic forms. Where such pools are absent or are largely depleted by a given disturbance, or where biological populations are prevented from utilizing these reserves (e.g., due to the accumulation of pollutants), disturbances may push the ecosystem into a new basin of attraction. Similarly, resilience is related to the rate at which essential elements are recycled and made available for uptake by regenerating biota following disturbance. The rapid recovery of the mixed hardwood forest on Coweeta WS 7 following clearcutting (Chapters 12, 16, 25) clearly illustrates this relationship. In contrast, where rates of element mobilization and uptake are severely limited by climatic or physicochemical constraints, where available elements are physically immobilized or lost from the ecosystem in solution or gaseous form, or where the disturbance itself inhibits rates of recycling and uptake, then the rate of ecosystem recovery will be reduced.

Summary and Conclusions

From the mosaic of living forms present on the earth, persistent units of structure or pattern are abstracted at certain scales of space and time for study as ecosystems. The ecosystem is thus viewed as a scale-dependent macroscopic unit of study, and as a persistent structure maintained in organized states far from thermal equilibrium by coupled processes of energy dissipation and biogeochemical cycling. The organization and dynamics of the ecosystem depend on functional processes operating at the scale of observation. Factors which disrupt the space-time coupling of the ecosystem to a specific physicochemical environment decouple the processes of energy dissipation and element cycling from the existing structure, and force the ecosystem into a new organizational state. The ecosystem may thus be viewed abstractly (in terms of its dynamic behaviors) as a conditionally stable attractor or (as a unit of ecological organization) as a basin of attraction in an environmentally defined state space.

Viewed in terms of the inverse relation between mass and metabolism inherent in biological organization, the original concepts of resistance and resilience are reinterpreted theoretically as scaling variables which reflect the space-time coupling of specific biotic assemblages to physicochemical environments. These terms should not be applied to the response of distinctly different ecosystems to exogenous disturbance. Such postdisturbance responses of functionally different ecosystems reflect dynamics inherent in scale-dependent ecosystem organization.

However, the concepts of resistance and resilience may be usefully applied to postdisturbance responses of different local realizations of a given ecosystem type exhibiting comparable levels of structure and function at similar scales of space and time. In this restricted context, it is important to determine factors which prevent the ecosystem from assuming a new organizational state following disturbance (ecosystem resistance), and which regulate the rate of recovery to the undisturbed or nominal state (ecosystem resilience). These two components of ecosystem relative stability are related to characteristics of biogeochemical element cycles, particularly the presence of large element storages both in abiotic pools and in living and slowly decaying organic pools (resistance) and the rate at which elements are mobilized and recycled by biota (resilience).

The macroscopic approach adopted here clarifies much of the existing confusion concerning ecosystem stability, and partially resolves conflicting approaches to stability among population and ecosystem ecologists. Implicit in the present theoretical treatment of ecosystems is the concept of conditional stability, suggesting that undue emphasis has been placed on this topic in the past. However, the fact that conditional stability is a direct and necessary consequence of the theoretical treatment of the ecosystem as a hierarchical biogeochemical system is not a trivial result. Rather, this theoretical approach provides the basis for going beyond past confusions and debates to address more useful questions in the future.

Among the most interesting of such questions is the relationship between the predictable and conservative properties of ecosystems viewed macroscopically, and the highly dynamic sorting of species populations in space and time. Although macroscopic requirements of matter and energy processing impose significant constraints on the dynamics and properties of species populations, sufficient behavioral flexibility results from individualistic adaptations to provide substantial functional redundancy in ecosystem composition. Thus, functional integrity at the macroscopic ecosystem level may appear as multiple stable states at the population level. Rigorous evidence supporting this view of ecosystem functional convergence is presently lacking. Determining relationships between the ecosystem viewed as a hierarchy of functional processes vs. as a hierarchy of species assemblages remains an exciting and pressing challenge for empirical and theoretical ecologists.

29. European Experiences in Long-Term Forest Hydrology Research

H.M. Keller

It is difficult to summarize European experiences in long-term forest hydrology research. The difficulties are due not only to the many languages spoken and written in Europe, but also because the term "forest hydrology" has been used in many different ways. In fact, the term is much younger than its science. Today, forest hydrology is a wide field. This chapter is restricted to a few aspects of European forest hydrology. On the occasion of the 50 year anniversary of hydrologic research on small forested watersheds at Coweeta, it seems appropriate to base this report on watershed studies. Also, the discussion will be limited to those studies which have a long-term record, and which show a continuous operation from its beginning to present.

Long-term basin studies do not just happen. The time and the conditions of the economic and political situation, as well as the standing of the scientific community, need to mature until all is right to start such costly operations on a long-term basis. In this report, historic review will be combined with activities and experience in this field of research. However, the picture given will be incomplete and the view and interpretation of the past 100 years in Europe will be given as it is seen by the author. The following discussion is incomplete and probably has some bias towards Swiss activities. It follows more or less the chronology of European developments, and concludes with an overview of current research problems.

Uncertainties About the Hydrologic Role of Forests in the Alps

Flood disasters in the Alps in the 1860s and 1870s caused much damage in addition to loss of lives and property. As in most other countries of central Europe, the local Swiss

mountain communities, sometimes with regional support, had to solve these problems
themselves. The Swiss federal government decided to send experienced specialists
from the freshly founded federal School of Technology in Zurich (technical University)
to the Alps to investigate causes of flood damage and to make proposals. The result was
obvious – large scale deforestations and past intensive use of mountain lands, together
with infrequent, but very high rainfall events, were responsible for the disastrous
floods. Mountain lands needed immediate protection and rehabilitation; hence,
reforestation in the mountains called for support by the federal government. Federal
forest legislation which included these items was introduced in 1876 and amended in
1902. Such opinion-based procedures were criticized and since no research results
were available, Professor Engler at the School of Forestry in Zurich introduced the idea
of comparing two watersheds with respect to flow regime and sediment transport. In
1902, both Emmental watersheds, the Rappengraben and Sperbelgraben, were
equipped with flow measuring stations and precipitation gauges (See Figures 29.1 and
29.2). In 1919, based on these observations, a scientific publication, 626 pages long,
reported on the beneficial effects of forests on the flow regime (Engler 1919). The
problem seemed solved. The Swiss politicians had what they needed, and in many
European countries the idea of protecting forests, particularly in the mountains, was
increasingly supported. However, many scientists remained critical and could not follow all of Engler's arguments about the hydrologic role of the forest (Bates and Henry
1928; Penmann 1959). The Emmental project was continued and subsequent studies

Figure 29.1. Figure 29.1.A recording raingauge as it was used in the
early years of the Emmental
watershed study (around 1915).

Figure 29.2. The Sperbelgraben gauging station during the first year of the study (taken in about 1915).

were published by Burger (Burger 1934, 1945, 1954). Even though many scientific problems seem solved, some debates still go on today; in particular, the question on extreme floods in forested areas (Zeller 1982).

Forest hydrology research activity in Switzerland offers an example which shows that so-called facts are often based on opinions rather than on results of carefully designed research.

The question at the turn of the century was: to what extent could the early deforestations be blamed for the flood disasters in the 1860s and later? A scientist's answer is always based on scientific investigations, with careful consideration of the many processes involved. Such an answer requires time. A politician's answer, however, must be keyed to the need to develop appropriate legislation; therefore, solutions must be relatively simple. Also, political actions must be taken in a relatively short period of time. Oversimplification, however, can result in misinterpretation of scientific conclusions. The discrepancy between the scientific answer and the politically expedient one seems inevitable.

In this example, political decisions (e.g., legislation on forest protection) were made, even though there was no base of scientific findings. The appropriate research activity started after the political decision had been made. This experience is almost a hundred years old. How much have we learned?

Research Opportunities Before and After World War II

Unemployment in the 1930s and after World War II left visible signs in the forested areas of Europe. On the one hand, many reforestation efforts were made to improve forest protection in the mountains using the manpower of the many people left unemployed. On the other hand, during the war and postwar years, large amounts of timber and fuel were needed, and forests were clearcut.

Such activities cannot be planned by researchers. Sometimes, however, they offer opportunities to introduce studies which would otherwise be difficult or impossible to initiate under other conditions.

The Swiss Federal Forest Service and the local state forest organizations had for a number of years favored reforestation to decrease erosion damage, to stabilize the slopes, and to improve and generally protect streamflow regime. Unemployment in the early 1930s resulted in federal employment programs such as tree planting in the mountains.

The small Melera watershed in southern Switzerland was instrumented in 1934, at the same time that Coweeta started its operation. The Melera watershed was chosen as a study site because large reforestation projects had been done in the watershed during previous years. A long-term monitoring program was set up to learn from field observations if and how the flow regime would be affected by reforestation.

It was a pity that the early plan to establish a paired watershed study in the Melera area failed. Melirolo, the neighboring watershed, had been left in pasture for that purpose. However, there was insufficient funding for a paired watershed study. Melirolo was never instrumented.

Melera has been kept operational since 1934, even though scientific publications on this watershed ended in 1955, shortly before the operation was handed over from the Forest Research Institute to the Swiss Federal Hydrological Service. Twice, floods covered the gauging station with debris, but both times researchers were able to convince the adminstration to pay for repairs and to continue funding for continued monitoring.

Also in 1934, the first cooperative forest hydrologic research convention was signed in Switzerland to study the effects of reforestation on streamflow and bed load transport in the Baye de Montreux basin (near Lake Geneva). Three federal agencies and the water supply agencies of the state and local community were involved. Such cooperation was rare at that time. The study went through many critical periods and ended after 42 years, because only a small percentage of the watershed had been reforested and the basin proved to lose too much water through underground seepage. The water balance approach could not be applied, and consequently some of the involved agencies lost interest. Even though the original goal was never reached, much knowledge was gained in measuring precipitation in rugged terrain and at high elevations (Sevruk 1970).

In 1948, shortly after the war in West Germany, the now well-known Harz mountain watershed studies were begun. Towards the end of World War II and in postwar years, the Harz forests were heavily logged. The local forest service (Wagenhoff and Von Wedel 1959) was able to start a multiple watershed study before reforestation was started. The Lange Bramke basin is now fully forested, and the long-term records show very clearly the effect of reforestation on the water balance (Liebscher and Wilke 1981). This project consists of a variety of research efforts. It currently incorporates a study which deals with influence of acid rain, a problem of strong political significance.

The Watershed Approach in Research

The general approach to studying forest influences on the basin scale changed very little from 1902 (Engler 1919) to 1948 (Liebscher and Wilke 1981) and even into the 1950s; one or more small watersheds were instrumented with precipitation and discharge measuring stations. Only rarely were sediment basins installed. Daily, monthly, and annual precipitation and streamflow were then calculated, tabulated, and interpreted.

Comparisons between watersheds were frequently made. The primary factor used for evaluation was percentage forest cover of the respective basins. The link to practical application was obvious, and favored this approach for a long time. It included many statistics, but little knowledge, from process studies. Later simulation modelling offered new interpretations; however, modelling has required a long time to gain acceptance. Measurements at a gauging station appeared as hard facts; printed simulation results appeared as less reliable. For a time, monitoring and simulation modelling were considered coequal alternatives. Today, however, long term monitoring has regained its primary position and support. Simulation modelling is no longer regarded as a quick, cheap substitute, but as a support procedure.

The American Influence

Many European forest hydrologists received some of their training in the United States. American schools teach large scale experiments on paired or multiple watersheds. These programs, which also include development and application of simulation models seem simple, clear, and effective. However, not all countries, and not all political and scientific systems, can produce such programs. In addition, the image of American science has not always been favored in Europe; in some aspects European experience differed from American research results. This created barriers of understanding at times and made it difficult to work with U.S. methods.

It is interesting to note that before the Krofdorf study started in 1971 (Brechtel, Balazs, and Kille 1982), no true paired forest watershed study incorporating large scale experiments had been started in Europe. The studies mentioned above (Engler 1919, Burger 1945, Sevruk 1970) were one, two, or multiple basin studies with the aim of comparing the differences between the basins as they exist and as they develop naturally or through management practices.

The Krofdorf Experiments near Hann. Munden (Hessia, FRG) were started in 1971 with two pairs of watersheds. After 12 years of calibration, the experimental cuts were made as planned in 1983 (Brechtel et al. 1982). In Sweden, near Kloten (Grip 1982), a true field experiment on the basin scale was performed to investigate the effects of forest harvesting and fertilization. These two studies show a clear experimental approach, with much emphasis on managing the forests in the watersheds.

Many other hydrologic research basins were put into operation with the aim of establishing baseline information. Long-term monitoring is needed for the early detection of future possible changes. Such research programs were established in Switzerland for the Rotenbach in 1952 (Naegeli 1959, Keller 1965) and the Alptal basins in 1968 (Keller 1970, Keller and Strobel 1982), the Busna in Italy in 1969 (Susmel 1981, Fattorelli 1982), the Plinlimon (1971) in Wales (Roberts et al. 1983), the Rila Mountain watersheds (1961) in Bulgaria (Raev et al. 1981), the Grosse Ohe (1978) in Bavaria, FRG (Baumgartner 1985), the Homerka basin (1969) in Poland (Froehlich, 1983), l'Avic (1978) in Spain (Escarre 1983), and Mont Lozere (1981) in France (Dupraz et al. 1982). All are still in operation today. These studies are of the traditional European descriptive school, avoiding large scale experiments and putting emphasis on simultaneous observation of the most relevant input and output processes.

The American School is also present in Europe through the application of simulation models. This opened the way in Europe to work with different approaches in long-term forest hydrology research. Consequently, new opportunities for the future have emerged. The fact that the hydrology of forested watersheds can be adequately simulated has served not as an alternative, but as a complementary tool to be combined with watershed field research. The many simple watershed studies and the many long-term data sets available from various regions all over Europe will eventually combine modelling and monitoring. This combination will proceed rapidly as research results are exchanged across political and language boundaries. The International Union of Forestry Research Organizations and other international organizations have already had a remarkable influence on bridging communication gaps.

From Water Balance to Ecosystem Studies

The influence of forest cover on streamflow was and still is a primary focal point of forest hydrology research in Europe. Increasing population size in many countries, greater need to use water for various purposes, the increasing pollution of rivers and lakes, and the need for improving the quality of our environment are all circumstances which in the 1960s created a need to incorporate water quality considerations into long-term forest hydrology projects. Within only a few years, many European programs with similar objectives were started. These programs were intended to study the influence of forests and their management on the quality (dissolved and suspended matter) of streamflow, and of groundwater. The following studies and their starting date should be mentioned:

Switzerland: Alptal (SZ) 1968
Sweden: Kloten 1968
Great Britain: Plinlimon (Wales) 1971
Fed. Rep. Germany: Krofdorf (Hessen) 1971, Grosse Ohe (Bavaria) 1978
Spain: l'AVIC (Barcelona) 1978
France: Mont Lozere (Toulouse) 1981
Italy: Cordon (Veneto) 1984

These few are only examples; there are many more operated by various agencies. Only the future, however, can tell which of these will be long-term study sites and which will be abandoned after only a few years of investigation.

The multitude of watershed information available has resulted in several attempts to summarize findings (e.g. Keller 1983). Why so many studies? There seems to be a need for local/regional information. Extrapolation in the field of nutrient behavior in forest ecosystems is obviously far more complex and much more difficult to extrapolate into areas with no measurements than was the case for water volumes.

This development is too rapid to make conclusions and interpretations at this time. It seems that the European public today is increasingly interested in the preservation and protection of the environment. The forest decline phenomenon and widespread acid rain discussions make it necessary that forest hydrologist rapidly produce information on the quality of forest ecosystems. Politicians cannot wait to obtain answers from scientists. The present situation seems in a way similar to what it was about 100 years ago.

Conclusions

In most European long-term watershed studies, names are often almost synonymous with the project (Engler, Burger, Wagenhoff). This is still valid today. We need scientists who are willing to conduct such studies over long periods, but we also need organizations which support such circumstances actively. Much of the operational experience necessary cannot be transmitted to a new person in one day. Understanding of the forest ecosystem depends on personal experience in the field, which is then combined with the results of monitoring work. Such combined experience has brought much expertise into the interpretation of forest hydrology research data.

In the early days, only a few people were involved in the operations of watershed research. Today, with the increasing interest of ecologists in such studies we find a much larger group involved. The watershed is not only the site of a hydrologic basin study, it is concurrently the location of many process-level studies; this combination brings about further understanding of processes in the basin. Hence, we depend on our ability to cooperate.

Some cooperative projects look interdisciplinary on paper; others list only one institute and one name. What is the actuality? Do publications tell the true story? It is a common experience that, among other things, successful cooperative research depends on:

1. the character of individual persons,
2. the number of persons involved in a project,
3. the continuity of the composition of the group,
4. moral support from administrators to engage in cooperative work.

Therefore, it is important that forest hydrologists make every effort to foster cooperation across language and political barriers. The quality of research will improve and scientists will receive much personal satisfaction.

Many of our forest watershed studies have been established based on the hypothesis that climate and the large scale environment change only very slowly, and that for practical purposes these parameters can be regarded as constant. However, the recent past has shown that these assumptions do not hold true anymore. It seems that we are rapidly losing track of baseline reference points. How can we detect reasons for changes in forest ecosystems if many environmental parameters change simultaneously? I'm convinced that data from long-term forest hydrology research basins, and in particular those with little direct human impact, will become increasingly important in the near future. I hope that the exchange of ideas and information between Europe and the United States maintains its lively pace and grows. The forests and our children will be thankful.

I'd like to conclude with the words of Professor Otto Luetschg, who ended his speech as president of IAHS during the General Assembly of August 11, 1939 in Washington, D. C. as follows: "May the labors of the hydrologists in the different countries not only throw clear light upon the complicated problems of high mountain hydrology, but also favor the bonds of friendship which are so necessary to peace and to better understanding among the nations."

30. Past and Future of Ecosystem Research – Contribution of Dedicated Experimental Sites

Most fields of science seem to profit disproportionately from the central role of one or a few outstanding institutions. This symposium recognizes one such institution, the Coweeta Hydrologic Laboratory. For 50 years, the laboratory has generated scientific information about the forested landscapes of the southern Appalachian Mountains and, particularly, about the relationship between land treatment and the water resource. Its contributions to applied sciences, such as forestry, have been inestimable. Of special relevance to this keynote address is Coweeta's leadership in the development of the young field of ecosystem science.

In many senses, Coweeta Hydrologic Laboratory is the mentor for numerous other ecosystem research sites and programs. Many have benefited from the concepts, materials, and data developed at, and from personnel based or trained at Coweeta. Dr. F. Herbert Bormann (personal communication) relates how the watershed research program at Coweeta inspired the program initiated at Hubbard Brook. As a student at Duke University and then as a young assistant professor at Emory University, Bormann became impressed with the whole ecosystem approach to research at Coweeta. Later, struggling with approaches to biogeochemical research as a professor at Dartmouth, Bormann recalled that, "years of experience with Coweeta quickly made it apparent that small watersheds with monitored hydrologic input and output could be used as vehicles to study whole system biogeochemistry." Subsequently, the Hubbard Brook ecosystem study was born. When watershed research was initiated at newly established H. J. Andrews Experimental Forest in Oregon, many methods and concepts were adopted from Coweeta. I used a Southeastern Forest Experiment Station paper from

Coweeta in analyzing Andrews stream hydrographs during my first summer with the
Forest Service.

In this chapter I will make some personal observations on ecological research during
the last 15 years, with particular reference to the role that has been played by Coweeta
and similar sites. I will consider accomplishments in ecosystem research, factors that
have contributed to these accomplishments, and implications for future progress in
ecological research. These are, of course, my personal views, with all of the implied
biases and experiences.

Accomplishments in Ecosystem Science

Recent years have seen astonishing progress in our understanding of ecosystems and
how they function. There have been numerous discoveries, many of them that were
clearly surprises. In the following sections I review some of my favorite examples.

The magnitude and dynamics of belowground processes in forest ecosystems has
been documented. A rule of thumb assigned approximately 20% of the live forest
biomass to belowground. Not appreciated were the turnover rates in fine roots and
mycorrhizae, nor the fact that highly disproportionate amounts of photosynthate, typically around 50%, must be directed to maintenance of these intensely dynamic belowground tissues. Further research is showing that the belowground energy requirements
are highly variable (from perhaps 10 to 70% of the photosynthate) and that this variation is related to site conditions. In more favorable environments, such as highly fertile
sites, the proportion devoted to belowground maintenance is less. The implications of
this emerging body of information for development and interpretation of silvicultural
practices, such as fertilization, is only now beginning to be developed.

The importance of interactions between forest canopies and the atmosphere has
emerged. Many such interactions have, of course, been appreciated for some time; for
example, the role of the canopy in interception and transpiration processes (Chapter 9,
this volume). Fog drip and cloud moisture have been viewed as significant in some
local ecosystems. But, the complexity and extent of these condensing/precipitating surfaces was not fully appreciated; good measures of the surface areas did not even exist.
A report by Swank and Helvey (1970; see also Chapter 22, this volume) was one of the
first watershed scale experiments illustrating the dynamic relationship between the
quantity of forest foliar surface, evapotranspiration, and streamflow. We know that
there may be an incredible 400,000 m²/ha of foliar surface in coniferous forests of the
Pacific Northwest, and that the canopy can dramatically increase water yields where
fog is common (Harr 1982). It is only recently that we have begun to understand the
importance of plant canopies in cycles of several elements and as a site for deposition
of pollutants. It is certainly no accident that much of the forest decline that we are see-
ing is in cloud forest zones (see, e.g., Lovett, Reiners and Olson 1982). The topic of
canopy-atmosphere interactions is one that we have only begun to understand.

Ecosystem research has enumerated the numerous and complex interactions between
forests and streams as illustrated throughout this volume. Vegetative influences on
stream chemistry, light, and temperature regimes have been quantified and incorporated into predictive models. We now recognize the importance of riparian vegetation

to aquatic organisms in providing energy resources diverse in quality and quantity (Chapter 19) as well as in their delivery in time and space. Our view of woody debris in aquatic ecosystems has reversed from a negative to a positive influence. The importance of wood in creating habitat diversity, contributing to spatial stability, and providing retentiveness are all recent (or, at least, rediscovered) concepts (Chapter 21).

Some of the ecological roles of coarse woody debris, standing dead trees, and down boles have been identified and quantified in terrestrial environments. The extent and importance of these structures simply was not appreciated by most ecologists. This was even true in the Pacific Northwest, where 150 Mt/ha of coarse woody debris is common in forests and where > 500 Mt/ha may occur in an intact stand. If you doubt the degree to which these structures were unappreciated, the International Biological Programme's (IBP) woodlands synthesis volume did not even have a data set category for down dead wood on the forest floor (Reichle 1981). Today we recognize the fact that wood structures are as important dead as alive—important to animals as food and habitat, to geomorphic processes, in carbon and nutrient cycles, and as sites for nitrogen fixation (Harmon et al. 1985).

Advances in our appreciation of the nitrogen dynamics in forest ecosystems have been substantial, particularly of the scale and sources of nitrogen additions. Early forest soil textbooks had little to say about nitrogen additions other than a general suggestion that lightning and freeliving blue green algae were probably primary sources. The role of woody leguminous and nonleguminous plants with nitrogen-fixing microbial associates has been identified during the last two decades. Many of these plants, such as *Alnus*, *Ceanothus*, and *Purshia* in the west and *Robinia* in the east (Chapter 12), appear to have critical roles in maintaining productivity of temperate forest sites because of the high levels of nitrogen fixation. Especially impressive has been the discovery of the numerous routes for nitrogen accretion—fixation in forest canopy lichens, in rotting wood, in decaying leaf litter, and in the root rhizosphere. No doubt many additional pathways of nitrogen additions and losses remain to be identified and quantified.

The ecological importance of nongrowing season photosynthesis has been recognized. This has particular importance in the forests of the Pacific Northwest, where over 50% of the net photosynthesis can occur outside of the growing season and partially explains the prodigious growth rates possible during the cool, short summers. This phenomenon is very important on a variety of habitats, including the extremely productive coastal *Picea sitchensis* and *Sequoia sempervirens* regions and sites, which suffer significant summer drought. Nongrowing season photosynthesis is obviously an important factor in the superiority of evergreen conifers over deciduous hardwoods in the western environments (Waring and Franklin 1979).

The importance of vegetative regrowth in preventing nutrient losses following catastrophic disturbances has been discovered. The Hubbard Brook experiment provided the most graphical demonstration of this phenomenon (Likens et al. 1978), which has been explored and quantified in a variety of other environments including Coweeta (Boring et al., Chapter 12). Indeed, several theoretical constructs have been developed around this phenomenon (Bormann and Likens 1979).

There have been many other ecological discoveries that are illustrated by papers in this symposium. There is the tight linkage between hydrologic and nutrient cycles

(Chapter 4). There have been discoveries about the paths and rates of sediment processing in stream ecosystems including the recognition of the long-lasting effects of delivering such materials to streams (Chapter 21). The denitrification process is emerging as important in some ecosystems. Differences in water usage between hardwoods and conifers have been quantified with results that are surprising in the magnitude of the differences (Chapter 22). The eastern white pine-hardwood comparison has been repeated in the Rocky Mountains with similar results; although responsible processes differ, quaking aspen has less impact on water yield than either Engelmann spruce-subalpine fir or lodgepole pine forests. Unsaturated soil water movement has been shown to be important in maintaining stream flow (Chapter 8). The fact and importance of biological regulation of the sulphur cycle has been demonstrated for forest soils (Swank et al. 1984; Chapter 18). The role of insect populations in regulating nutrient cycles has been demonstrated at an ecosystem scale (Swank et al. 1981).

The list of important discoveries is long and would dramatically expand if we moved beyond considerations of temperate forests to grasslands, deserts, tropical forests, taiga, and tundra. We have progressed in our understanding of ecosystems. This knowledge is having a major impact upon our ability to predict system responses and modifying the way that forest ecosystems are being managed. And, we have had numerous surprises along the way, which has several implications for the science. While surprises have made the research fun, they are also reminders of the need to expand our scientific frameworks, to deal with problems of faulty logic and inadequate observations, and to appreciate the importance of research at discipline boundaries.

Factors Contributing to Advances in Ecosystem Research

What are some factors contributing to these discoveries? Doubtless each of us would have our own list of favorites. Here are some of the factors from mine.

The existence of long-term data bases has contributed significantly to ecosystem research. For example, the data on baseline stream chemistry at Coweeta made it possible for an alert research team to discover the effects of the fall cankerworm defoliation on nitrogen losses from a watershed (Swank et al. 1981). Other long-term stream records indicate that southern Appalachian forests may be in the early stages of acidification (Swank and Waide, Chapter 4). Similarly, the long-term watershed records confirmed and quantified the reduced water yields that followed conversion of hardwood to white pine forest. Note the serendipitous nature of the first two examples and the unexpected magnitude of the third.

All of the discoveries mentioned in the preceding section came from research programs with a holistic point of view—a whole ecosystem perspective—and most were generated by interdisciplinary research teams. Pioneering programs such as Hubbard Brook made a significant contribution, not just of scientific findings, but in helping to make ecosystem science a respectable undertaking. New concepts and methods were contributed by creative scientists, such as Eugene Odum with his textbooks and stimulating treatises; especially "Strategies of Ecosystem Development" (Odum 1969).

The International Biological Programme (IBP) made a special contribution to the advances of the last 15 years. The Biome projects were grandiose in concept and unrealistic in some of their goals, but they were holistic. These projects cracked many black boxes and forced many groups to at least attempt to address whole ecosystems. IBP created a new appreciation of the importance of decomposition processes and organisms. Belowground processes, land and water interactions, and coarse woody debris have surfaced as major topics because of Biome programs. What young assistant professor, on his or her own, would have selected the labor-intensive, methodologically difficult area of root production and turnover to build a career? What National Science Foundation panel would have funded such work? There had to be the impetus and support of a large, holistic research effort to stimulate such topical areas. The IBP set the stage for subsequent generations of more tightly structured ecosystem proposals, experimental and hypothesis-based efforts, which exploit and generate fertile topics.

Few advances in ecosystem research have been the products of isolated scientists or results of laboratory studies. Almost none are the consequence of null hypotheses or outgrowths of ecological theory.

Musings on Ecological Theory

I would like to diverge momentarily with some of my thoughts on ecological theory. Let me say at the outset that I do believe that the development of theoretical constructs has been important in the progress of ecology. They have helped ecologists to move from a pedantic and dominantly descriptive science to one with greater rigor and structure; as Paul Risser (personal communication) has commented, at least the theories have prompted ecologists to reconcile their observations with those made elsewhere.

My criticisms concern primarily the relatively narrow body of work increasingly (and unfortunately) referred to as "community ecology": the world of broken-stick models, r- and k-selection, Lotka-Volterra equations, and competition as the structuring angel of all communities (or not, depending on your point of view). For many years I was in awe of these theoreticians, who so effectively dominated what has been known as ecological theory. Although much of the theory seemed nonsensical and failed when I attempted to apply it to communities of plants or to ecosystems, I assumed the fault lay with my perception of the real world and not with the theory.

In fact this body of ecological constructs has proved of marginal value, a fact increasingly acknowledged by theoreticians themselves. I attended a conference on application of ecological theory to societal problems several years ago, at which an eminent theoretical ecologist acknowledged a reliance on ecological knowledge of organisms and communities, and not on theory, when giving advice on solutions for real world problems. Much current ecological theory addresses special cases and lacks broader application. Other ecological theory addresses biological phenomena which are interesting, but of limited consequence in the behavior of ecological systems. I think that we may be doing our students and ecological science a disservice with an excessive emphasis on such theory.

Indeed, we may have done a disservice to our science with several other of our recent emphases. Use of the null hypothesis in ecology can be useful; it has, for example, encouraged rigor in the design and exposition of research. But, an exclusive reliance on this approach conditions one to think in terms of absolutes instead of probabilities or proportional contributions of several processor factors. The disparagement of natural history research, which is the ultimate source material in ecology, has been unfortunate. Mediocrity is not a necessary condition of natural history research, nor brilliance uniformly represented in theoretical efforts. Species do make a difference, particularly where they fill a unique structural or functional role within an ecosystem. Consider the character of Sierran mixed conifer forest with and without giant sequoia, as one example.

I wish that we would teach ecological students to think more broadly. We have often created a view of the natural world that is strongly deterministic, filled with the best adapted and highly coevolved organisms. We provide a world that is almost devoid of surprises. I view the world as having a lot of randomness, one in which there are many successful strategies, many unfilled niches and, perhaps, organisms more capable of utilizing resources than those currently present. Anyone doubting the last point should review the status of cheatgrass in the shrub-steppe of the Pacific Northwest or the aggressiveness of lodgepole pine above the original timberline in New Zealand.

In my view, the most useful ecological paradigms will have to accommodate the great diversity inherent in the world's ecosystems. Key processes and controlling factors will vary across ecosystems, and this must be recognized in our paradigms. MacMahon (1981) illustrates one approach in his contrasts of successional processes in different biomes, from desert to rainforest. Processes of denudation, migration, establishment, competition, and site modification differ in patterns, rates, and intensity in each biome.

Tree mortality in mature forests of the Pacific Northwest provides a regional example of how dominant processes vary on a probabilistic basis. In coastal Sitka spruce-western hemlock forests, wind accounts for about 80% of the mortality. In the Douglas fir-western hemlock forests of the Cascade Range, wind accounts for 35 to 50% of the mortality. In further contrast, only 15 to 20% of the mortality is attributable to wind in interior ponderosa pine forests. Mortality due to insects and disease has a reverse relationship.

Almost any successional scenario seems possible depending upon the physical setting and the organisms. Productivity and system complexity can decline with time, as when muskeg succeeds forests in boreal regions. Disturbances can accelerate succession as well as retard it, as illustrated by the effects of catastrophic wind in eliminating overstories of a shade-intolerant species, such as Douglas fir, and releasing understories of shade-tolerant species, such as western hemlock and Pacific silver fir. Species diversity may peak early in a sere, or late, or at some intermediate point.

Some of the existing ecological models provide another approach to ecological paradigms. Models such as JABOWA and FORET allow us to use our knowledge of species ecology and of environment to generate probabilistic outputs. They allow us to synthesize our natural history information into a usable form. Note that such models can integrate the fields of physiological ecology, population biology, and ecosystem science.

We need to get on with the process of developing more robust ecological theory. I suspect that some of the greater advances will come from collaborations between various subdisciplines in ecology, such as the ecosystem and population scientists. I doubt that much usable ecological theory will consist of simple mathematical formulae. Ecological science is not physics, and it is time to stop trying to fit it into that mold. Ecological theory must incorporate the realities and unpredictability of the natural world.

Future Directions in Ecosystem Research

What about future directions in ecosystem research? We need to continue to encourage development of tightly structured hypotheses and definitive experiments, particularly as we look in detail at subsystems and processes. We see such trends in the proposals submitted to the National Science Foundation, such as in increased numbers of exciting proposals dealing with soil chemistry as it affects nutrient cycling, productivity, and ecosystem response to pollutants.

We need to continue to adapt new technologies that allow us to do new things or old things better. Opportunities are being created at both the micro- and macro-levels, as illustrated by the use of optical fibers to study root growth, and remote sensing to study productivity at the global level. It is also important that we continue to incorporate other disciplines, with their fresh perspectives and exotic technologies, into ecosystem research programs.

We need to move up in our scale of study, both temporally and spatially. Heterogeneity and spatial patterns at all scales up to the landscape are an exciting topic. Larger spatial and temporal scales are presently very critical issues in terms of the relevance of ecological research to societal problems, as illustrated by concerns over cumulative effects and long-term site productivity. Some of this probably involves scaling up some of the modeling efforts and incorporating spatial patterns; juxtaposition of pieces of an ecosystem or landscape can be critical to outcomes. Much of our science has not effectively made the jump from stands or small watersheds to drainage basins. Are there thresholds at which point a landscape or a stream drainage may unravel?

Despite my harsh words about the past contributions of ecological theory, I believe that we do need to work hard at organizing our knowledge as general principles or theory as it is more broadly interpreted. This theoretical framework would include use of model structures and syntheses of major systems and processes. To be successful, a much broader and more broadly experienced ecological community will need to take an active part in formulation of ecological theory than seems to have been the case.

I am sure that we will do all of these things. They are logical, they are in our self interest, and they fit traditional biases or current trends. But, there are some other emphases that I view as equally critical and that I fear may lack sufficient attention or support.

Intensive, whole ecosystem studies must continue. Many types of ecosystems, particularly in tropical regions, still lack detailed examinations. Additional processes and compartments will be recognized as the result of such efforts. As I recall how long it took us to "see" coarse woody debris, I have no doubt that many fundamental structural and functional features of ecosystems remain to be identified; we have not even recognized all of the basic physical and chemical processes operating in these systems. The selection of additional sites for comprehensive studies will have to be done very carefully, as we cannot afford many such studies.

Natural history research needs to be returned to its rightful place as a respected element of ecological science. Such research provides us with our ultimate source material. Careful review is important in the selection of projects for support, but so is the need to nurture this type of study.

Comparative analyses of ecosystems are essential so that we can systematically examine system-to-system variability in the types and relative importance of compartments, processes, and controlling factors. Such analyses are essential to developing truly general predictive capabilities. I do not mean comparisons involving a random selection of sites; I do mean logical sets, within and between biomes that will allow us to look at responses along gradients or across environmental fields. Vitousek et al. (1979) have illustrated the power of such an approach. We must minimize the tendency to impose our local view on the world and join other scientists in developing response surfaces for regions and continents.

Collaborations of interdisciplinary teams will continue to be important – if ecosystem science is going to advance. There are too many interesting and critical questions that simply cannot be tackled individually or even by small groups. Prime examples include questions about ecosystem level responses to acid rain and other pollutants, cumulative effects of human activities on watershed- and landscape-level responses, effects of management practices on long-term site productivity, and the structure and function of old-growth forests. We must continue to train our students in the merit and art of cooperative research, and work to modify scientific institutions to recognize and reward such efforts.

Long-term data bases must be developed and maintained. They can only increase in importance with their critical role in providing critical tests of hypotheses, providing the "raw meat" for formulation of hypotheses, measuring the rates of long-term processes, providing baselines and identifying trends, and developing an appreciation of and information on episodic phenomena. Long-term data bases mean long-term experimental manipulations and permanent sample plots.

Scientific properties like Coweeta are essential to the three needs of comparison, collaboration, and long-term data bases. We must have locales dedicated to science and education with experimental capabilities, whole ecosystems available for study, and logistical support. We must have sites with long-term measurements, superior scientific leadership, and diverse scientific teams.

Many of the best locales for such research are federal properties under the jurisdiction of U.S.D.A. Forest Service, Department of Energy, and U.S.D.I. National Park Service. Coweeta illustrates the power of the governmental-academic linkage very well. Most of the highly productive ecological research programs have evolved where government and academic personnel have combined efforts at a site. Such sites can and do operate along a continuum of concept, experimentation, and application. With a good basic research program, management issues and societal problems can be addressed with understanding – essentially all research proves relevant!

Networking of the sites and research groups dedicated to long-term ecosystem research is critical. I think that suprasite synthesis is where the next generation of ecological paradigms will arise, and it needs greater attention and some fresh approaches.

Need for Stewardship

The continuity of Coweeta and comparable facilities is a shared concern and responsibility. Their existence depends upon a combined stewardship of institutions and of the scientific community. I want to express my concern about potential failures in this stewardship. Too often, the continued existence of facilities such as Coweeta must be credited to one or a few individuals. Maintaining sites and data bases have cost individuals and programs, sometimes dearly. All of the federal sites are operated out of research funds, whether such funds have come from Forest Service, National Science Foundation, or other sources. In effect, someone's research budget is paying the overhead of maintaining these sites. Someone's curriculum vitae is going to be the thinner for the time and cost involved.

The institutions responsible for the sites and programs need to lift some of the burden from individual research programs. Some of the long-term monitoring programs need to be supported "off the top", not out of local research projects. National Science Foundation needs to reconsider its unwillingness to provide support for monitoring and operating facilities. Forest Service, National Park Service, and Department of Energy need to provide support from the national level.

The scientific community must see that dedicated research sites are protected and maintained. We can actively support such properties by utilizing them and by encouraging institutional financial support.

Agencies and institutions have a particular responsibility since they provide the funds, protect the lands, maintain the facilities, and archive the data sets. Even the most valuable of facilities have been threatened with closure or disestablishment-- during times of tight budgets, tenures of unsympathetic administrators, or changes in program emphasis. Many old hands in Forest Service research remember the pendulum from field to laboratory studies in the 1960s. Coweeta was repeatedly scheduled for closure in the last fifteen years. H. J. Andrews was almost dissolved in the early 1960s on the assumption that we knew all that we needed to about old-growth Douglas fir forests. San Joaquin Experimental Range, a biosphere reserve, was proposed for disposal as excess federal property in 1980.

It is to the credit of the Forest Service that none of these things happened, but they could have. In lesser instances, they did, such as the disposal of numerous structures that should have remained available as research facilities. Nor is the problem unique to the Forest Service, as scientists involved with the National Environmental Research Parks can attest.

As individuals and as organizations we need to be vigilant in the protection of these scientific resources. We must insure that the scientific community and institutions take the longer view, especially in unstable times.

Conclusion

Ecosystem science is healthy. There is a sense of excitement with many important discoveries during the last two decades and the promise of many more. There are a diversity of approaches and scales of study and a tolerance of this diversity. There is a sense of community that facilitates cooperation and collaboration. The science is relevant; findings are being utilized in the solution of societal problems.

Coweeta Hydrological Laboratory has been a major leader in this progression of ecosystem science. The scientific community looks forward to its next 50 years of contribution.

References

Abbott DT (1980) Woody litter decomposition at Coweeta Hydrologic Laboratory, North Carolina. Ph.D. Dissertation, University of Georgia, Athens.

Abbott DT and Crossley DA Jr (1982) Woody litter decomposition following clearcutting. Ecology 63:35–42.

Abbott DT, Seastedt TR, and Crossley DA Jr (1980) The abundance, distribution and effects of clear-cutting on Cryptostigmata in the Southern Appalachians. Environ. Entomol. 9:618–623.

Aber JD and Mellilo JM (1980) Litter decomposition: measuring state of decay and percent transfer into forest soils. Can. J. Bot. 58:416–421.

Aber JD, Mellilo JM, Nadelhoffer JM, McClaugherty KJ, and Pastor J (1985) Fine root turnover in forest ecosystems in relation to quantity and form of nitrogen availability: A comparison of two methods. Oecologia 66:317–321.

Allen TFH and Iltis HH (1980) Overconnected collapse to higher levels: Urban and agricultural origins, a case study. pp. 96–103. In Banathy BH (editor), Systems Science and Science, Proceedings of the Twenty-Fourth Annual North American Meeting of the Society for the General Systems Research.

Allen TFH and Starr TB (1982) Hierarchy: Perspectives for Ecological Complexity. University of Chicago Press, Chicago.

Allen TFH and Wyleto EP (1983) A hierarchical model for the complexity of plant communities. J. Theor. Biol. 101:529–540.

Allen TFH, O'Neill RV, and Hoekstra TW (1984) Interlevel relations in ecological research and management: Some working principles from hierarchy theory. General Technical Report RM-110, USDA Forest Service, Rocky Mountain Forest and Range Experiment Station, Fort Collins, Colorado.

Anderson JM (1973) The breakdown and decomposition of sweet chestnut (*Castanea sativa* Mill) and beech (*Fagus sylvatica* L.) leaf litter in two deciduous woodland soils. I. Breakdown, leaching and decomposition. Oecologia 12:251–274.

Anderson JM and Ineson P (1983) Interactions between soil arthropods and microorganisms in carbon, nitrogen and mineral element fluxes from decomposing leaf litter. pp. 413–432. In Alee J, McNeill S, and Rorison IH (editors), Nitrogen as an Ecological Factor. Blackwell Scientific Publishers, Oxford.

Anderson MG and Burt TP (1978) Toward a more detailed field monitoring of variable source areas. Water Resour. Res. 14:1123–1131.

Anderson RV, Coleman DC, and Cole CV (1981) Effects of saprotrophic grazing on net mineralization. pp. 201–216. In Clark FE and Rosswall T (editors), Terrestrial Nitrogen Cycles. Ecol. Bull., Stockholm.

Andresen AM, Johnson AH, and Siccama TG (1980) Levels of lead, copper, and zinc in the forest floor in the northeastern United States. J. Environ. Qual. 9:293–296.

APHA (1975) Standard Methods for the Analysis of Water and Wastewater. Fourteenth Edition.

Armond PA and Mooney HA (1978) Correlation of photosynthetic unit size and density with photosynthetic capacity. Carnegie Inst. Yearb. 77:234–237.

Arney K and Pugh R (1983) Showcase for wildlife: Catoosa. Tenn. Wildl. 6:20–25.

Arnon DI, Fratzke WE, and Johnson CM (1942) Hydrogen ion concentration in relation to absorption of inorganic nutrients by higher plants. Plant Physiol. 17:515–524.

Aubuchon RR, Thompson DR, and Hinckley TM (1978) Environmental influences on photosynthesis within the crown of a white oak. Oecologia 35:295–306.

Baes CF III and Ragsdale HL (1981) Age-specific lead distribution in xylem rings of three tree genera in Atlanta, Georgia. Environ. Pollut. 2:21–36.

Baes CF III and McLaughlin SB (1984) Trace elements in tree rings, evidence of recent and historical air pollution. Science 224:494–497.

Bandy OL and Marincovich L (1973) Rates of late Cenozoic uplift, Baldwin Hills, Los Angeles, California. Science 181:653–654.

Barber SA (1984) Soil Nutrient Bioavailability: A Mechanistic Approach. Wiley-Interscience, New York.

Barden LW (1981) Forest development in canopy gaps of a diverse hardwood forest of the southern Appalachian Mountains. Oikos 37:205–209.

Barnes JR and Minshall GW (editors) (1983) Stream Ecology: Application and Testing of General Ecological Theory. Plenum Press, New York.

Barrett JW (1980) Regional silviculture of the United States. John Wiley and Sons, New York.

Baskerville GL (1972) Use of logarithmic regression in the estimation of plant biomass. Can. J. For. Res. 2:49–53.

Bates CG and Henry AJ (1928) Forest and streamflow experiments at Wagon Wheel Gap, Colorado. United States Department of Agriculture, Weather Bureau Monthly Weather Review, Suppl. No. 30, Washington, DC, USA.

Baumgartner A (1984) Das Wassereinzugsgebiet Grosse Ohe. Proceedings Symposium Wald und Wasser, 2–5.9.84 in Grafenau BRD.

Beadle NCW (1966) Soil phosphate and its role in moulding segments of the Australian flora and vegetation, with special references to xeromorphy and sclerophylly. Ecology 47:992–1007.

Beard JS (1955) The classification of tropical American vegetation types. Ecology 36:89–100.

Beasley RS (1976) Contribution of subsurface flow from the upper slopes of forested watersheds to channel flow. Soil Sci. Am. Proc. 40:955–957.

Beaton JD, Burns GR, and Platou J (1968) Determination of sulphur in soils and plant materials. Tech. Bull. 14. The Sulphur Institute, Washington DC.

Beck DA and McGee CE (1974) Locust sprouts reduce growth of yellow poplar seedlings. United States Forest Service Research Note SE-201.

Beck DE and Della-Bianca L (1981) Yellow-poplar: characteristics and management. United States Forest Service Agricultural Handbook No. 583.

Becker GF (1895) A reconnaissance of the goldfields of the southern Appalachians. U.S. Geological Survey, 16th Annual Report, Pt. 3, 251–331.

Belser LW and Mays EL (1980) Specific inhibition of nitrite oxidation by chlorate and its use in assessing nitrification in soils and sediments. Appl. Environ. Microbiol. 39:505–510.

Benke AC (1979) A modification of the Hynes method for secondary production with particular significance for multivoltine populations. Limnol. Oceanogr. 24:168–174.

Benke AC (1984) Secondary production of aquatic insects. In: Ecology of Aquatic Insects. pp. 289–322. In Resh VH and Rosenberg DM (editors). Praeger Publishers, New York.

Benke AC and Wallace JB (1980) Trophic basis of production among net-spinning caddisflies in a southern Appalachian stream. Ecology 61:108–118.

Benninger LK, Lewis DM, and Turekian KK (1975) The use of natural Pb-210 as a heavy metal tracer in the river-estuarine system. In: Marine Chemistry and the Coastal Environment. American Chemical Society Symposium Series No. 18. Church TM (editor).

Berish CW and Ragsdale HL (1984) Comparative trace metal analysis of tree rings from remote and urban southeastern forests. Abstract. 35th Annual AIBS Meeting.

Berish CW and Ragsdale HL (1985) Chronological sequence of element concentrations in wood of *Carya* spp. in the Southern Appalachian Mountains. Can. J. For. Res.

Berish CW and Ragsdale HL (1986) Metals in low-elevation southern Appalachian forest and soil. J. Environ. Qual. 15:183–187.

Bernier PY (1982) VSAS2: A revised source area simulator for small forested basins. Ph.D. Dissertation, University of Georgia, Athens.

Berry JL (1977) Chemical weathering and geomorphological processes at Coweeta, North Carolina. Geol. Soc. Am. Abstr. 9:120.

Berryman AA and Wright LC (1978) Defoliation, tree condition and bark beetles. pp. 81–87. In Brookes MH, Stark RW, and Campbell RW (editors), The Douglas-fir Tussock Moth: A Synthesis. USDA Tech. Bull. 1585. U.S. Dept. of Agriculture, Washington.

Best GR (1971) Potassium, sodium, calcium and magnesium flux in a mature hardwood forest watershed and eastern white pine forest watershed at Coweeta. M.S. Thesis, University of Georgia, Athens.

Best GR (1976) Treatment and biota of an ecosystem affect nutrient cycling. Ph.D. Dissertation, University of Georgia, Athens.

Best GR and Monk CD (1975) Cation flux in hardwood and white pine watersheds. pp. 847–861. In Howell FG, Gentry JB, and Smith MH (editors), Mineral Cycling in Southeastern Ecosystems. Energy Research and Development Administration Symposium Series. CONF-740513.

Betson RP (1964) What is watershed runoff? J. Geophys. Res. 69:1541–1552.

Betson RP and Marius JB (1969) Source area of storm runoff. Water Resour. Res. 5:574–582.

Beven KT and Kirkby MJ (1979) A physically based, variable contributing area model of basin hydrology. Hydrol. Sci. Bull. 24:43–69.

Bicknell SM (1979) Pattern and process of plant succession in a revegetating northern hardwood ecosystem. Ph.D. Dissertation, Yale University, New Haven.

Bilby RE (1981) Role of organic debris dams in regulating export of dissolved and particulate matter from a forested watershed. Ecology 62:1234–1243.

Bilby RE and Likens GE (1980) Importance of organic debris dams in the structure and function of stream ecosystems. Ecology 61:1107–1113.

Black CA (1968) Soil-plant relationships. Second Edition, John Wiley & Sons, New York.

Black PE (1957) Interception in a hardwood stand. M.F. Thesis, University of Michigan, Ann Arbor.

Black PE and Clark PM (1958) Timber, water, and Stamp Creek. United States Department of Agriculture, Forest Service, Southeastern Forest Experiment Station and Region 8, Asheville, North Carolina, USA.

Blackburn TR (1973) Information and the ecology of scholars. Science 181:1141–1146.

Boardman NK (1977) Comparative photosynthesis of sun and shade plants. Annu. Rev. Plant Physiol. 28:355–377.

Bogenschutz H and Konig E (1976) Relationships between fertilization and tree resistance to forest insect pests. pp. 281–289. In Fertilizer Use and Plant Health. Proc. 12th Colloquium of the International Potash Institute held at Izmir, Turkey, 1976. International Potash Institute, Bern, Switzerland.

Bollinger GA (1973) Seismicity and uplift. Am. J. Sci. 273A:397–408.

Bonnel M and Gilmour DA (1978) The development of overland flow in a tropical rain forest. Amsterdam, Elsevier Scientific. J. Hydrol. 39:365–382.

Boring LR (1979) Early forest regeneration and nutrient conservation on a clearcut southern Appalachian watershed. M.S. Thesis, University of Georgia, Athens.

Boring LR (1982) The role of black locust (*Robinia pseudo-acacia* L.) in forest regeneration and nitrogen fixation in the southern Appalachians. Ph.D. Dissertation, University of Georgia, Athens.

Boring LR, Monk CD, and Swank WT (1981) Early regeneration of a clear-cut southern Appalachian forest. Ecology 62:1244-1253.

Boring LR and Swank WT (1984a) The role of black locust (*Robinia pseudoacacia*) in forest succession. J. Ecol. 72:749-766.

Boring LR and Swank WT (1984b) Symbiotic nitrogen fixation in regenerating black locust (*Robinia pseudoacacia* L.) stands. For. Sci. 30:528-537.

Boring LR, Swank WT, Waide JB, and Henderson GS (1986) The relative importance of nitrogen fixation and atmospheric deposition as nitrogen inputs to terrestrial ecosystems. Biogeochem. in review.

Bormann FH (1982) The effects of air pollution on the New England landscape. Ambio 11:338-346.

Bormann FH and Graham BF Jr (1959) The occurrence of natural root grafting in eastern white pine, *Pinus strobus* L., and its ecological implications. Ecology 40:677-691.

Bormann FH and Likens GE (1967) Nutrient cycling. Science 155:424-429.

Bormann FH and Likens GE (1979) Pattern and process in a forested ecosystem. Springer-Verlag, New York.

Bosch JM and Hewlett JD (1982) A review of catchment experiments to determine the effect of vegetation changes on water yield and evapotranspiration. J. Hydrol. (Amsterdam) 55:3-23.

Botkin DB (1980) A grandfather clock down the staircase: stability and disturbance in natural ecosystems. pp. 1-10. In Waring RH (editor), Forests: Fresh Perspectives from Ecosystem Analysis. Oregon State University Press, Corvallis.

Bowman RA and Focht DD (1974) The influence of glucose and nitrate concentrations upon denitrification rates in a sandy soil. Soil Biol. Biochem. 6:297-301.

Bradford BH (1977) Precipitation frequency in mountainous terrain. School of Engineering, Georgia Institute of Technology cooperative agreement Progress Report. Coweeta Hydrologic Laboratory, United States Department of Agriculture, Forest Service, Otto, North Carolina, USA.

Braun EL (1972) Deciduous Forests of Eastern North America. Hafner Publishing Company, New York.

Bray JR (1961) Measurement of leaf utilization as an index of minimum level of primary consumption. Oikos 12:70-74.

Bray JR and Gorham E (1964) Litter production in forests of the world. Adv. Ecol. Res. 2:101-157.

Brechtel HM, Balazs A, and Kille K (1982) Natural correlation of streamflow characteristics from small watersheds in the forest research area of Krofdorf—Results of a paired watershed calibration. Proc. Symp. Hydrol. Research Basins, Sonderheft Landeshydrologie Bern, 1982,291-300.

Bremner JM (1965) Total nitrogen. In Black CA (editor), Methods of Soil Analysis. American Society of Agronomy, Madison, Wisconsin.

Bremner JM and Mulvaney CS (1982) Nitrogen—Total. pp. 595-624. In Page AL (editor), Methods of Soil Analysis, Part 2. Chemical and Microbiological Properties. American Society of Agronomy, Madison, Wisconsin.

Bren LJ and Turner AK (1979) Overland flow on a steep, forested infiltrating slope. Aust. J. Soil. Res. 30:43-52.

Brezonik PL, Edgerton ES, Hendry CD (1980) Acid precipitation and sulfate deposition in Florida. Science 208:1027-1029.

Bronowski J (1972) New concepts in the evolution of complexity, Part II. Am. Schol. 42:110-122.

Brown BJ (1982) Productivity and herbivory in high and low diversity tropical successional ecosystems in Costa Rica. Ph.D. Dissertation, University of Florida, Gainesville.

Brown GW and Krygier JT (1970) Effects of clear-cutting on stream temperatures. Water Resour. Res. 6:1135–1139.

Brown GW and Krygier JT (1971) Clearcut logging and sediment production in the Oregon Coast Range. Water Resour. Res. 7:1189–1199.

Buckner E and McCracken W (1978) Yellow poplar: a component of climax forests? J. For. 76:421–423.

Burger H (1934) Einfluss des Waldes auf den Stand der Gewasser. II. Mitteilung. Der Wasserhaushalt im Sperbel – und Rappengraben von 1915/16 bis 1926/27. Mitt. Eidg. Anst. Forstl. Versuchswes. 18:311–416.

Burger H (1945) Der Wasserhaushalt im Valle die Melera von 1934/35 bis 1943/44. Mitt. Eidg. Anst. Forstl. Versuchswes. 24:133–218.

Burger H (1954) Einfluss des Waldes auf den Stand der Gewasser. Mitteilung 5: Der Wasserhaushalt im Sperbel – und Rappengraben von 1942/43 bis 1951/52. Mitteilungen des Schweiz. Eidg. Anst. Forstl. Versuchswes. 31:9–58.

Burrell DC (1974) Atomic spectrometric analysis of heavy-metal pollutants in water. Arba Science Publishing, Inc., Ann Arbor.

Burton KW and John E (1977) A study of heavy metal contamination in the Rhonda Fawr, South Wales. Water Air Soil Pollut. 7:45–68.

Cameron EN (1951) Feldspar deposits of the Bryson City District, North Carolina. North Carolina Division of Mineral Resources Bulletin 62.

Carson WS and Green WT (1977) Soil Survey of Rabun and Towns Counties, Georgia, 97.

Caskey WH and Schepers JS (1985) Modelling of microbial activity: Mineralization, immobilization, nitrification, and denitrification. Proceedings USDA/ARS Natural Resources Modelling Symposium. Pingree Park, CO.

Cataneo R and Stout GE (1968) Raindrop-size distributions in humid continental climates, and associated rainfall rate-radar reflectivity relationships. J. Appl. Meteorol. 7:901–907.

Chabot BF and Hicks DJ (1982) The ecology of leaf life spans. Annu. Rev. Ecol. Syst. 13:229–259.

Chapin FS III (1980) The mineral nutrition of wild plants. Annu. Rev. Ecol. Syst. 11:233–260.

Chapin FS III and Kadrowski RA (1983) Seasonal changes in nitrogen and phosphorus fractions and autumn retranslocation in evergreen and deciduous taiga species. Ecology 64:376–391.

Chapman AG (1935) The effect of black locust on associated species with special reference to forest trees. Ecol. Monogr. 5:37–60.

Chapman CA (1958) Control of Jointing by Topography. J. Geol. 66:552–558.

Chapman SB (1967) Nutrient budgets for a dry heath ecosystem in the south of England. J. Ecol. 55:677–689.

Chow VT (1959) Open channel hydraulics. McGraw-Hill, New York.

Clayton JL (1979) Nutrient supply to soil by rock weathering. pp. 75–96. In Impact of Intensive Harvesting on Forest Nutrient Cycling, State University of New York, College of Environmental Science and Forestry, Syracuse, New York.

Cleaves ET and Costa JE (1979) Equilibrium, cyclicity, and problems of scale – Maryland's Piedmont landscape. Maryland Geol. Surv. Inf. Circ. 29:32.

Cleaves ET, Fisher DW, and Bricker OP (1974) Chemical weathering of serpentinite in the eastern Piedmont of Maryland. Geol. Soc. Am. Bull. 85:437–444.

Cleaves ET, Godfrey AE, and Bricker OP (1970) Geochemical balance of a small watershed and its geomorphic implications. Geol. Soc. Am. Bull. 81:3015–3032.

Cleland WW (1967) The statistical analysis of enzyme kinetic data. Adv. Enzymol. Relat. Areas Mol. Biol. 29:1–32.

Clifford HF (1966) The ecology of invertebrates in an intermittent stream. Invest. Indiana Lakes Streams 7:57–98.

Cogbill CV and Likens GE (1974) Acid precipitation in the Northeastern United States. Water Resour. Res. 10:1133–1137.

Cole DW and Rapp MR (1980) Elemental cycling in forest ecosystems. pp. 341–409. In Reichle D (editor), Dynamic Properties of Forest Ecosystems. Cambridge Univ. Press, Cambridge.

Coleman DC, Reid CP, and Cole CV (1983) Biological strategies of nutrient cycling in soil systems. Adv. Ecol. Res. 13:1–55.

Connell JH and Sousa WP (1983) On the evidence needed to judge ecological stability or persistence. Am. Nat. 121:789–824.

Conrad M (1973) Thermodynamic extremal principles in evolution. Biophysik 9:191–196.

Conrad SG, Wilson WF, Allen EP, and Wright TJ (1963) Anthophyllite asbestos in North Carolina. North Carolina Divison of Mineral Resources Bulletin 77.

Cornaby BW (1973) Population parameters and systems models of litter fauna in a white pine ecosystem. Ph.D. Dissertation, University of Georgia, Athens.

Cornaby BW, Gist CS, and Crossley DA Jr (1975) Resource partitioning in leaf-litter faunas from hardwood and hardwood-converted-to-pine forests. pp. 588–597. In Howell FG, Gentry JB, and Smith MH (editors), Mineral Cycling in Southeastern Ecosystems. ERDA Symposium Series. (CONF-740513).

Cornaby BW and Waide JB (1973) Nitrogen fixation in decaying chestnut logs. Plant Soil 39:445–448.

Costa JE (1974) Response and recovery of a Piedmont watershed from tropical storm Agnes, June 1972. Water Resour. Res. 10:106–112.

Costa JE and Cleaves ET (1984) The Piedmont landscape of Maryland: a new look at an old problem. Earth Surf. Processes Landforms 9:59–74.

Coulson RN and Witter JA (1984) Forest Entomology. John Wiley & Sons, New York.

Covington WW (1981) Changes in forest floor organic matter and nutrient content following clear cutting in northern hardwoods. Ecology 62:41–48.

Craighead FC (1937) Locust borer and drought. J. For. 35:792–793.

Critchfield HJ (1966) General Climatology. Second Edition. Prentice Hall, Englewood Cliffs, New Jersey, USA.

Crocker MT (1986) Interstitial dissolved organic carbon at a spring seep. M.S. Thesis, University of Georgia, Athens.

Cromack J Jr, Sollins P, Todd RL, Crossley DA Jr, Fender WM, Fogel R, and Todd AW (1977) Soil Microorganism—Arthropod Interactions: Fungi as Major Calcium and Sodium Sources. pp. 78–84. In Mattson WJ (editor), The Role of Arthropods in Forest Ecosystems. Springer-Verlag, New York.

Cromack K Jr (1973) Litter production and decomposition in a mixed hardwood watershed at Coweeta hydrologic Station, North Carolina. Ph.D. Thesis, University of Georgia, Athens.

Cromack K Jr and Monk CD (1975) Litter production and decomposition and nutrient cycling in a mixed hardwood watershed and a white pine plantation. pp. 609–624. In Howell FG, Gentry JR, and Smith MH (editors), Mineral Cycling in Southeastern Ecosystems. ERDA Symposium Series. (CONF-740513).

Crossley DA Jr (1977a) The role of terrestrial saprophagous arthropods in forest soils: Current status of concepts. pp. 49–56. In Mattson WJ (editor), The Role of Arthropods in Forest Ecosystems. Springer-Verlag, New York.

Crossley DA Jr (1977b) Oribatid mites and nutrient cycling. pp. 71–86. In Dindal DL (editor), Biology of Oribatid Mites. Syracuse, NY: State University NY at Syracuse, Coll. Environ. Sci. For.

Crossley DA Jr, Callahan JT, Gist CS, Maudsley JR, and Waide JB (1976) Compartmentalization of arthropod communities in forest canopies at Coweeta. J. Georgia Entomol. Soc. 11:44–49.

Crossley DA Jr, Duke KM, and Waide JB (1975) Fallout Cesium-137 and mineral element distribution in food chains of granitic outcrop ecosystems. pp. 580–587. In Howell FG and Smith MH (editors), Mineral Cycling in Southeastern Ecosystems. ERDA Symposium Series.

Cuffney TF, Wallace JB, and Webster JR (1984) Pesticide manipulation of a headwater stream ecosystem: significance for ecosystem processes. Freshwater Invertebr. Biol. 3:153–171.

Cummins KW (1974) Structure and function of stream ecosystems. BioScience 24:631–641.

Cummins KW, Klug MJ, Wetzel RG, Petersen RC, Suberkropp KF, Manny BA, Wuycheck JC, and Howard FO (1972) Organic enrichment with leaf leachate in experimental lotic ecosystems. BioScience 22:719–722.

Cunningham GL and Syvertsen JD (1977) The effect of nonstructural carbohydrate levels on dark CO_2 release in Creosotebush. Photosynthetica 11:291–295.

Dahm CN (1981) Pathways and mechanisms for removal of dissolved organic carbon from leaf leachate in streams. Can. J. Fish. Aquat. Sci. 38:68–76.

Dale TN (1923) The Commercial Granites of New England, USGS Bull. 780, p. 488.

David MD, Mitchell MJ, Nakes JP (1982) Organic and inorganic sulfur constitutents of a forest soil and their relationship to microbial activity. Soil Sci. Soc. Am. J. 46:847–852.

Davidson E (1986) Gaseous nitrogen losses from two forested watersheds via nitrification and denitrification. Ph.D. Dissertation, North Carolina State University, Raleigh.

Davis MB (1976) Pleistocene biography of temperate deciduous forests. Geosci. Man. 13:113–126.

Day FP Jr (1971) Vegetation structure of a hardwood watershed at Coweeta. M.S. Thesis, University of Georgia, Athens.

Day FP Jr (1974) Primary production and nutrient pools in the vegetation on a southern Appalachian watershed. Ph.D. Dissertation, University of Georgia, Athens.

Day FP Jr and McGinty DT (1975) Mineral cycling strategies of two deciduous and two evergreen tree species on a southern Appalachian watershed. pp. 736–743. In Howell FG, Gentry JR, and Smith MH (editors), Mineral Cycling in Southeastern Ecosystems. ERDA Symposium Series. (CONF-740513).

Day FP Jr and Monk CD (1974) Vegetation patterns on a southern Appalachian watershed. Ecology 55:1064–1074.

Day FP Jr and Monk CD (1977a) Net primary production and phenology on a southern Appalachian watershed. Am. J. Bot. 64:1117–1125.

Day FP Jr and Monk CD (1977b) Seasonal nutrient dynamics in the vegetation on a southern Appalachian watershed. Am. J. Bot. 64:1126–1139.

Dick WA and Tabatabai MA (1979) Ion chromatographic determination of sulfate and nitrate in soils. Soil Sci. Soc. Am. J. 43:899–904.

Dils RE (1952) Changes in some vegetation, surface soil and surface runoff characteristics of a watershed brought about by forest cutting and subsequent mountain farming. Thesis, Michigan State College of Agriculture and Applied Science, East Lansing.

Dils RE (1953) Influence of forest cutting and mountain farming on some vegetation, surface soil and surface runoff characteristics. United States Department of Agriculture, Forest Service, Southeastern Forest Experiment Station Paper 24, Asheville, North Carolina, USA.

Dils RE (1957) A guide to the Coweeta Hydrologic Laboratory. United States Department of Agriculture, Forest Service, Southeastern Forest Experiment Station, Asheville, North Carolina, USA.

D'Itri FM (ed.) (1982) Acid precipitation: effects on ecological systems. Ann Arbor Science, Ann Arbor, MI, 506.

Dominski A (1971) Accelerated nitrate production and loss in the northern hardwood forest ecosystem underlain by podzol soils following clear cutting and addition of herbicides. Ph.D. Dissertation, Yale University, New Haven.

Douglass JE (1962) A method for determining the slope of neutron moisture meter calibration curves. SEFES Paper 143. U.S. Department of Agriculture, Forest Service, Southeastern Forest Experiment Station, 11.

Douglass JE (1966) Volumetric calibration of neutron moisture probes. Soil Sci. Soc. Am. Proc. 30:541–544.

Douglass JE (1974) Flood frequencies and bridge and culvert sizes for forested mountains of North Carolina. United States Department of Agriculture, Forest Service, Southeastern Forest Experiment Station General Technical Report SE-4, Asheville, North Carolina, USA.

Douglass JE (1983) A summary of some results from the Coweeta Hydrologic Laboratory. Appendix B. pp. 137–141. In Hamilton LS and King PN (editors), Tropical forested watersheds: hydrologic and soil response to major uses or conversions. Westview Press, Boulder.

Douglass JE (1983) The potential for water yield augmentation from forest management in the eastern United States. Water Resour. Bull. 19:351–358.

Douglass JE, Cochrane DR, Bailey GW, Teasley JI, and Hill DW (1969) Low herbicide concentration found in streamflow after a grass cover is killed. Southeastern Forest Experiment Station, Research Note SE-108, 3.

Douglass JE and Swank WT (1972) Streamflow modification through management of eastern forests. United States Department of Agriculture, Forest Service, Southeastern Forest Experiment Station, Research Paper SE-94, Asheville, North Carolina.

Douglass JE and Swank WT (1975) Effects of management practices on water quality and quantity: Coweeta Hydrologic Laboratory, North Carolina. pp. 1–13. In Municipal Watershed Management Proceedings, United States Department of Agriculture, Forest Service General Technical Report NE-13.

Douglass JE and Swank WT (1976) Multiple use in southern Appalachian hardwoods — a ten-year case history. pp. 425–436. In Proceedings of the XVI International Union of Forestry Research Organizations World Congress, Oslo, Norway. IUFRO Secretariat, Schonbrunn-Triolergarten, A-1131 Vienna, Austria.

Douglass JE and Swift LW Jr (1977) Forest Service studies of soil and nutrient losses caused by roads, logging, mechanical site preparation, and prescribed burning in the Southeast. pp. 489–503. In Correll DL (editor), Watershed research in eastern North America. Volume II. Smithsonian Institution, Chesapeake Bay Center for Environmental Studies, Edgewater, Maryland.

Drablos D and Tollan A (1980) Ecological impact of acid precipitation. Proc. Intern. Confer., Sandefjord, Norway 11–14 March 1980. SNSF, Oslo-As, NLH, Norway, 383.

Dunford EG and Fletcher PW (1947) The effect of removal of streambank vegetation upon water yield. Trans. Am. Geophys. Union. 28:105–110.

Dunne T (1978) Field studies of hillslope flow processes. pp. 227–293. In Kirkby MJ (editor), Hillslope Hydrology. Wiley-Interscience, New York.

Dunne T and Black RD (1970a) An experimental investigation of runoff production in permeable soils. Water Resour. Res. 6:478–490.

Dunne T and Black RD (1970b) Partial area contributions to storm runoff in a small New England watershed. Water Resour. Res. 6:1296.

Dupraz C, Lelong F, Troy JP, Dumazet B (1982) Comparative study of the effects of vegetation on the hydrological and hydrochemical flows in three minor catchments of Mont Lozere (France). Proceedings Symposium Hydrologic Research Basins, Sonderheft Landeshydrologie, Bern 1982:671–681.

Duvigneaud P and Denaeyer-DeSmet S (1969) Biological cycling of minerals in temperate deciduous forests. In Reichle DE (editor), Analysis of Temperature Forest Ecosystems. Springer-Verlag, New York.

Edlefsen NF and Bodman GB (1941) Field measurements of water movement through silt loam soil. Am. Soc. Agron. J. 33:713–731.

Edwards NT and Ross-Todd BM (1983) Soil carbon dynamics in a mixed deciduous forest following clear-cutting with and without residue removal. Soil Sci. Soc. Am. J. 47:1014–1021.

Edwards RT (1985) The role of seston bacteria in the metabolism and secondary production dynamics of southeastern blackwater rivers. Ph.D. Dissertation, University of Georgia, Athens.

Egerton FN (1973) Changing concepts of the balance of nature. Q. Rev. Biol. 48:322–350.

Ehrenfeld JG (1980) Understory response to canopy gaps of varying size in a mature oak forest. Bull. Torrey Bot. Club 107:29–41.

Elias RW, Hirao Y, and Patterson CC (1975) Impact of present levels of aerosol Pb concentrations on both natural ecosystems and humans. pp. 257–271. In Hutchinson TC (editor), Proceedings of the International Conference on Heavy Metals in the Environment, Vol. 2, Part I. Institute of Environmental Sciences, University of Toronto, Toronto.

Engler A (1919) Untersuchungen uber den Einfluss des Waldes auf den Stand der Gewasser. Mitteilungen. Eidg. Anst. Forstl. Versuchswes. 12:1–626.

EPA (1977) Quality Assurance Handbook for Air Pollution Measurement Systems. Vol. II EPA-600/4/77/0261. USEPA. Office of Research and Development. Research Triangle Park, North Carolina.

Escarre A et al. (1983) Nutrient budgets in a forested mediterranean watershed. Paper presented at the MAB 5-workshop in Budapest, Oct. 1983, 15.

Eschner AR and Patric JH (1982) Debris avalanches in eastern upland forests. J. For. 80:342–347.

Evans LS (1984) Acid preparation effects on terrestrial vegetation. Annu. Rev. Phytopathol. 22:397–420.

Faeth SE (1980) Invertebrate predation of leaf-miners at low densities. Ecol. Entomol. 5:111–114.

Fanning DS and Keramidas VZ (1977) Micas. pp. 195–258. In Dixon JB and Weed SB (editors), Minerals in Soil Environments. Soil Science Society of America, Madison, Wisconsin.

Fassel VA (1978) Quantitative elemental analyses by plasma emission spectroscopy. Science 202:183–191.

Fattorelli S (1982) Riclerche idrologiche in tre piccoli bacini della Valli Gindicarie. Quad Idro Mont. 1:11–30.

Findlay S and Meyer JL (In press) Significance of bacterial biomass and production as an organic carbon source in lotic detrital systems. Bull. Mar. Sci.

Findlay S, Meyer JL, and Edwards RT (1984) Measuring bacterial production via rate of incorporation of [^{3}H] thymidine into DNA. J. Microbiol. Methods 2:57–72.

Findlay S, Meyer JL, and Risley R (1986) Benthic bacterial biomass and production in two blackwater rivers. Can. J. Fish. Aquat. Sci. 43:1271–1276.

Fisher SG and Gray LJ (1983) Secondary production and organic matter processing by collector macroinvertebrates in a desert stream. Ecology 64:1217–1224.

Fisher SG and Likens GE (1973) Energy flow in Bear Brook, New Hampshire: an integrative approach to stream ecosystem metabolism. Ecol. Monogr. 43:421–439.

Fitter AN and Hay RKM (1981) Environmental physiology of plants. Academic Press, London.

Fitzgerald JW (1976) Sulfate ester formation and hydrolysis: a potentially important yet ofter ignored aspect of the sulfur cycle of aerobic soils. Bacteriol. Rev. 40:698–721.

Fitzgerald JW and Andrew TL (1984) Mineralization of methionine sulphur in soils and forest floor layers. Soil Biol. Biochem. 16:565–570.

Fitzgerald JW, Andrew TL, and Swank WT (1984) Availability of carbon bonded sulfur for mineralization in forest soils. Can. J. For. Res. 14:839–843.

Fitzgerald JW, Ash JT, Strickland TC, and Swank WT (1983) Formation of organic sulfur in forest soils: a biologically mediated process. Can. J. For. Res. 13:1077–1082.

Fitzgerald JW and Johnson DW (1982) Transformations of sulphate in forested and agricultural lands. pp. 411–426. In More AI (editor), Sulphur 82, Vol I., British Sulphur Corp., London.

Fitzgerald JW, Strickland TC, Ashe JT (1985) Isolation and partial characterization of forest floor and soil organic sulfur. Biogeochemistry 1:155–167.

Fitzgerald JW, Strickland TC, and Swank WT (1982) Metabolic fate of inorganic sulfate in soil samples from undisturbed and managed forest ecosystems. Soil Biol. Biochem. 14:529–536.

Flaccus E (1959) Revegetation of landslides in the White Mountains of New Hampshire. Ecology 40:692–703.

Force ER (1976) Metamorphic source rocks of titanium placer deposits—A geochemical cycle: U.S. Geological Survey Professional Paper 959-B: B1-B13.

Fowler MC (1977) Laboratory trials of a new triazine herbicide (DPX 3674) on various aquatic species of macrophytes and algae. Wood Res. 17:191.

Freney JR (1961) Some observations on the nature of organic sulphur compounds in soil. Aust. J. Agric. Res. 12:424–432.

Freney JR, Melville GE, and Williams CH (1970) The determination of carbon bonded sulfur in soil. Soil Sci. 109:310–318.

Fritz RS (1983) Ant protection of a host plant's defoliator: consequence of an ant-membracid mutualism. Ecology 64:789–797.

Froehlich W (1975) Dynamika Transportu fluwialnego kamiency Nawojowskiej. Prace Geograficzne nr. 114,112p Polska akademia Nauk, Warsawa.

Froehlich W (1983) The mechanism of dissolved solids transport in flysch drainage basins. IAHS publ. Nr 141:99–108.

Fuhrman J and Azam F (1982) Thymidine incorporation as a measure of heterotrophic bacterio-plankton production in marine surface waters: Evaluation and field results. Mar. Biol. 66:109–120.

Fullagar PD, Hatcher RD Jr, and Merschat CE (1979) 1200 million year old gneisses in the Blue Ridge Province of North and South Carolina. Southeast Geol. 20:69–77.

Galloway JN, Likens GE, Keene WC, and Miller JM (1982) The composition of precipitation in remote areas of the world. J. Geophys. Res. 87:8771–8786.

Garrels RM (1967) Genesis of some ground waters from igneous rocks. pp. 405–420. In Abelson PH (editor), Researches in Geochemistry, V. II. Wiley.

Garrels RM and Mackenzie FT (1967) Origin of the chemical compositions of some springs and lakes 67:222–242. In Stumm W (editor), Equilibrium Concepts in Natural Water Systems, American Chemical Society Advances in Chemistry Series.

Gash JHC (1979) An analytical model of rainfall interception by forest. Q. J. R. Meteorol. Soc. 105:43–55.

Gates DM (1980) Biophysical ecology. Springer-Verlag, New York.

Geesey GG, Mutch R, Costerton JW, and Green RB (1978) Sessile bacteria: an important component of the microbial population in small mountain streams. Limnol. Oceanogr. 23:1214–1223.

Georgian TJ Jr (1982) The seasonal and spatial organization of a guild of periphyton grazing stream insects. Ph.D. Dissertation, University of Georgia, Athens.

Georgian TJ Jr and Wallace JB (1981) A model of seston capture by net-spinning caddisflies. Oikos 36:147–157.

Georgian TJ Jr and Wallace JB (1983) Seasonal production dynamics in a guild of periphyton grazing insects in a southern Appalachian stream. Ecology 64:1236–1248.

Gerritsen J and Patten BC (1985) System theory formulation of ecological disturbance. Ecol. Model. 29:383–397.

Gholz HL, Hawk GM, Campbell AG, Cromack K, and Brown AT (1985) Early vegetation recovery and element cycles on a clearcut watershed in western Oregon. Can. J. For. Res. 15:400–409.

Gilbert GK (1904) Domes and dome structures in the high sierra. Bull. Geol. Soc. Am. 15:26–36.

Gist CS (1972) Analysis of mineral pathways in a cryptozoan food web. Ph.D. Dissertation, University of Georgia, Athens.

Gist CS and Crossley DA Jr (1975a) The litter arthropod community in a Southern Appalachian hardwood forest: Numbers, biomass and mineral element content. Am. Midl. Nat. 93:107–122.

Gist CS and Crossley DA Jr (1975b) A model of mineral cycling for an arthropod foodweb in a Southeastern hardwood forest litter community. pp. 84–106. In Howell FG, Gentry JB, and Smith MH (editors), Mineral Cycling in Southeastern Ecosystems. ERDA Symposium Series. (CONF-740513).

Gist CS and Swank WT (1974) An optical planimeter for leaf area determination. Am. Midl. Nat. 92:213–217.

Givnish TJ and Vermeij GJ (1976) Sizes and shapes of lianae leaves. Am. Nat. 110:743–779.

Goldberg DE (1982) The distribution of evergreen and deciduous trees relative to soil type: An example from the Sierra Madre, Mexico and a general model. Ecology 63:942–951.

Goldich SS (1938) A study in rock weathering. J. Geol. 46:17–38.

Goldstein RA and Grigal DF (1972) Computer programs for the ordination and classification of ecosystems. Environmental Science Division Publication Number 417, ORNL-IBP-71-10, 125.

Golladay SW, Webster JR, and Benfield EF (1983) Factors affecting food utilization by a leaf shredding aquatic insect: leaf species and conditioning time. Holarct. Ecol. 6:157–162.

Gosz JR (1981) Nitrogen cycling in coniferous ecosystems. pp. 405–426. In Clark FE and Rosswall T (editors), Terrestrial Nitrogen Cycles: Processes, Ecosystem Strategies, and Management Impacts. Ecol. Bull. (Stockholm).

Goodman D (1975) The theory of diversity-stability relationships in ecology. Q. Rev. Biol. 50:237–266.

Gould SJ (1982) Darwinism and the expansion of evolutionary theory. Science 216:380–387.

Grafius E and Anderson NH (1979) Production dynamics, bioenergetics, and the role of *Lepidostoma quercina* Ross (Trichoptera: Lepidostomatidae) in an Oregon woodland stream. Ecology 60:433–441.

Granat L (1972) On the relation between pH and the chemical composition in atmospheric precipitation. Tellus 24:550–560.

Grant WH (1958) Geology of Hart County, Georgia. Georgia Geol. Surv. Bull. 67, p. 7.

Grant WH (1983) Debris avalanching and slow alluviation, a mechanism for rapid valley growth. Geological Society of America, Abstracts 15,55.

Grant WH (1958) Geology of Hart County, Georgia. Georgia Geol. Surv. Bull. 67.

Grime JP (1977) Evidence for the existence of three primary strategies in plants and its relevance to ecological and evolutionary theory. Am. Nat. 111:1169–1194.

Grip H (1982) Water chemistry and runoff in forest streams at Kloten (Sweden). Uppsala University, UNGI Report Nr. 58,114.

Groet SS (1976) Regional and local variations in heavy metal concentrations of bryophytes in the northeastern United States. Oikos 27:445–456.

Grzenda AR, Nicholson HP, Teasley JI, and Patric JH (1964) DDT residues in mountain stream water as influenced by treatment practices. J. Econ. Entomol. 57:615–618.

Gurtz ME (1981) Ecology of stream invertebrates in a forested and a commercially clear-cut watershed. Ph.D. Dissertation, University of Georgia, Athens.

Gurtz ME and Wallace JB (1984) Substrate-mediated response of stream invertebrates to disturbance. Ecology 65:1556–1569.

Gurtz ME, Webster JR, and Wallace JB (1980) Seston dynamics in southern Appalachian streams: effects of clearcutting. Can. J. Fish. Aquat. Sci. 37:624–631.

Hack JT (1966) Circular patterns and exfoliation in crystalline terrance, grandfather mountain area, North Carolina. Bull. Geol. Soc. Am. 77:975–986.

Hack JT (1980) Rock Control and Tectonism—Their importance in shaping the Appalachian highlands. U.S. Geological Survey Professional Paper, No. 1125-E, 17.

Haefner JD (1980) The effects of old field succession on stream insects in the southern Appalachians and production of two net-spinning caddisflies. M.S. Thesis, University of Georgia, Athens.

Haefner JD and Wallace JB (1981a) Shifts in aquatic insect populations in a first-order southern Appalachian stream following a decade of old field succession. Can. J. Fish. Aquat. Sci. 38:353–359.

Haefner JD and Wallace JB (1981b) Production and potential seston utilization by *Parapsyche cardis* and *Diplectrona modesta* in two streams draining contrasting southern Appalachian watersheds. Environ. Entomol. 10:433–441.

Hadley JB (1949) Preliminary report on corundum deposits in the Buck Creek peridotite, Clay County, North Carolina. U.S. Geological Survey Bulletin, 948-E:103–128.

Hadley JB and Goldsmith R (1963) Geology of the eastern Great Smoky Mountains, North Carolina and Tennessee. U.S. Geological Survey Professional Papers 349-B.

Hadley JB and Nelson AE (1971) Geologic map of the Knoxville quadrangle, North Carolina, Tennessee, and South Carolina. U.S. Geological Survey Miscellaneous Geologic Investigations Map I-654, 1: 250,000.

Hagvar S and Kjondal BR (1981) Succession, diversity and feeding habits of microarthropods in decomposing birch leaves. Pedobiologia 22:385–408.

Haines BL (1981) The effects of simulated acid rain on element leaching from leaves, leaf litter, and from soil and upon calcium uptake by tulip poplar roots and phosphorus uptake by mycorrhizal pine roots. Technical Report, U.S. Environmental Protection Agency, Corvallis, Oregon.

Haines BL (1986) Calcium uptake kinetics of plants from a southern Appalachian forest succession. Bot. Gaz. (in press).

Haines BL, Chapman J, and Monk C (1985) Rates of mineral element leaching from leaves of nine plant species from a southern Appalachian forest succession subjected to simulated acid rain. Bull. Torrey Bot. Club. 112:258–264.

Haines BL, Jernstedt JA, and Neufeld HS (1985) Direct foliar effects of simulated acid rain. II. Leaf surface characteristics. New Phytol 99:407–416.

Haines BL, Jordan C, Clark H, and Clark K (1983) Acid rain in an Amazon rainforest. Tellus 35B:77–80.

Haines BL, Stefani M, and Hendrix F (1980) Acid rain: threshold of leaf damage in eight plant species from a southern Appalachian forest succession. Water Air Soil Pollut. 14:403–407.

Haines BL and Waide J (1980) Predicting potential impacts of acid rain on element cycling in a southern Appalachian deciduous forest at Coweeta. pp. 335–340. In Hutchinson TC and Havas M (editors), Effects of acid precipitation on Terrestrial Ecosystems. Plenum, New York.

Haines BL, Waide JB, and Todd RL (1982) Soil solution nutrient concentrations sampled with tension and zero-tension lysimeters: report of discrepancies. Soil Sci. Soc. Am. J. 46:658–661.

Hains JJ Jr (1981) The response of stream flora to watershed perturbation. M.S. Thesis, Clemson University, Clemson.

Hall RJ, Likens GE, Fiance SB, and Hendrey GR (1980) Experimental acidification of a stream in the Hubbard Brook Experimental Forest, New Hampshire. Ecology 61:976–989.

Hamilton WB (1961) Geology of Richardson and Jones Coves quadrangles, Tennessee. U.S. Geological Survey Professional Paper 349-A.

Hanlon RDG and Anderson JM (1979) The effects of Collembola grazing on microbial activity in decomposing leaf litter. Oecologia 32:93–99.

Harding DJL and Stuttard RA (1974) Microarthropods 27:489.

Hardy RW, Burns RC, and Holsten RD (1973) Applications of the acetylene-ethylene assay for measurement of nitrogen fixation. Soil Biol. Biochem. 5:47–81.

Hargrove WW Jr (1983) Forest Canopy Consumption by arthropod herbivores: An average availability model. M.S. Thesis, University of Georgia, Athens.

Hargrove WW Jr, Crossley DA Jr, and Seastedt TR (1984) Shifts in insect herbivory in the canopy of black locust. *Robinia pseudo-acacia* L., following fertilization. Oecologia 43:322–328.

Harmon ME, Franklin JF, and Swanson FJ, et al. (1986) Ecology of coarse woody debris in temperate ecosystems. Adv. Ecol. Res. 15:133–302.

Harper RM (1914) The "pocosin" of Pike County, Alabama and its bearing on certain problems of succession. Bull. Torrey Bot. Club. 41:209–220.

Harr RD (1977) Water flux in soil and subsoil on a steep forested slope. J. Hydrol. 33:37–58.

Harr RD (1982) Fog drip in the Bull Run Municipal Watershed, Oregon. Water Res. Bull. 18:785–789.

Harris WF, Santantonio D, and McGinty D (1979) The dynamic below-ground ecosystem. pp. 119–129. In Waring RH (editor), Forests: fresh perspectives from ecosystem analysis. Oregon State University Press, Corvallis.

Harwell MA, Cropper WP Jr, and Ragsdale HL (1977) Nutrient cycling and stability: A reevaluation. Ecology 58:660–666.

Harwood JL and Nicholls RG (1979) The plant sulpholipid—a major component of the sulphur cycle. Biochem. Soc. Trans. 7:440–447.

Hassler WW and Tebo LB Jr (1958) Project completion report for Project F-4-R. Fish management investigations on trout streams. North Carolina Wildlife Resources Commission, Raleigh, North Carolina.

Hatcher RD Jr (1971) Geology of Rabun and Habersham Counties, Georgia: A reconnaissance study. Georgia Department of Mines, Min. Geol. Bull. 83.

Hatcher RD Jr (1974) Introduction to the Blue Ridge tectonic history of northeast Georgia. Georgia Geological Society Guidebook 13-A.

Hatcher RD Jr (1976) Introduction to the geology of the eastern Blue Ridge of the Carolinas and nearby Georgia. South Carolina Division of Geology, Carolina Geological Society Guidebook.

Hatcher RD Jr (1977) Macroscopic polyphase folding illustrated by the Toxaway Dome, eastern Blue Ridge, South Carolina-North Carolina. Geol. Soc. Am. Bull. 89:1678–1688.

Hatcher RD Jr (1978) Tectonics of the western Piedmont and Blue Ridge, southern Appalachians: Review and speculation. Am. J. Sci. 278:276–304.

Hatcher RD Jr (1979) The Coweeta Group and Coweeta syncline: major features of the North Carolina-Georgia Blue Ridge. Southeast. Geol. 21:17–29.

Hatcher RD Jr (1980) Geologic Map of the Prentiss Quadrangle, North Carolina Department of Natural Resources and Community Development.

Hatcher RD Jr and Butler JR (1979) Guidebook for southern Appalachian field trip in the Carolinas, Tennessee and northeastern Georgia. International Geological Correlation Program – Caledonide Orogen Project 27.

Heichel GH and Turner NC (1983) CO_2 assimilation of primary and regrowth foliage of red maple (*Acer rubrum* L.) and red oak (*Quercus rubra* L.): response to defoliation. Oecologia, (Berlin) 57:14–19.

Heinrichs H and Mayer R (1980) The role of forest vegetation in the biogeochemistry of cycle of heavy metals. J. Environ. Qual. 9:111–118.

Helvey JD (1964) Rainfall interception by hardwood forest litter in the Southern Appalachians. Research Paper SE-8. U.S. Department of Agriculture, Forest Service, Southeastern Forest Experiment Station.

Helvey JD (1967) Interception by eastern white pine. Water Resour. Res. 3:723–729.

Helvey JD (1970) Interception of rain. pp. 89–93. In Toebes C and Ouryvaev V (editors), Representative and experimental basins; an international guide for research and practice. United Nations Educational Scientific and Cultural Organization. Henkes-Holland, Haarlam.

Helvey JD (1971) A summary of rainfall interception by certain conifers of North America. In Monke EJ (editor), Biological effects in the hydrological cycle-terrestrial phase. Proceedings of the third international seminar for hydrology professors; July 18–30, 1971, in West Lafayette, Purdue University, Department of Agricultural Engineering, Agricultural Experiment Station.

Helvey JD (1981) Flood frequency and culvert sizes needed for small watersheds in the central Appalachians. United States Department of Agriculture, Forest Service, Northeastern Forest Experiment Station General Technical Report NE-62, Broomall, Pennsylvania, USA.

Helvey JD and Hewlett JD (1962) The annual range of soil moisture under high rainfall in the southern Appalachians. J. For. 60:485–486.

Helvey JD, Hewlett JD, and Douglass JE (1972) Predicting soil moisture in the southern Appalachians. Soil Sci. Soc. Am. Proc. 36:954–959.

Helvey JD and Patric JH (1965) Canopy and litter interception of rainfall by hardwoods of eastern United States. Water Resour. Res. 1:193–206.

Helvey JD and Patric JH (1966) Design criteria for interception studies. Symposium Design of Hydrological Networks. Int. Assoc. Sci. Hydrol. Bull. 67:131–137.

Helwig JT and Council KA (1979) SAS user's guide. SAS Institute Inc., Cary, North Carolina.

Henderson GS and Harris WF (1975) An ecosystem approach to characterization of the nitrogen cycle in a deciduous forest watershed. In Berneir B and Wignet CH (editors), Forest Soils and Forest Land Management. Les Presses de l'Univ. Laval, Quebec.

Henderson GS, Swank WT, Waide JB, and Grier CC (1978) Nutrient budgets of Appalachian and Cascade region watersheds: a comparison. For. Sci. 24:385–397.

Hendry CD, Berish CW, and Edgerton ES (1984) Chemistry of precipitation at Turrialba, Costa Rica. Water Resour. Res. 20:1677–1684.

Hendry GR, Baalsrud K, Traaen TS, Laske M, and Raddman G (1976) Acid precipitation: some hydrobiological changes. Ambio 5:224–227.

Herlitzius H (1983) Biological decomposition efficiency in different woodland soils. Oceologia 57:78–97.

Hertzler RA (1936) History of the Coweeta Experimental Forest. Historical Report, unpublished report on file at Coweeta Hydrologic Laboratory, United States Department of Agriculture, Forest Service, Otto, North Carolina.

Hertzler RA (1938) Determination of a formula for the 120° V-notch weir. Civ. Eng. 8:756–757.

Hewlett JD (1961a) Watershed Management. pp. 62–66. In Report for 1961 Southeastern Forest Expt. Sta. Asheville, N.C. USDA Forest Serv., Southeastern Forest Expt. Sta.

Hewlett JD (1961b) Soil moisture as a source of base flow from steep mountain watersheds. United States Department of Agriculture, Forest Service, Southeastern Forest Experiment Station Paper 132.

Hewlett JD (1964a) Research in hydrology of forested headwaters of the Coweeta Hydrologic Laboratory. Transactions of the twenty-ninth North American Wildlife and Natural Resource Conference, March 9–11, 1964. Wildl. Manage. Inst. 103–112.

Hewlett JD (1964b) Coweeta Hydrologic Laboratory Plans for the research program. (Period 1964–1969). Unpublished report on file at Coweeta Hydrologic Laboratory, United States Department of Agriculture, Forest Service, Otto, North Carolina.

Hewlett JD (1967) A hydrologic response map for the State of Georgia. Water Resour. Bull. 3:4–20.

Hewlett JD (1971) Comments on the catchment experiment to determine vegetal effects on water yield. Water Resour. Bull. 7:376–381.

Hewlett JD, Cunningham GB, Troendle CA (1977) Predicting storm flow and peakflow from small basins in humid areas by the R-index method. Water Resour. Bull. 13:231–254.

Hewlett JD and Douglass JE (1961) A method of calculating error of soil moisture volumes in gravimetric sampling. For. Sci. 7:265–272.

Hewlett JD and Douglass JE (1968) Blending forest uses. United Department of Agriculture, Forest Service, Southeastern Forest Experiment Station Research Paper SE-37, Asheville, North Carolina.

Hewlett JD, Douglass JE, and Clutter JL (1964) Instrumental and soil moisture variance using the neutron scattering method. Soil Sci. 97:19–24.

Hewlett JD, Fortson JC, and Cunningham GB (1984) Additional tests on the effect of rainfall intensity on stormflow and peak flow from wild-land basins. Water Resour. Res. 20:985–989.

Hewlett JD and Helvey JD (1970) Effects of forest clear-felling on the storm hydrograph. Water Resour. Res. 6:768–782.

Hewlett JD and Hibbert AR (1961) Increases in water yield after several types of forest cutting. Int. Assoc. Sci. Hydrol. Bull. 6:5–17.

Hewlett JD and Hibbert AR (1963) Moisture and energy conditions within a sloping soil mass during drainage. J. Geophys. Res. 68:1080–1087.

Hewlett JD and Hibbert AR (1966) Factors affecting the response of small watersheds to precipitation in humid areas. pp. 275–290. In International Symposium on Forest Hydrology, Proceedings of a National Science Foundation Advanced Science Seminar. Pergamon Press, New York.

Hewlett JD and Nutter WL (1970) The varying source area of streamflow from upland basins. Interdisciplinary Aspects of Watershed Management, Proc. Symp., Bozeman, Montana, Am. Soc. Civil Eng., 65–83.

Hewlett JD and Troendle CA (1975) Non-point and diffused water sources: A variable source area problem. pp. 21–45. In Watershed Management. American Society of Civil Engineers, Logan, Utah.

Hibbert AR (1966) Forest treatment effects on water yield. pp. 527–543. In Proceedings of a National Science Foundation advanced science seminar, international symposium on forest hydrology. Pergamon Press, New York.

Hibbert AR (1969) Water yield changes after converting a forested catchment to grass. Water Resour. Res. 5:634–640.

Hibbert AR and Cunningham GB (1966) Streamflow data processing opportunities and application. pp. 725–736. In Sopper WB and Lull HW (editors), Forest Hydrology, Proceedings of the International Symposium. Pergamon Press, New York.

Hobbie J, Daley R, and Jasper S (1977) Use of nucleopore filters for counting bacteria by fluorescence microscopy. Appl. Environ. Microbiol. 33:1225–1228.

Hodges TK (1973) Ion absorption by plant roots. pp. 163–207. In Brady NC (editor), Advances in Agronomy. Academic Press, New York.

Holdren GR Jr and Berner RA (1979) Mechanisms of feldspar weathering—I. Experimental studies. Geochim. Cosmochim. Acta 43:1161–1171.

Holdridge L (1967) Life Zone Ecology. Tropical Science Center, San Jose, Costa Rica.

Holling CS (1973) Resilience and stability of ecological systems. Annu. Rev. Ecol. Syst. 4:1–24.

Holling CS (1981) Forest insects, forest fire, and resilience. pp. 445–464. In Mooney HA, Bonnickson TM, Christensen NL, Lotan JE, and Reiners WE (Technical Coordinators), Proceedings of the Conference, Fire Regimes and Ecosystem Properties. General Technical Report, WO-26. United States Department of Agriculture Forest Service, Washington, DC.

Hoover MD (1944) Effect of removal of forest vegetation upon water-yields. Trans. Am. Geophys. Union Part 6:969–977.

Hoover MD and Hursh CR (1943) Influence of topography and soil depth on runoff from forest land. Trans. Am. Geophys. Union Part 2:693–698.

Hoppe E (1896) Precipitation measurement under tree crowns. Translated from German by AH Krappe, Division of Silvics, U.S. Forest Service, 1935, Translation No. 291, 50.

Hopson CA (1958) Exfoliation and weathering at Stone Mountain, Georgia and their bearing on the disfigurement of the confederate memorial. Georgia Miner. Newsl. 11:65–79.

Horn H (1976) The adaptive geometry of trees. Princeton University Press, Princeton.

Horton JW (1982) Geology and mineral resources summary of the Rosman quadrangle, North Carolina. North Carolina Geological Survey GM 185-NE, scale 1/24,000.

Horton RE (1933) The role of infiltration in the hydrologic cycle. Trans. Am. Geophys. Union 14:446–460.

Horton RE (1943) Discussion of "Infiltration capacities of some plant-soil complexes on Utah range watershed lands" by Lowell Woodward. Trans. Am. Geophys. Union 24:473–475.

Horton RE (1945) Erosional development of streams and their drainage basins: Hydrophysical approach to quantitative geomorphology. Bull. Geol. Soc. Am. 56:275–370.

Howard PJA and Howard DM (1974) Microbial decomposition of tree and shrub leaf litter. Oikos 25:341–352.

Howell FG, Gentry JB, and Smith MH (editors) (1975) Mineral cycling in Southeastern Ecosystems, Proceedings Symposium at Augusta, GA, National Technical Information Service, U.S. Department of Commerce, Springfield, VA, 898.

Huff DD and Swank WT (1985) Modelling changes in forest evapotranspiration. pp. 125–151. In Anderson MG and Burt TP (editors), Hydrological Forecasting. John Wiley and Sons, Ltd., Chichester.

Hunter CE (1941) Forsterite olivine deposits of North Carolina and Georgia. Georgia Department of Mines, Mining and Geology Bulletin 47.

Hurlbert SH (1984) Pseudoreplication and the design of ecological field experiments. Ecol. Monogr. 54:187–211.

Hursh CR (1932a) Statement of problem—General discussion of streamflow and erosion studies in the southern Appalachian Mountains. Unpublished Report. Coweeta Hydrologic Laboratory, United States Department of Agriculture, Forest Service, Otto, North Carolina.

Hursh CR (1932b) Flood conditions of the southern Appalachian Mountain Region. Unpublished Report. Coweeta Hydrologic Laboratory, United States Department of Agriculture, Forest Service, Otto, North Carolina.

Hursh CR (1935) Control of exposed soil on road banks. United States Department of Agriculture, Forest Service, Appalachian Forest Experiment Station Technical Note 12, Asheville, North Carolina.

Hursh CR (1936) Storm water and absorption. Trans. Am. Geophys. Union, Part 2, 301–302.

Hursh CR (1938) Mulching for road bank fixation. United States Department of Agriculture, Forest Service, Appalachian Forest Experiment Station Technical Note 31, Asheville, North Carolina.

Hursh CR (1939) Roadbank stabilization at low cost. United States Department of Agriculture, Forest Service, Appalachian Forest Experiment Station Technical Note 38, Asheville, North Carolina.

Hursh CR (1941) The geomorphic aspects of mudflows as a type of accelerated erosion in the Southern Appalachians. Am. Geophys. Union Report and Papers part II:253–254.

Hursh CR (1942a) The Coweeta Experimental Forest. Unpublished Report. Coweeta Hydrologic Laboratory, United States Department of Agriculture, Forest Service, Otto, North Carolina.

Hursh CR (1942b) Naturalization of roadbanks – an integral part of highway construction. Roads Bridges (Canada) 80(7):22–26, 131–134.

Hursh CR (1942c) Naturalized roadbanks. Better Roads 12(6):13–15, 24–25; 12(7):17–20.

Hursh CR (1942d) The naturalization of roadbanks. United States Department of Agriculture, Forest Service, Appalachian Forest Experiment Station Technical Note 51, Asheville, North Carolina.

Hursh CR (1943) The Coweeta Experiment Forest Progress Report. Historical Report. Unpublished report on file at Coweeta Hydrologic Laboratory, United States Department of Agriculture, Forest Service, Otto, North Carolina.

Hursh CR (1944) Appendix B – Report of the subcommittee on subsurface flow. Trans. Am. Geophys. Union 25:743–746.

Hursh CR (1945) Plants, shrubs, trees in slope stabilization. Contract. Eng. Mon. 42:26–27.

Hursh CR (1949) Climatic factors controlling roadside design and development. In Proceedings of the Highway Research Board, National Research Council, Washington, DC, 9–19.

Hursh CR and Brater EF (1941) Separating storm hydrographs into surface- and subsurface-flow. Trans. Am. Geophys. Union 22:863–871.

Hursh CR and Hoover MD (1942) Soil profile characteristics pertinent to hydrologic studies in the southern Appalachians. Soil Sci. Soc. Am. Proc. 6:414–422.

Hursh CR, Hoover MD, and Fletcher PW (1942) Studies in the balanced water economy of experimental drainage areas. Trans. Am. Geophys. Union, Part 2 509–517.

Hursh RA and Fletcher PW (1943) The soil profile as a natural reservoir. Soil Sci. Soc. Am. Proc. 7:480–486.

Hurst VJ (1955) Stratigraphy, structure and mineral resources of the Mineral Bluff quadrangle, Georgia. Georgia Department of Mines, Min. Geol. Bull. 63.

Huryn AD and Wallace JB (1985) Life history and production of *Goerita semata* Ross (Trichoptera: Limnephilidae) in the southern Appalachian mountains. Can. J. Zoo. 63:2604–2611.

Huryn AD and Wallace JB (1986). A method for obtaining in situ growth rates of larval Chironomidae (Diptera) and its application to studies of secondary production. Limnol. Oceanogr. 31:216–222.

Hutchinson BA and Matt DR (1976) Beam enrichment of diffuse radiation in a deciduous forest. Agric. Meteorol. 17:93–110.

Hutchinson GE (1965) The Ecological Theater and the Evolutionary Play. Yale University Press, New Haven.

Hutchinson KJ and King KL (1980) The effect of sheep stocking level on invertebrate abundance, biomass and energy utilization in a temperate, sown grassland. J. Appl. Ecol. 17:369–387.

Hutchinson TC and Havas M (1980) Effects of acid precipitation on terrestrial ecosystems. Plenum Press, New York, 654.

Hynes HBN (1983) Groundwater and stream ecology. Hydrobiologia 100:93–99.

Iglich EM (1975) Population age structures and dynamics of six hardwood species in a forested watershed ecosystem in the southern Appalachians. M.S. Thesis, University of Georgia, Athens.

Ingham RE, Trofymow JA, Ingham ER, and Coleman DC (1985) Interactions of bacteria, fungi and their nematode grazers: Effects on nutrient cycling and plant growth. Ecol. Monogr. 55:119–140.

Inners JD and Wilshusen JP (1983) Anatomy of a landslide in Luzerne County, Pennsylvania. Pennsylvania Geol. Surv. 14:12–16.

Jackson LWR (1967) Effect of shade on leaf structure of deciduous tree species. Ecology 48:498–499.

Jackson ML (1958) Soil chemical analysis. Prentice-Hall, Englewood Cliffs, New Jersey.

Jackson NM Jr (1976) Magnitude and frequency of floods in North Carolina. United States Geological Survey, Water Resources Investigations 76-17, Raleigh, North Carolina, USA.

Jacobs WP (1979) Plant Hormones and Plant Development. Cambridge University Press, Cambridge.

Jaworoski Z (1975) Stable and radioactive pollutants in a Scandinavian glacier. Environ. Pollut. 9:305–315.

John KR (1964) Survival of fish in intermittent streams of the Chiricahua Mountains, Arizona. Ecology 45:112–119.

Johnson AH and Siccama TG (1983) Acid deposition and forest decline. Environ. Sci. Technol. 17:294–305.

Johnson AH, Siccama TG, and Friedland AJ (1982) Spatial and temporal patterns of lead accumulation in the forest floor in the northeastern United States. J. Environ. Qual. 11:577–580.

Johnson AH, Siccama TG, Wang D, Turner RS, and Barringer TH (1981) Recent changes in patterns of tree growth rate in New Jersey Pinelands: a possible effect of acid rain. J. Environ. Qual. 10:427–430.

Johnson DW (1984) The effects of harvesting intensity on nutrient depletion in forests. pp. 157–166. In Ballard P and Gessel SP (editors), Proceedings of the IUFRO Symposium on Forest Site and Continuous Productivity. Gen. Tech. Rept. PNW-163, Pacific Northwest Forest and Range Experiment Station, USDA Forest Service, Portland, Ore.

Johnson DW, Hornbeck JW, Kelly JM, Swank WT, and Todd DE (1980) Regional patterns of soil sulfate accumulation: relevance to ecosystem sulfur budgets. pp. 507–520. In Shriner DS, Richmond CR, and Lindberg SE (editors), Atmospheric Sulfur Deposition: Environmental Impact and Health Effects. Ann Arbor Press, Ann Arbor.

Johnson EA (1952) Effect of farm woodland grazing on watershed values in the southern Appalachian Mountains. J. For. 50:109–113.

Johnson PL and Swank WT (1973) Studies on cation budgets in the southern Appalachians on four experimental watersheds with contrasting vegetation. Ecology 54:70–80.

Johnson PS (1975) Growth and structural development of red oak sprout clumps. For. Sci. 21:413–417.

Jones L (1955) A watershed study in putting a hardwood forest at the Coweeta Hydrologic Laboratory in the southern Appalachian Mountains under intensive management. Graduate Problem, University of Georgia, Athens.

Kaczmarek W (1967) Elements of organization in the energy flow of forest ecosystems (preliminary notes). pp. 683–685. In Petrusewicz K (editor), Secondary Productivity in Terrestrial Ecosystems. Panstwowe Wydawnictwo Naukowe, Warsaw, Poland.

Kaczmarek M and Wasilewski A (1977) Dynamics of numbers of the leaf-eating insects and its effect on foliage production in the "Grabowy" Reserve in the Kampinos National Park. Ekol. Pol. 25:653–673.

Kaplan LA and Bott TL (1982) Algal generated DOC diel fluctuations in a piedmont stream. Limnol. Oceanogr. 27:1091–1100.

Kaplan LA and Bott TL (1983) Microbial heterotrophic utilization of dissolved organic matter in a piedmont stream. Freshwater Biol. 13:363–377.

Kaplan LA, Larson RA, and Bott TL (1980) Patterns of dissolved organic carbon in transport. Limnol. Oceanogr. 25:1034–1043.

Kazmierczak RF Jr, Webster JR, and Benfield EF (In press) Characteristics of seston in a regulated Appalachian mountain river, USA. In Regulated Rivers.

Keefer RK (1984) Rock avalanches caused by earthquakes; source characteristics. Science 223:1288–1289.

Keeney DR (1980) Prediction of soil nitrogen availability in forest ecosystems: A literature review. For. Sci. 26:1591–171.

Keeney DR (1982) Nitrogen-availability indices. pp. 711–734. In Page AL (editor), Methods of Soil Analysis, Part 2 – Chemical and Microbiological Properties. American Society of Agronomy, Madison.

Keeney DR and Nelson DW (1982) Nitrogen-inorganic forms. pp. 643–698. In Page AL (editor), Methods of Soil Analysis, Part 2 – Chemical and Microbiological Properties. American Society of Agronomy, Madison.

Keller KM (1965) Hydrologische Beobachtungen im Flyschgebiet beim Schwarzsee (Kt. Freiburg). Mitt. Schweiz. Anst. Forstl. Versuchswes. 41:23–60.

Keller HM (1970) Der Chemismus kleiner Bache in teilweise bewaldeten Einzugsgebieten in der Flyschzone eines Voralpentales. Mitt. Schweiz. Anst. Forstl. Versuchswes. 46:113–155.

Keller HM (In press) The export of nutrients from forest lands, a literature review. Proceedings MAB 5-workshop on in Budapest, September 1983.

Keller HM and Strobel T (1982) Water balance and nutrient budgets in subalpine basins of different forest cover. pp. 683–694. In Proceedings of Symposium on Hydrological Research Basins, Bern.

Keith A (1970a) Description of the Pisgah quadrangle, North Carolina-South Carolina. U.S. Geological Survey Atlas Folio 147.

Keith A (1970b) Description of the Nantahala quadrangle, North Carolina-Tennessee. U.S. Geological Survey Atlas Folio 143.

Keith A (1952) Geologic map of the Coweeta quadrangle, North Carolina-South Carolina. U.S. Geological Survey Open File Map, scale 1/125,000.

Kimmins JP (1977) Evaluation of the consequences for future tree productivity of the loss of nutrients in whole-tree harvesting. For. Ecol. Manage. 1:169–183.

King PB (1964) Geology of the central Great Smoky Mountains, Tennessee. U.S. Geological Survey Professional Paper 349-C.

Kirkby MJ and Chorley RJ (1967) Throughflow, overland flow and erosion. Bull. Int. Assoc. Sci. Hydrol. 12:5–21.

Kitchell JF, O'Neill RV, Webb D, Gallepp GW, Bartell SM, Koonce JF, and Ausmus BS (1979) Consumer regulation of nutrient cycling. Bioscience 29:28–34.

Knowles R (1982) Free-living dinitrogen-fixing bacteria. pp. 1071–1092. In Page AL (editor), Methods of Soil Analysis, Part 2 – Chemical and Microbiological Properties. American Society of Agronomy, Madison.

Kochenderfer JN (1970) Erosion control on logging roads in the Appalachians. United States Department of Agriculture, Forest Service, Northeastern Forest Experiment Station Research Paper NE-158, Broomall, Pennsylvania.

Kochenderfer JN, Wendel GW, and Smith HC (1984) Cost of and soil loss on "minimum-standard" forest truck roads constructed in the central Appalachians. United States Department of Agriculture, Forest Service, Northeastern Forest Experiment Station Research Paper NE-544, Broomall, Pennsylvania.

Kovner JL (1955) Changes in streamflow and vegetation characteristics brought about by a forest cutting and subsequent natural regrowth. Ph.D. Dissertation, State College of New York, Syracuse.

Kovner JL (1956) Evapotranspiration and water yields following forest cutting and natural regrowth. Proc. Soc. Am. For. 1956:106–110.

Krynine DP and Judd WR (1957) Principles of engineering geology and geotechnics. McGraw-Hill, New York.

Kuserk FT, Kaplan LA, Bott TL (1984) In situ measures of dissolved organic carbon flux in a rural stream. Can. J. Fish. Aquat. Sci. 41:964–973.

Ladd TI, Ventullo RM, Wallis PM, and Costerton JW (1982) Heterotrophic activity and biodegradation of labile and refractory compounds by groundwater and stream microbial populations. Appl. Environ. Microbiol. 44:321–329.

Langbein WB (1974) Topographic characteristics of drainage basins. U.S. Geological Survey Water-Supply Paper 968-C, 127–157.

Larcher W (1975) Physiological plant ecology. Springer-Verlag, Berlin.

Larson RA (1978) Dissolved organic matter of a low-coloured stream. Freshwater Biol. 8:91–104.

Lee R (1970) Theoretical estimates versus forest water yield. Water Resour. Rec. 6:1327–1334.

Lee R (1978) Forest Microclimatology. Columbia University Press, New York.

Lefcoff LJ (1981) Application of the variable source area simulator to a forested Piedmont watershed. M.S. Thesis, University of Georgia, Athens.

Legenre L and Demers S (1984) Towards dynamic biological oceanography and limnology. Can. J. Fish. Aquat. Sci. 41:2–19.

Leitch CJ and Flinn DW (1983) Residues of hexazinone in streamwater after aerial application to an experimental catchment planted with radiata pine. Aust. For. 46:126–131.

Leopold DJ and Parker GR (1985) Vegetation patterns on a southern Appalachian watershed after two clearcuts. Castanea 50:164–186.

Leopold LB, Wolman MG, and Miller JP (1964) Fluvial processes in geomorphology. W.H. Freeman and Company, San Francisco.

Lesure FG (1968) Mica deposits of the Blue Ridge in North Carolina. U.S. Geological Survey Professional Paper 577.

Leyton L, Reynolds ERC, and Thompson FB (1967) Rainfall interception in forest and moorland. pp. 163–178. In Sopper WE and Lull HW (editors), International Symposium on Forest Hydrology. Pergamon Press, Oxford.

Lieberman JA and Hoover MD (1948a) The effect of uncontrolled logging on stream turbidity. Water Sewage Works 95:255–258.

Lieberman JA and Hoover MD (1948b) Protecting quality of streamflow by better logging. South Lumberman 177:236–240.

Liebscher HJ and Wagenhoff A (1978) Abschlussbericht zum Forschungsvorhaben Niederschlag, Abfluss und Verdunstung in bewaldeten Einzugsgebieten im Mittelgebirge. Interner Bericht, 114 S.

Liebscher JH and Wilke K (1981) Simulation of runoff by means of statistical processes in the experimental basins in the upper Harz mountains. Proceedings of the IUFRO workshop on water and nutrient simulation models. Swiss Federal Institute for Forestry Research Birmensdorf 87–110.

Likens GE and Bilby RE (1982) Development, maintenance, and role of organic-debris dams in New England streams. pp. 122–128. In Swanson FJ, Janda RJ, Dunne T, and Swanson DN (editors), Sediment budgets and routing in forested drainage basins. U.S. Forest Service Pacific Northwest Forest and Range Experiment Station General Technical Report PNW-141.

Likens GE and Bormann FH (1974) Acid rain: a serious regional environmental problem. Science 184:1176–1179.

Likens GE, Bormann FH, and Johnson NM (1968) Nitrification: Importance to nutrient losses from a cutover forested ecosystem. Science 163:1205–1206.

Likens GE, Bormann FH, Johnson NM, Fisher DW, and Pierce RS (1970) The effects of forest cutting and herbicide treatment on nutrient budgets in the Hubbard Brook watershed-ecosystem. Ecol. Monogr. 40:23–47.

Likens GE, Bormann FH, Pierce RS, Eaton JS, and Johnson NM (1977) Biogeochemistry of a Forested Ecosystem. Springer-Verlag, New York.

Likens GE, Bormann FH, Pierce RS, and Reiners WA (1978) Recovery of a deforested ecosystem. Science 199:492–496.

Likens GE and Butler TJ (1981) Recent acidification of precipitation in North America. Atmos. Environ. 15:1103–1109.

Lindberg SE and Harris RC (1981) The role of atmospheric decomposition in an eastern U.S. deciduous forest. Water Air Soil Pollut. 16:13–31.

Lindberg SE and Harris RC (1983) Water and acid-soluble trace metals in atmospheric particles. J. Geophys. Res. 88:5091–5100.

Lindberg SE and Turner RR (1983) Trace metals in rain at forested sites in the Eastern United States. pp. 107–114. In Proceedings of the International Conference on Heavy Metals in the Environment. CEP Consultants, Edinburgh.

Linthurst RA and Altschuller AP (1984) The acidic deposition phenomenon and its effects: Critical assessment review papers Vols I & II. EPA-600/8-83-016BF. U.S. Environmental Protection Agency, Office of Research and Development, Washington, DC.

Livingston JL (1966) Geology of the Brevard zone and Blue Ridge in southwestrn Transylvania County, North Carolina. Dissertation, Rice University, Houston.

Lock MA (1981) River epilithon—a light and energy transducer. pp. 3–40. In Lock MA and Williams DD (editors), Perspectives in Running Water Ecology. Plenum Press, New York.

Lock MA and Hynes HBN (1975) The disappearance of four leaf leachates in a hard and soft water stream in southwestern Ontario, Canada. Internationale Revue der Gesamten Hydrobiologie 60:847–855.

Lock MA and Hynes HBN (1976) The fate of "dissolved" organic carbon derived from autumn-shed maple leaves (*Acer saccharum*) in a temperate hardwater stream. Limnol. Oceanogr. 21:436–443.

Lorimer CG (1980) Age structure and disturbance history of a southern Appalachian virgin forest. Ecology 61:1169–1185.

Lorimer CG (1984) Development of the red maple understory in northeastern oak forests. For. Sci. 30:3–22.

Loucks OL (1970) Evolution of diversity, efficiency, and community stability. Am. Zoo. 10:17–25.

Loveless AR (1961) A nutritional interpretation of sclerophylly based on differences in the chemical composition of sclerophyllous and mesophytic leaves. Ann. Bot. 25:168–184.

Loveless AR (1962) Further evidence to support a nutritional interpretation of sclerophylly. Ann. Bot. 26:551–561.

Lovett GM, Reiners WA, and Olson RK (1982) Cloud droplet deposition in subalpine balsam fir forests: hydrological and chemical inputs. Science 218:1303–1304.

Lowdermilk WC (1933) Forests and streamflow: A discussion of the Hoyt-Troxell report. J. For. 31:296–307.

Lowe RL, Golladay SW, and Webster JR (1986) Periphyton response to nutrient manipulation in streams draining clearcut and forested watersheds. J. North Amer. Benth. Soc. 5:221–229.

Lush DL and Hynes HBN (1973) The formation of particles in freshwater leachates of dead leaves. Limnol. Oceanogr. 18:968–977.

Lush DL and Hynes HBN (1978) The uptake of dissolved organic matter by a small spring stream. Hydrobiologia 60:271–275.

Lyell C (1875) Principles of Geology, 12th ed., v. I, John Murray, London.

MacMahon JA (1981) Successional processes: comparisons among biomes with special reference to probable roles of and influences on animals. pp. 277–304. In West DC, Shugart HH, and Botkin DB (editors), Forest Succession: Concepts and Applications. Springer-Verlag, New York.

Malas D and Wallace JB (1977) Strategies for coexistence in three species of net-spinning caddisflies (Trichoptera) in a second order southern Appalachian stream. Can. J. Zoo. 55:1829–1840.

March WJ, Wallace JR, and Swift LW Jr (1979) An investigation into the effect of storm type on precipitation in a small mountain watershed. Water Resour. Res. 15:298–304.

Marks PL (1974) The role of pin cherry in the maintenance of stability in northern hardwood ecosystems. Ecol. Monogr. 44:73–88.

Marks PL and Bormann FH (1972) Revegetation following forest cutting: Mechanisms for return to steady-state nutrient cycling. Science 176:914–915.

Martinez JA (1975) Seasonal trends of CO_2 exchange in two understory evergreen shrubs in the eastern deciduous forest. M.S. Thesis, University of Georgia, Athens.

Mattson KG (1986) Soil organic carbon cycling following clearcutting. Ph.D. Dissertation, University of Georgia, Athens.

Mattson WJ (1980) Herbivory in relation to plant nitrogen content. Annu. Rev. Ecol. Syst. 11:119–161.

Mattson WJ and Addy ND (1975) Phytophagous insects as regulators of forest primary production. Science 190:515–522.

May RM (1977) Thresholds and breakpoints in ecosystems with a multiplicity of stable states. Nature 269:471–477.

Mayack DT, Bush PB, Neary DG, and Douglass JE (1982) Impact of hexazinone on invertebrates after application to forested watersheds. Arch. Environ. Contam. Toxicol. 11:209–217.

McClaughtery CA, Aber JD, and Mellilo JM (1982) The role of fine roots in organic matter and nitrogen budgets of two forest ecosystems. Ecology 63:1481–1490.

McConathy RK (1983) Tulip-poplar leaf diffusion resistance calculated from stomatal dimensions and varying environmental parameters. For. Sci. 29:139–148.

McDiffett WF (1970) The transformation of energy by a stream detritivore, *Pteronarcys scotti* (Plecoptera). Ecology 51:975–988.

McGee CE (1978) Size and age of tree affect white oak stump sprouting. United States Forest Service Research Note SO-239.

McGee CE and Hooper RM (1970) Regeneration after clearcutting in the southern Appalachians. United States Forest Service Research Note SE-70.

McGee CE and Hooper RM (1975) Regeneration trends ten years after clearcutting of an Appalachian hardwood stand. United States Forest Service Research Note SE-227.

McGinty DT (1972) The ecological role of *Kalmia latifolia* L. and *Rhododendron maximum* L. in the hardwood forest at Coweeta. M.S. Thesis, University of Georgia, Athens.

McGinty DT (1976) Comparative root and soil dynamics on a white pine watershed in the hardwood forest in the Coweeta Basin. Ph.D. Dissertation, University of Georgia, Athens.

McKniff JM (1967) Geology of the Highlands-Cashiers area, North Carolina, South Carolina and Georgia. Dissertation, Rice University, Houston.

McLaughlin SB, McConathy RK, and Dinger BE (1978) Seasonal changes in respiratory metabolism of yellow-poplar (*Liriodendron tulipifera*) branches. J. Appl. Ecol. 15:327–334.

McNaughton SJ (1983) Compensatory plant growth as a response to herbivory. Oikos 40:329–336.

McNaughton SJ (1985) The ecology of a grazing ecosystem: The Serengeti. Ecol. Monogr. 55:259–294.

McNaughton SJ and Wolf LL (1979) General Ecology. 2nd Edition. Holt, Rinehart and Winston, New York.

Meade BK (1971) Report of the subcommission on recent crustal movements in North America. See Bollinger GA 1973 (op. cit).

Meentemeyer V (1978) Macroclimate and lignin control of litter decomposition rates. Ecology 59:465–472.

Meginnis HG (1964) Watershed Management Research (Research Programs, Problem Selection–Analysis, Project 1601, Water Yield Improvement.) 4300 Memo for the record. Unpublished report on file at Coweeta Hydrologic Laboratory, United States Department of Agriculture, Forest Service, Otto, North Carolina, USA.

Meinzer OE (1942) Hydrology. Dover Publications, New York. 712.

Mellilo JM (1981) Nitrogen cycling in deciduous forests. pp. 427–442. In Clark FE and Rosswall T (editors), Terrestrial Nitrogen Cycles: Processes, Ecosystem Strategies and Management Impacts. Ecological Bulletin, (Stockholm) 33.

Mellilo JM, Aber JD, and Muratore JF (1982) Nitrogen and lignin control of hardwood leaf litter decomposition dynamics. Ecology 63:621–626.

Merritt RW and Cummins KW (editors) (1984) An introduction to the aquatic insects of North America. 2nd edition. Kendall/Hunt Publishing Company, Dubuque.

Messina FJ (1981) Plant protection as a consequence of an antmembracid mutualism: interactions on goldenrod (*Solidago* spp.). Ecology 62:1433–1440.

Meyer JL and O'Hop J (1983) Leaf-shredding insects as a source of dissolved organic carbon in headwater streams. Am. Midl. Nat. 109:175–183.

Meyer JL and Tate CM (1983) The effects of watershed disturbance on dissolved organic carbon dynamics of a stream. Ecology 64:33–44.

Milewski AV (1983) A comparison of ecosystems in Mediterranean Australia and southern Africa: Nutrient-poor sites at the barrens and the caleden coast. Annu. Rev. Ecol. Syst. 14:57–76.

Millot G and Bonifas M (1955) Transformations isovolumetriques dans les phenomens de lateritisation et de bauxitisation. Bull. Ser. Carte. Geol. Alsace Lorraine 8:3–20.

Mitchell JE, Waide JB, and Todd RL (1975) A preliminary compartment model of the nitrogen cycle in a deciduous forest ecosystem. pp. 41–57. In Howell FG (editor), Mineral Cycling in Southeastern Ecosystems. ERDA Symposium Series. (CONF-740513).

Moeller JR, Minshall GW, Cummins KW, Petersen RC, Cushing CE, Sedell JR, Larson RA, and Vannote RL (1979) Transport of dissolved organic carbon in streams of differing physiographic characteristics. Organ. Geochem. 1:139–150.

Mohr DW (1975) Geology of the Noland Creek quadrangle, North Carolina. North Carolina Geological Survey GM 158-NE, scale 1/24,000.

Molles MC Jr (1982) Trichoptera communities of streams associated with aspen and conifer forests: long term structural change. Ecology 63:1–6.

Monk CD (1966) An ecological significance of evergreenness. Ecology 47:504–505.

Monk CD (1971) Leaf decomposition and loss of ^{45}Ca from deciduous and evergreen trees. Am. Midl. Nat. 86:379–384.

Monk CD (1975) Nutrient losses in particulate form as weir pond sediments from four unit watersheds in the southern Appalachians. pp. 862–876. In Howell FG, Gentry JB, and Smith MH (editors), Mineral Cycling in Southeastern Ecosystems. ERDA Symposium Series. (CONF-740513).

Monk CD, Child G, and Nicholson S (1970) Biomass, litter and leaf surface area estimates of an oak-hickory forest. Oikos 21:138–141.

Monk CD, Crossley DA Jr, Todd RL, Swank WT, Waide JB, and Webster JR (1977) An overview of nutrient cycling research at Coweeta Hydrologic Laboratory. pp. 35–50. In Correll DL (editor), Watershed research in eastern North America: a workshop to compare results. Smithsonian Institution.

Monk CD and Day FP Jr (1984) Vegetation analysis, primary production and selected nutrient budgets for a southern Appalachian oak forest: A synthesis of IBP studies at Coweeta. For. Ecol. Manage. 10:87–113.

Monk CD, McGinty DT, and Day FP Jr (1985) The ecological importance of *Kalmia latifolia* and *Rhododendron maximum* in the deciduous forest of the southern Appalachians. Bull. Torrey Bot. Club 112:187–193.

Montagnini F (1985) Nitrogen turnover and leaching from successional and mature ecosystems in the Southern Appalachians. Ph.D. Dissertation, University of Georgia, Athens.

Mooney HA (1983) Carbon-gaining capacity and allocation patterns of Mediterranean-climate plants. In Kruger FG, Mitchell DT, and Jarvis JVM (editors), Mediterranean-type ecosystems. Springer-Verlag, New York.

Moore A and Swank WT (1975) A model of water content and evaporation for hardwood leaf litter. pp. 58–69. In Howell FG, Gentry JB, and Smith MH (editors), Mineral cycling in southeastern ecosystems. Energy Research and Development Agency Symposium Series. (CONF-740513).

Mopper K (1978a) Improved chromatographic separations on anion-exchange resins. I. Partition chromatography of sugars in ethanol. Anal. Biochem. 85:528–532.

Mopper K (1978b) Improved chromatographic separations on anion-exchange resins. III. Sugars in borate medium. Anal. Biochem. 87:162–168.

Morowitz HJ (1978) Foundations of Bioenergetics. Academic Press, New York.

Morrison IK (1984) Acid rain, a review of literature on acid deposition effects in forest ecosystems. For. Abstr. 45:483–506.

Mosley MP (1981) The influence of organic debris on channel morphology and bedload transport in a New Zealand forest stream. Earth Surf. Processed Landforms 6:571–579.

Mosley MP (1982) Subsurface flow velocities through selected forest soils, South Island, New Zealand. J. Hydrol. 55:65–92.

Mowbray TB and Oosting HJ (1968) Vegetation gradients in relation to environment and phenology in a southern Blue Ridge gorge. Ecol. Monogr. 38:309–344.

Mowry FL (1980) Measurements of boundary layer flow over a forested mountain slope. Ph.D Dissertation, Duke University, Durham.

Mulholland PJ (1981) Formation of particulate organic carbon in water from a southeastern swamp-stream. Limnol. Oceanogr. 26:790–795.

Murphy CE Jr (1970) Energy sources for the evaporation of precipitation intercepted by tree canopies. Ph.D. Dissertation, Duke University, Durham.

Naegeli W (1959) Die Wassermesstationen im Flyschgebiet beim Schwarzsee (Kt. Freiburg). Mitt. Schweiz. Anst. Forstl. Versuchswes. 35:225–241.

Naiman RJ (1982) Characteristics of sediment and organic carbon export from pristine boreal forest watersheds. Can. J. Fish. Aquat. Sci. 39:1699–1718.

Naiman RJ and Sedell JR (1979) Benthic organic matter as a function of stream order in Oregon. Arch. Hydrobiol. 87:404–422.

National Academy of Sciences (NAS) (1981) Atmospheric-Biosphere Interactions. National Research Council. National Academy Press, Washington, DC.

National Academy of Sciences (NAS) (1983) Atmospheric Processes in Atmospheric Deposition. National Research Council. National Academy Press, Washington, DC.

NBS, Standard Reference Materials Catalog (1981) NBS Special Publication 260. Washington, DC, 114.

Neary DG, Bush PB, and Douglass JE (1981) 2-, 4-, and 14-month efficacy of hexazinone for site preparation. Proc. South Weed Sci. Soc. 34:181–191.

Neary DG, Bush PB, and Douglass JE (1983) Off-site movement of hexazinone in stormflow and baseflow from forest watersheds. Weed Sci. 31:543–551.

Neary DG, Bush PB, Douglass JE, and Todd RL (1985) Picloram movement in an Appalachian hardwood forest watershed. J. Environ. Qual. 14:585–592.

Neary DB, Bush PB, and Grant MA (1986) Water quality of ephemeral forest streams after site preparation with hexazinone. For. Ecol. Manage. 14:23–40.

Neary DG, Douglass JE, and Fox W (1979) Low picloram concentrations in streamflow resulting from forest application of Tordon 10K. Proc. South Weed Sci Soc. 32:182–197.

Neary DG, Douglass JE, Ruehle JL, and Fox W (1984) Converting rhododendron-laurel thickets to white pine using a picloram herbicide and mycorrhizae-inoculated seedlings. South. J. Appl. For. 8:164–169.

Needleman HL (1980) Lead exposure and human health: recent data on an ancient problem. Tech. Rev., March/April, 39–45.

Neet KE (1953) Cooperativity in enzyme function: Equilibrium and kinetic aspects. pp. 267–320. In Purich DL (editor), Contemporary Enzyme Kinetics and Mechanisms. Academic Press, New York.

Nelson DW and Sommers LE (1982) Total carbon, organic carbon, and organic matter. pp. 539–580. In Page AL (editor), Methods of Soil Analysis, Part 2 – Chemical and Microbiological Properties. American Society of Agronomy, Madison.

Nelson TC (1955) Chestnut replacement in the southern highlands. Ecology 39:352–353.

Neufeld HS, Jernstedt JA, and Haines BL (1985) Direct foliar effects of simulated acid rain. I. Damage, growth and gas exchange. New Phytol. 99:389–405.

Neumann RB, Nelson WH (1965) Geology of the western Great Smoky Mountains, Tennessee. U.S. Geological Survey Professional Paper 349-D.

Newell K (1984a) Interaction between two decomposed basidiomycetes and a collembolan under sitka spruce: distribution abundance and selective grazing. Soil Biol. Biochem. 16:227–234.

Newell K (1984b) Interaction between two decomposed basidiomycetes and a collembolan under sitka spruce: grazing and its potential effects on fungal distribution and litter decomposition. Soil Biol. Biochem. 16:235–240.

Nicholson PB, Borock KL, and Heal OWH (1966) Studies on the decomposition of the fecal pellets of a millipede (*Glomeris marginata* Villers). J. Ecol. 54:755–766.

Nielsen DR, Kirkham D, and Van Wyck WK (1959) Measuring water stored temporarily above the field moisture capacity. Soil Sci. Soc. Am. Proc. 23:408–412.

Nobel PS, Zaragoza LJ, and Smith WK (1975) Relation between mesophyll surface area, photosynthetic rate and illumination level during development for leaves of *Plectranthus parviflorus* Henckel. Plant Physiol. 55:1067–1070.

Norris LA (1981) The behavior of herbicides in the forest environment and risk assessment. p. 305. In Holt HH and Fischer BC (editors), Weed Control in Forest Management. Proceedings of the 1981 J.S. Wright Forestry Conference, Purdue University, Feb. 3–5, 1981.

Nriagu JO (1978) Lead in the Atmosphere. pp. 137–183. In Nriagu JO (editor), The Bio-geochemistry of Lead in the Environment. Elsevier, Amsterdam.

O'Doherty EC (1982) The life history of a stream-dwelling harpacticoid copepod. M.S. Thesis, University of Georgia, Athens.

O'Doherty EC (1985) Stream-dwelling copepods: their life history and ecological significance. Limnol. Oceanogr. 30:554–564.

Odom AL, Kish SA, and Leggo PJ (1973) Extension of "Grenville basement" to the southern extremity of the Appalachians: U-Pb ages of zircons. Geol. Soc. Am. Abstr. Progr. 5:425.

Odum EP (1953) Fundamentals of Ecology. W.B. Saunders, Philadelphia.

Odum EP (1969) The strategy of ecosystem development. Science 164:262–270.

Ogata G and Richards LA (1957) Water content changes following irrigation of bare field soil that is protected from evaporation. Soil Sci. Soc. Am. Proc. 21:355–356.

O'Hop JR Jr, Wallace JB, and Haefner JD (1984) Production of a stream shredder, *Peltoperia maria* (Plecoptera:Peltoperlidae) in disturbed and undisturbed hardwood catchments. Freshwater Biol. 14:13–21.

Olson JS, Watts JA, and Allison LJ (1983) Carbon in live vegetation of major world ecosystems. ORNL-5862. Environmental Sciences Division, Oak Ridge National Laboratory, Oak Ridge.

O'Neill RV and Waide JB (1981) Ecosystem theory and the unexpected: Implications for environmental toxicology. pp. 43–73. In Cornaby BW (editor), Management of Toxic Substances in Our Ecosystems: Taming the Medusa. Ann Arbor Science Publishers, Inc., Ann Arbor.

O'Neill RV, DeAngelis DL, Waide JB, and Allen THF (1986) A Hierarchical Concept of Ecosystems. Princeton University Press, Princeton.

Onstad CA and Brakensiek DL (1968) Watershed simulation by stream path analogy. Water Resour. Res. 4:965–971.

Ovington JD (1965) Organic production, turnover and mineral cycling in woodlands. Biol. Rev. 40:295–336.

Page FS (1927) Living stumps. J. For. 25:687–690.

Pačes T (1983) Rate constants of dissolution derived from the measurements of mass balance in hydrological catchments. Geochim Cosmochim Acta 47:1855–1863.

Parker GR and Swank WT (1982) Tree species response to clearcutting a Southern Appalachian watershed. Amer. Midl. Nat. 108:304–310.

Parkhurst D and Loucks OL (1972) Optimal leaf size in relation to environment. J. Ecol. 60:505–537.

Parkinson D (1980) Aspects of microbial ecology of forest ecosystems. pp. 109–117. In Waring RH (editor), Forests: Fresh Perspectives from Ecosystem Analysis. Corvallis, Oregon, Oregon State University.

Patric JH and Douglass JE (1965) Soil moisture flux following mid-season defoliation on plastic-covered forest plots. Trans. Am. Geophys. Union 46:57.

Patric JH, Douglass JE, and Hewlett JD (1965) Soil water absorption by mountain and piedmont forests. Soil Sci. Soc. Am. Proc. 29:303–308.

Patten BC and Odum EP (1981) The cybernetic nature of ecosystems. Am. Nat. 118:886–895.

Patten BC and Witkamp M (1967) Systems analysis of ^{134}cesium kinetics in terrestrial microcosms. Ecology 48:814–824.

Penman HL (1959) Notes on the water balance of the Sperbelgraben and Rappengraben. Eidg. Anst. Forstl. Versuchswes. Mitt., Birmensdorf 35:99–109.

Penman HL (1963) Vegetation and hydrology. Technical Communication No. 53. Commonwealth Bureau of Soils, Harpendon. 124.

Petelle M (1980) Aphids and melezitose: a test of Owen's 1978 hypothesis. Oikos 35:127–128.

Petersen H and Luxton M (1982) A comparative analysis of soil fauna populations and their role in decomposition processes. Oikos 39:287–388.

Phillips DL and Murdy WH (1985) Effects of rhododendron (*Rhododendron maximum* L.) on regeneration of southern Appalachian hardwoods. For. Sci. 31:226–233.

Phillips DR and Saucier JR (1979) A test of prediction equations for estimating hardwood understory and total stand biomass. Georgia Forestry Commission Research Paper 7.

Phillips SO, Skelly JM, and Burkhart HE (1977) Growth fluctuations of loblolly pine due to periodic air pollution levels: interaction of rainfall and age. Phytopathology 67:716–720.

Pilgrim DH, Huff DD, and Steel TD (1978) A field evaluation of subsurface and surface runoff, II Runoff Processes. J. Hydrol. 38:319–341.

Pimm SL (1980) The complexity and stability of ecosystems. Nature 307:321–326.

Pittillo JD and Lee M (1984) Reference plant collection of the Coweeta Hydrologic Laboratory. USDA Forest Service General Technical Report SE-29.

Prigogine I (1980) From Being to Becoming: Time and Complexity in the Physical Sciences. W.H. Freeman, San Francisco.

Radford AE, Ahles HE, and Bell CR (1964) Manual of the Vascular Flora of the Carolinas. University of North Carolina Press, Chapel Hill.

Ragsdale HL, Berish CW, and Swank WT (1984) Chemical element comparisons between control watersheds using LTER reference plots at the Coweeta Hydrologic Laboratory. Abstract. 35th Annual AIBS meeting.

Reav I, Shipovenski D, and Zyapkov L (1981) Forest Hydrology Research in Bulgaria. J. Hydrol. Sci. (Bulgaria) 8:11–16.

Record FA, Bubenick DV, and Kindya RJ (1982) Acid Rain Information Book. Noyes Data Corp., Park Ridge, NY.

Reichle DE (editor) (1981) Dynamic properties of forest ecosystems. Cambridge University Press, Cambridge.

Reichle DE and Crossley DA Jr (1969) Trophic level concentrations of cesium-137, sodium and potassium in forest arthropods. pp. 687–695. In Nelson DJ and Evans FC (editors), Symposium on Radioecology. USAEC, Washington, DC.

Reichle DE, Goldstein RA, Van Hook RI Jr, and Dodson GJ (1973) Analysis of insect consumption in a forest canopy. Ecology 54:1076–1084.

Reichle DE, O'Neill RV, and Harris WF (1975) Principles of energy and material exchange in ecosystems. pp. 27–43. In van Dobben WH and Lowe-McConnell RH (editors), Unifying Concepts in Ecology. Dr. W. Junk Pub., The Hague.

Reiners WA (1973) Terrestrial detritus and the carbon cycle. pp. 303–327. In Woodwell GM and Pecan EV (editors), ERDA Symposium Series. (CONF-720510).

Remsen I, Randolf JR, and Barksdale HC (1960) The zone of aeration and ground water recharge in sandy sediments at Seabrook, New Jersey. Soil Sci. 89:145–156.

Resh VH and Rosenberg DM (1984) Ecology of aquatic insects. Praeger Publishers, New York.

Reynolds BC, McSwain MR, and Beale RH (1986) Procedures for chemical analyses at the Coweeta Hydrologic Laboratory. Unpublished Report, Coweeta Hydrologic Laboratory, Otto, North Carolina.

Reynolds LJ (1976) Relative abundance of forest floor Collembola as influenced by leaf litter decomposition rate. M.S. Thesis, University of Georgia, Athens.

Rhoades DF (1983) Herbivore population dynamics and plant chemistry. pp. 155–220. In Denno RF and McClure MS (editors), Variable Plants and Herbivores in Natural and Managed Systems. Academic Press, New York.

Richards PW (1952) The Tropical Rain Forest. Cambridge University Press.

Rickman RW, Ramig RE, and Allmaras RR (1975) Modeling dry matter accumulation in dryland winter wheat. J. Agron. 67:283–289.

Rieper M (1978) Bacteria as food for marine harpacticoid copepods. Mar. Biol. 45:337–345.

Rieper M and Flotow C (1981) Feeding experiments with bacteria, ciliates and harpacticoid copepods. Kiel. Meeresforsch., Sonderheft 5:370–375.

Risley LS (1983) Canopy Arthropod Dynamics in Successional and Mature Deciduous Forest Ecosystems. M.S. Thesis, University of Tennessee, Knoxville.

Robbins JS, Pruitt WO, and Gardner GH (1954) Unsaturated flow in field soils and its effect on soil moisture investigations. Soil Sci. Soc. Am. Proc. 18:344–347.

Roberts G, Hudson JA, and Blackie JR (1983) Nutrient cycling in the Wye and Severn at Plinlimon. Report 86, Institute of Hydrology, Wallingford.

Roberts SW, Strain BR, and Knoerr KR (1980) Seasonal patterns of leaf water relations in four co-occurring forest tree species: parameters from pressure-volume curves. Oecologia 46:330–337.

Robertson GP and Tiedje JM (1984) Denitrification and nitrous oxide production in successional and old-growth Michigan forests. Soil Sci. Soc. Am. J. 48:383–389.

Rodin LY and Brazilevich NI (1967) Production and Mineral Cycling in Terrestrial Vegetation. Oliver and Boyd, London.

Roessel BWP (1950) Hydrologic problems concerning the runoff in headwater regions. Trans. Am. Geophys. Union 31:431–442.

Root RB (1967) The niche exploitation pattern of the blue-gray flycatcher. Ecol. Monogr. 37:317–350.

Ross DH and Wallace JB (1981) Production of *Brachycentrus spinae* Ross (Trichoptera: Brachycentridae) and its role in seston dynamics of a southern Appalachian stream (U.S.A.). Environ. Entomol. 10:240–246.

Ross DH and Wallace JB (1982) Factors influencing the longitudinal distribution of larval Hydropsychidae (Trichoptera) in a southern Appalachian stream system (USA). Hydrobiologia 96:185–199.

Ross DH and Wallace JB (1983) Longitudinal patterns of production, food consumption and seston utilization by net-spinning caddisflies (Trichoptera) in a southern Appalachian stream (U.S.A.). Holarct. Ecol. 6:270–284.

Rounick JS and Winterbourn MJ (1983) Leaf processing in two contrasting beech forest streams: effects of physical abiotic factors on litter breakdown. Arch. Hydrobiol. 96:448–477.

Rowe R, Todd RL, and Waide JB (1976) A micro-technique for MPN analysis. Appl. Environ. Microbiol. 33:675–680.

Ryan DF and Bormann FH (1982) Nutrient resorption in a northern hardwood forest. Bioscience 32:29–32.

Ryan CA (1983) Insect-induced chemical signals regulating natural plant protection processes. p. 717. In Denno RF and McClure MS (editors), Variable Plants and Herbivores in Natural and Managed Systems. Academic Press, New York.

Sander IL (1972) Size of oak advance reproduction: key to growth following harvest cutting. United States Forest Service Research Paper NC-79.

Santos PF and Whitford WG (1981) The effects of microarthropods on litter decomposition in a Chihuahuan desert ecosystem. Ecology 62:654–663.

Schaeffer MF, Stevens RE, and Hatcher RD (1979) In situ stress and its relationship to joint formation in the Toxaway Gneiss, northwestern South Carolina. Southeast. Geol. 20:129–143.

Schaill HA (1936) Parshall Flumes—Coweeta Testing Station. Unpublished Report on file at Coweeta Hydrologic Laboratory, United States Department of Agriculture, Forest Service, Otto, North Carolina.

Schindler JE, Waide JB, Waldron MC, Hains JJ, Schreiner SP, Freeman ML, Benz SL, Pettigrew DR, Schissel LA, and Clarke PJ (1980) A microcosm approach to the study of biogeochemical systems: 1. Theoretical Rationale. pp. 192–203. In Geisy JP (editor), Microcosms in Ecological Research. DOE Symposium Series. (CONF-781101).

Schlesinger WH and Reiners WA (1974) Deposition of water and cations on artificial foliar collectors in fir krumholz of New England mountains. Ecology 55:378–386.

Schlesinger WH (1977) Carbon balance in terrestrial detritus. Annu. Rev. Ecol. Syst. 8:51–82.

Schnitzer M and Skinner SIM (1967) Organometallic interactions in soils: 7. Stability constants of Pb^{2+-}, Ni^{2+-}, Mn^{2+-}, Co^{2+-}, Ca^{2+-}, and Mg^{2+-} fulvic acid complexes. Soil Sci. 103:247–252.

Scholl DG and Hibbert AR (1973) Unsaturated flow properties used to predict outflow and evapotranspiration from a sloping lysimeter. Water Resour. Res. 9:1645–1655.

Schowalter TD (1981) Insect herbivore relationship to the state of the host plant: biotic regulation of ecosystem nutrient cycling through ecological succession. Oikos 37:126–130.

Schowalter TD (1985) Adaptations of Insects to Disturbance. pp. 235–252. In Pickett STA and White PS (editors), The Ecology of Natural Disturbance and Patch Dynamics. Academic Press, New York.

Schowalter TD, Coulson RN, and Crossley DA Jr. (1981a) Role of southern pine beetle and fire in maintenance of structure and function of the southeastern coniferous forest. Environ. Entomol. 10:821–825.

Schowalter TD and Crossley DA Jr (1983) Forest canopy arthropods as sodium, potassium, magnesium and calcium pools in forests. For. Ecol. Manage. 7:143–148.

Schowalter TD, Pope DN, Coulson RN, and Fargo WS (1981b) Patterns of southern pine beetle (*Dendroctonus frontalis* Zimm.) infestation enlargement. For. Sci. 27:837–849.

Schowalter TD, Webb JW, and Crossley DA Jr. (1981c) Community structure and nutrient content of canopy arthropods in clearcut and uncut forest ecosystems. Ecology 62:1010–1019.

Schultz JC and Baldwin IT (1982) Oak leaf quality declines in response to defoliation by Gypsy moth larvae. Science 217:149–151.

Sealy JL, Dick D., Arvik GH, Zindahl RL, and Skogerboe RK. (1972) Determination of lead in soil. Appl. Spectrosc. 26:456–460.

Seastedt TR (1979) Microarthropod response to clearcutting in the southern Appalachians: effects on decomposition and mineralization of litter. Ph.D. Dissertation. University of Georgia, Athens.

Seastedt TR (1984) The role of microarthropods in the decomposition and mineralization of litter. Annu Rev. Ecol. Syst. 29:25–46.

Seastedt TR and Crossley DA Jr (1980) Effects of microarthropods on the nutrient dynamics of forest litter. Soil Biol. Biochem. 12:337–342.

Seastedt TR and Crossley DA Jr (1981) Microarthropod response following cable logging and clearcutting in the southern Appalachians. Ecology 62:126–135.

Seastedt TR and Crossley DA Jr (1983) Nutrients in forest litter treated with naphthalene and simulated throughfall: A field microcosm study. Soil Biol. Biochem. 15:159–165.

Seastedt TR and Crossley DA Jr (1984) The influence of arthropods on ecosystems. Bioscience 34:157–161.

Seastedt TR, Crossley DA Jr, and Hargrove WW (1983) The effects of low-level consumption by canopy arthropods on the growth and nutrient dynamics of black locust and red maple trees in the southern Appalachians. Ecology 64:1040–1048.

Seastedt TR, Mameli L, and Gridley K (1981) Arthropod use of invertebrate carrion. Amer. Midl. Nat. 105:124–129.

Seastedt TR, Crossley DA Jr, Meentemeyer V, and Waide JB (1983) A two-year study of leaf litter decomposition as related to macroclimatic factors and microarthropod abundance in the Southern Appalachians. Holarct. Ecol. 6:11–16.

Seastedt TR and Tate CM (1981) Decomposition rates and nutrient contents of arthropod remains in forest litter. Ecology 62:13–19.

Sedell JR, Naiman RJ, Cummins KW, Minshall GW, and Vannote RL (1978) Transport of particulate organic matter in streams as a function of physical processes. Verh. Int. Ver. Limnol. 20:1366–1375.

Sevruk B (1970) Vergleichende Niederschlagsmessungen im Gebiet der Baye de Montreux. Mitteilungen Versuchsanstalt fur Wasserbau, Hydrologie und Glaziologie, Nr. 85:8.

Sharpe CFS (1938) see Coates DR (editor) (1977) Landslides. Geological Society of America, Reviews in Engineering Geology. 3:6.

Shimwell DW (1971) Description and Classification of Vegetation. Sidgwick and Jackson, Ltd., London.

Shugart HH Jr and West DC (1977) Development of an Appalachian deciduous forest succession model and its application to assessment of the impact of the chestnut blight. J. Environ. Manage. 5:161–179.

Siccama TG, Smith WH, and Mader DL (1980) Changes in lead, zinc, copper, dry weight, and organic matter content of the forest floor of white pine stands in Central Massachusetts over 16 years. Environ. Sci. Technol. 14:54–56.

Siegel DI (1984) The effect of kinetics on the geochemical balance of a small watershed on mafic terrain. EOS, Trans. Am. Geophys. Union 65:211.

Silsbee DG and Larson GL (1983) A comparison of streams in logged and unlogged areas of Great Smoky Mountains National Park. Hydrobiologia 102:99–111.

Simon HA (1962) The architecture of complexity. Proc. Am. Philos. Soc. 106:467–482.

Sluder ER (1958) Mountain farm woodland grazing doesn't pay. United States Department of Agriculture, Forest Service, Southeastern Forest Experiment Station Research Note 119, Asheville, North Carolina.

Smith DM (1962) The Practice of Silviculture. John Wiley and Sons, Inc., New York.

Smith HW (1963) Establishment of yellow-poplar in canopy openings. Ph.D. Dissertation, Yale University, New Haven.

Smith MS and Tiedje JM (1979) Phases of denitrification following oxygen depletion in soil. Soil Biol. Biochem. 11:261–267.

Smith WH (1973) Metal contamination of urban woody plants. Environ. Sci. Technol. 7:631–636.

Smith WH and Siccama TG (1981) The Hubbard Brook ecosystem study: Biogeochemistry of lead in the northern hardwood forest. J. Environ. Qual. 10:323–333.

Smith-Cuffney FL and Wallace JB (1987) The influence of microhabitat on availability of drifting invertebrate prey to a net-spinning caddisfly. Freshwater Biol. (in press).

Smolik JR and Dodd JL (1983) Effect of water, nitrogen, and grazing on nematodes in a shortgrass prairie. J. Range Manage. 36:744–748.

Sokal RR and Rohlf FJ (1969) Biometry. W.H. Freeman, San Francisco.

Sollins P. (1972) Organic matter budget and model for a southern Appalachian *Liriodendron* forest. Ph.D. Dissertation, University of Tennessee, Knoxville.

Spring PE (1973) Population dynamics of *Quercus prinus* L., *Acer rubrum* L., and *Cornus florida* L. in a forested watershed ecosystem. M.S. Thesis, University of Georgia, Athens.

Sprugel GD and Bormann FH (1981) Natural disturbance and the steady state in high-altitude forests. Science 211:390–393.

Steinberger Y, Freckman DW, Parker LW, and Whitford WG (1984) Effects of simulated rainfall and litter quantities on desert soil biota: nematodes and microarthropods. Pedobiologia 26:267–274.

Stephensen GR and Freeze RA (1974) Mathematical simulations of subsurface flow contributions to snowmelt runoff, Reynolds Creek Watershed, Idaho. Water Resour. Res. 10:284–298.

Strickland TC and Fitzgerald JW (1983) Mineralization of sulphur in sulphoquinovose by forest soils. Soil Biol. Biochem. 15:347–349.

Strickland TC and Fitzgerald JW (1984) Formation and mineralization of organic sulfur in forest soils. Biogeochemistry 1:79–95.

Strickland TC, Fitzgerald JW, and Swank WT (1984) Mobilization of recently formed forest soil organic sulfur. Can. J. For. Res. 14:63–67.

Stone LL and Skelly JM (1974) The growth of two forest tree species adjacent to a periodic source of air pollution. Phytopathology 64:773–778.

Susmel L (editor) (1981) Ricerche idrologiche comparate in ecosistemi di foresta e di prateria. Atti dell Instituto di ecologia e silvicoltura. Universita degli studi Padova II, 4:96–238.

Sutherland JP (1974) Multiple stable points in natural communities. Am. Nat. 108:859–873.

Swank WT and Caskey WH (1982) Nitrate depletion in a second-order mountain stream. J. Environ. Qual. 11:581–584.

Swank WT and Douglass JE (1974) Streamflow greatly reduced by converting deciduous hardwood stands to white pine. Science 185:857–859.

Swank WT and Douglass JE (1975) Nutrient flux in undisturbed and manipulated forest ecosystems in the southern Appalachian Mountains. Publication No. 117 de l'Association Internationale des Sciences Hydrologiques Symposium de Tokyo (Decembre 1975), 445–456.

Swank WT and Douglass JE (1977) Nutrient budgets for undisturbed and manipulated hardwood forest ecosystems in the mountains of North Carolina. pp. 343–364. In Correll DL (editor), Watershed Research in Eastern North America: a Workshop to Compare Results. Smithsonian Institution, Edgewater, MD.

Swank WT, Douglass JE, and Cunningham GB (1982) Changes in water yield and storm hydrographs following commercial clearcutting on a southern Appalachian catchment. pp. 583–594. In Proceedings of Symposium on Hydrological Research Basins, Bern.

Swank WT, Fitzgerald JW, and Ash JT (1984) Microbial transformation of sulfate in forest soils. Science 223:182–184.

Swank WT, Fitzgerald JW, and Strickland TC (1985) Transformations of sulfur in forest floor and soil of a forest ecosystem. pp. 137–145. In Johansson I (editor), Hydrological and hydrogeochemical mechanisms and model approaches to the acidification of ecological systems: International Hydrological Programme NHP-Report No. 10; Uppsala, Sweden; Oslo, Norway.

Swank WT, Goebel NB, and Helvey JD (1972) Interception loss in loblolly pine stands of the South Carolina piedmont. J. Soil Water Conserv. 27:160–164.

Swank WT and Helvey JD (1970) Reduction of streamflow increases following regrowth of clearcut hardwood forests. pp. 346–360. In Symposium on the results of research on representative and experimental basins. International Association of Scientific Hydrology Publication.

Swank WT and Henderson GS (1976) Atmospheric inputs of some cations and anions to forest ecosystems in North Carolina and Tennessee. Water Resour. Res. 12:541–546.

Swank WT and Miner NH (1968) Conversion of hardwood-covered watersheds to white pine reduces water yield. Water Resour. Res. 4:947–954.

Swank WT and Swank WTS (1984) Dynamics of water chemistry in hardwood and pine ecosystems. pp. 335–346. In Burt TP and Walling DE (editors), Catchment Experiments in Fluvial Geomorphology: Proceedings of a Meeting of the International Geophysical Union Commission on Field Experiments in Geomorphology. Exeter and Huddersfield, United Kingdom, Geo Books, Norwich, UK.

Swank WT and Waide JB (1980) Interpretation of nutrient cycling research in a management context: evaluating potential effects of alternative management strategies on site productivity. pp. 137–158. In Waring RH (editor), Forests: Fresh Perspectives From Ecosystem Analysis. Oregon State University Press, Corvallis.

Swank WT, Waide JB, Crossley DA Jr, and Todd RL (1981) Insect defoliation enhances nitrate export from forest ecosystems. Oecologia 51:297–299.

Swanson FJ, Fredriksen RL, and McCorison FM (1982) Material transfer in a western Oregon forested watershed. pp. 233–266. In Edmonds RL (editor), Analysis of coniferous forest ecosystems in the western United States. Hutchinson Ross, Stroudsburg.

Swanson FJ, Gregory SV, Sedell JR, and Campbell AG (1982) Land-water interactions: the riparian zone. pp. 267–291. In Edmonds RL (editor), Analysis of coniferous forest ecosystems in the western United States. Hutchinson Ross, Stroudsburg.

Swanson KA and Johnson AH (1980) Trace metal budgets for a forested watershed in the New Jersey Pine Barrens. Water Resour. Res. 16:373–376.

Swift LW Jr (1960) The effect of mountain topography upon solar energy theoretically available for evapotranspiration. Thesis, North Carolina State University, Raleigh.

Swift LW Jr (1964) A review of the relation of rain gage catch to actual precipitation on steep forested watersheds. Unpublished Report. Coweeta Hydrologic Laboratory, United States Department of Agriculture, Forest Service, Otto, North Carolina.

Swift LW Jr (1968) Comparison of methods for estimating areal precipitation totals for a mountain watershed. Am. Meteorol. Soc. Bull. 49:782–783.

Swift LW Jr (1972) Effect of forest cover and mountain physiography on the radiant energy balance. Dissertation, Duke University, Durham.

Swift LW Jr (1976) Algorithm for solar radiation on mountain slopes. Water Resour. Res. 12:108–112.

Swift LW Jr (1982) Duration of stream temperature increases following forest cutting in the southern Appalachian mountains. Proc. Int. Symp. Hydrometeorol. 273–275.

Swift LW Jr (1984a) Soil losses from roadbeds and cut and fill slopes in the southern Appalachian Mountains. South. J. Appl. For. 8:209–216.

Swift LW Jr (1984b) Gravel and grass surfacing reduce soil loss from mountain roads. For. Sci. 30:657–670.

Swift LW Jr (1986) Filter strip widths for forest roads in the southern Appalachians. South. J. Appl. For. 10:27–34.

Swift LW Jr and Cunningham GB (1986) Routines for collecting and summarizing hydrometrological data at Coweeta Hydrologic Laboratory. pp. 301–320. In Michener WK (editor), Research Data Management in the Ecological Sciences. Belle W. Baruch Library in Marine Science Number 16. University of South Carolina Press, Columbia.

Swift LW Jr and Knoerr KR (1973) Estimating solar radiation on mountain slopes. Agric. Meteorol. 12:329–336.

Swift LW Jr and Messer JB (1971) Forest cuttings raise temperatures of small streams in the southern Appalachians. J. Soil Water Conserv. 26:111–116.

Swift LW Jr and Schreuder HT (1981) Fitting daily precipitation amounts using the S_B distribution. Mon. Weather Rev. 109:2535–2541.

Swift LW Jr and Swank WT (1981) Long term responses of streamflow following clearcutting and regrowth. Hydrol. Sci. Bull. 26:245–256.

Swift LW Jr, Swank WT, Mankin JB, Luxmoore RJ, and Goldstein RA (1975) Simulation of evapotranspiration and drainage from mature and clearcut deciduous forests and young pine plantation. Water Resour. Res. 11:667–673.

Swift MJ, Heal OW, and Anderson JM (1979) Decomposition in Terrestrial Ecosystems. Blackwell Science, Oxford.

Switzer GL and Nelson LE (1972) Nutrient accumulation and cycling in a loblolly pine (*Pinus taeda* L.) plantation ecosystem, the first twenty years. Soil Sci. Soc. Am. Proc. 36:143–147.

Tabatabai JA and Bremner JM (1970) An alkaline oxidation method for determination of total sulfur in soils. Soil Sci. Soc. Am. Proc. 34:62–65.

Teasley JI (1984) Acid precipitation Series. Vol. 1–9. Butterworth Publishers, Boston.

Tebo LB Jr (1955) Effects of siltation, resulting from improper logging, on the bottom fauna of a small trout stream in the southern Appalachians. Prog. Fish Cult. 17:64–70.

Technicon (1971a) Low level ammonia in fresh and estuarine water. Industrial Method Number 108-71W/AAII. Technicon Industrial Systems, Tarrytown, New York.

Technicon (1971b) Nitrate and nitrite in water and waste water. Industrial Method Number 100-70W/AAII. Technicon Industrial Systems, Tarrytown, New York.

Thomas WA (1969) Accumulation and cycling of calcium by dogwood trees. Ecol. Monogr. 39:101–120.

Thornthwaite CW (1948) An approach toward a rational classification of climate. Geogr. Rev. 38:55–94.

Tichendorf WG (1969) Tracing storm flow to varying source areas in a small forested watershed in the southeastern Piedmont. Ph.D. Dissertation, University of Georgia, Athens.

Tilman D (1978) Cherries, ants and tent caterpillars: timing of nectar production in relation to susceptibility of caterpillars to ant predation. Ecology 59:686–692.

Todd RL, Meyer RD, and Waide JB (1978) Nitrogen fixation in a deciduous forest in the southeastern United States. Ecol. Bull., Stockholm 26:172–177.

Todd RL, Swank WT, Douglass JE, Kerr PC, Brockway DL, and Monk CD (1975a) The relationship between nitrate concentration in the southern Appalachian mountain streams and terrestrial nitrifiers. Agric. Ecosyst. Environ. 2:127–132.

Todd RL, Waide JB, and Cornaby BC (1975b) Significance of biological nitrogen fixation and denitrification in a deciduous forest ecosystem. pp. 729–735. In Howell FG, Gentry JB, and Smith MH (editors), Mineral Cycling in Southeastern Ecosystems. ERDA Symposium Series. (CONF-740513).

Tomlinson GH (1983) Air pollutants and forest decline. Environ. Sci. Technol.17:246–305.

Trewartha GT (1954) An introduction to climate. McGraw-Hill, New York.

Traaen TS (1980) Effects of acidity on decomposition of organic matter in aquatic environments. pp. 340–341. In Drablos D and Tollan A (editors), Proceedings of the International Conference on Ecological Impact of Acid Precipitation.

Trimble GR (1973) The regeneration of central Appalachian hardwoods, with emphasis on the effects of site quality and harvesting practice. U.S. Forest Service Research Note SE 282.

Triska FJ and Cromack K Jr (1980) The role of wood debris in forests and streams. p. 198. In Waring RH (editor), Forests: fresh perspectives from ecosystem analysis. Oregon State University Press, Corvallis.

Troendle CA (1979) A variable source area model for storm flow prediction on first-order forested watersheds. Ph.D. Dissertation, University of Georgia, Athens.

Troendle CA (1985) Variable source area models. pp. 347–403. In Anderson MG and Burt TP (editors), Hydrological Forecasting. John Wiley & Sons, New York.

Troendle CA and Homeyer JW (1971) Storm flow related to measured physical parameters on a small forested watershed in West Virginia. Trans. Am. Geophys. Union 52:204.

Tyler G (1972) Heavy metals pollute nature, may reduce productivity. Ambio 1:52–59.

Ulrich B, Mayer R, and Khanna PK (1980) Chemical changes due to acid precipitation in a loess-derived soil in Central Europe. Soil Sci. 130:193–199.

United States Department of Agriculture (1935) Drainage practice in the California Region, A supplement to the Forest Truck Trail Handbook. Forest Service California Region, San Francisco, California.

United States Department of Agriculture, Forest Service (1940) Superintendent's report. Unpublished report on file at Coweeta Hydrologic Laboratory, United States Department of Agriculture, Forest Service, Otto, North Carolina.

United States Department of Agriculture, Forest Service (1945) Superintendent's annual report. Unpublished report on file at Coweeta Hydrologic Laboratory, United States Department of Agriculture, Forest Service, Otto, North Carolina.

United States Department of Agriculture, Forest Service (1961) Some ideas about storm runoff and base flow. Southeast Forest Experiment Station, Annual Report, 62–66.

United States Department of Agriculture (1984a) Final environmental impact statement standards and guidelines for the southern regional guide. Forest Service Southern Region, Atlanta.

United States Department of Agriculture (1984b) Regional guide for the South. Forest Service Southern Region, Atlanta.

VanHassel JH, Ney JJ, and Garling DL (1979) Seasonal variation in the heavy metal concentrations of sediments influenced by highways of different traffic volumes. Bull. Environ. Contam. Toxicol. 23:592–596.

Van Hook RI Jr, Nielsen MG, and Shugart HH (1980) Energy and nitrogen relations for a *Macrosiphum liriodendri* (Homoptera: Aphididae) population in an east Tennessee *Liriodendron tulipifera* stand. Ecology 61:960–975.

Van Horn EC (1948) Talc deposits of the Murphy marble belt. North Carolina Geological Survey Bulletin 56.

Vannote RL (1978) A geometric model describing a quasi-equilibrium of energy flow in populations of stream insects. Proc. Nat. Acad. Sci. 75:381–384.

Vannote RL, Minshall GW, Cummins KW, Sedell JR, and Cushing CE (1980) The river continuum concept. Can. J. Fish. Aquat. Sci. 37:130–137.

Velbel MA (1982) Weathering and saprolitization in the southern Blue Ridge. XIth International Congress on Sedimentology, Abstracts, 171–172.

Velbel MA (1983) A dissolution-reprecipitation mechanism for the pseudomorphous replacement of plagioclase feldspar by clay minerals during weathering. pp. 139–147. In Hanon D and Noack Y (editors), Petrologie des Alterations et des Sols, Vol. 1. Sciences Geologiques, Memoires (Strasbourg).

Velbel MA (1984a) Weathering processes of rock-forming minerals. Chapter 4. pp. 67–111. In Fleet ME (editor), Environmental Geochemistry. Mineralogical Association of Canada Short Course Notes, vol. 10.

Velbel MA (1984b) Mineral transformations during rock weathering, and geochemical mass-balances in forested watersheds of the southern Appalachians. Ph.D. Dissertation, Yale University, New Haven.

Velbel MA (1984c) Natural weathering mechanisms of almandine garnet. Geology 12:631–634.

Velbel MA (1985a) Hydrogeochemical constraints on mass balances in forested watersheds of the Southern Appalachians. pp. 231–247. In Drever JI (editor), The Chemistry of Weathering. NATO Advanced Research Workshop, July, 1984 (Rodez, France), D. Reidel, Holland.

Velbel MA (1985b) Geochemical mass balances and weathering rates in forested watersheds of the southern Blue Ridge. Am. J. Sci. 285:904–930.

Velbel MA (1985c) Clay mineral formation and water-rock interactions during rock weathering in the southern Blue Ridge Mountains, USA. 8th International Clay Conference. Abstracts, p. 245.

Velbel MA (1986a) Chapter 18: The mathematical basis for determining rates of geochemical and geomorphic processes in small forested watersheds by mass balance: Examples and implications. pp. 439–451. In Coleman S and Dethier D (editors), Rates of Chemical Weathering of Rocks and Minerals. Academic Press, Orlando.

Velbel MA (1986b) Influence of surface area, surface characteristics, and solution composition on feldspar weathering rates. In Davis JA and Hayes KF (editors), Geochemical Processes at Mineral Surfaces. American Chem. Soc. Symposium Series, No. 323:615–634.

Vitousek PM (1981) Clear-cutting and the nitrogen cycle. pp. 631–642. In Clarke FE and Rosswall T (editors), Terrestrial Nitrogen Cycles: Processes, Ecosystem Strategies, and Management Impacts. Ecol. Bull., Stockholm, vol. 33.

Vitousek PM, Gosz JR, Grier CC, Melillo JM, Reiners WA, and Todd RL (1979) Nitrate losses from disturbed ecosystems. Science 204:469–474.

Vitousek PM and Matson PA (1984) Mechanisms of nitrogen retention in forest ecosystems: A field experiment. Science 225:51–52.

Vitousek PM and Melillo JM (1979) Nitrate losses from disturbed forests: Patterns and mechanisms. For. Sci. 25:605–619.

Vitousek PM and Reiners WA (1975) Ecosystem succession and nutrient retention: a hypothesis. BioScience 25:376–381.

Vogel DS (1984) Invertebrate consumers and leaf breakdown rates two years following pesticide treatment of a headwater stream. M.S. Thesis, University of Georgia, Athens.

Vogelmann HW (1982) Catastrophe on Camels Hump. Nat. Hist. 91:8–14.

Voight B (editor) (1978) Rockslides and Avalanches, 1. Natural Phenomena. Elsevier Scientific, New York.

Wagenhoff A and Von Wedel K (1959) Vergleich der bisherigen Ergebnisse der Untersuchungen uber den Wasserhaushalt und der Beobachtung im Harz mit den langfristigen Beobachtungen im Emmental in der Schweiz. Mitt. Schweiz Anst. Forstl. Versuchswes. 35:127–138.

Waide JB and Jager Y (1981) Methods for investigating the stability of wetland ecosystems. Report to the Institute of Water Resources, U.S. Army Corps of Engineers, Fort Belvoir.

Waide JB and Swank WT (1976) Nutrient recycling and the stability of ecosystems: Implications for forest management in the southeastern United States. pp. 404–424. Proc. Soc. Am. For.

Waide JB and Swank WT (1977) Simulation of potential effects of forest utilization on the nitrogen cycle in different southeastern ecosystems. pp. 767–789. In Correll DL (editor), Watershed Research in Eastern North America. Smithsonian Institution.

Waide JB, Schindler JE, Waldron MC, Hains JJ, Schreiner SP, Freedman ML, Benz SL, Pettigrew DR, Schissel LA, and Clark PJ (1980) A microcosm approach to the study of biogeochemical systems: responses of aquatic laboratory microcosms to physical, chemical, and biological perturbations. pp. 204–223. In Giesy JP (editor), Microcosms in Ecological Research. U.S. Department of Energy Symposium Series (CONF-781101), National Technical Information Service, Springfield, Va.

Waide JB, Swank WT, and Reynolds LJ (1986) Variation in space and time of element concentrations in soils and soil solutions in forested watershed ecosystems in the southern Appalachian mountains of North Carolina. In the IUFRO Symposium on Water and Nutrient Movement in Forest Soils: Spatial and Temporal Variation, Hampton Beach, NH (manuscript in preparation).

Waughman GJ, French RJ, and Jones K (1981) Nitrogen fixation in some terrestrial environments. pp. 135–192. In Broughton WJ (editor), Nitrogen Fixation, Volume I: Ecology. Clarendon Press, Oxford.

Walker LC (1957) Integration of forest and watershed management—Watersheds 40 and 41. Unpublished report on file at Coweeta Hydrologic Laboratory, United States Department of Agriculture, Forest Service, Otto, North Carolina.

Wallace JB and Gurtz ME (1986) Response of baetid mayflies (Ephemeroptera) to catchment logging. Am. Midl. Nat. 115:25–41.

Wallace JB, Ross DH, and Meyer JL (1982a) Seston and dissolved organic carbon dynamics in a southern Appalachian stream. Ecology 63:824–838.

Wallace JB, Webster JR, and Cuffney TF (1982b) Stream detritus dynamics: regulation by invertebrate consumers. Oecologia 53:197–200.

Wallace JB, Webster JR, and Woodall WR (1977) The role of filter feeders in stream ecosystems. Arch. Hydrobiol. 79:506–532.

Wallace LL (1978) Comparative photosynthesis of three gap phase successional tree species. Ph.D. Dissertation, University of Georgia, Athens.

Wallace LL and Dunn EL (1980) Comparative photosynthesis of three gap phase successional tree species. Oecologia 45:331–340.

Wallis PM, Hynes HBN, and Telang SA (1981) The importance of groundwater in the transportation of allochthonous dissolved organic matter to the streams draining a small mountain basin. Hydrobiologia 79:77–90.

Wallwork JA (1976) The Distribution and Diversity of Soil Fauna. Academic Press, London. 355.

Waring RH and Franklin JF (1979) Evergreen coniferous forests of the Pacific Northwest. Science 204:1380–1386.

Warren CF, Wales JH, Davis GE, and Doudoroff P (1964) Trout production in an experimental stream enriched with sucrose. J. Wildl. Manage. 28:617–660.

Waters TF (1979) Influence of benthos life history upon the estimation of secondary production. J. Fish. Res. Board Can. 36:1425–1430.

Weather Bureau (1961) Rainfall frequency atlas of the United States for durations from 30 minutes to 24 hours and return periods from 1 to 100 years. Technical Paper 40, United States Department of Commerce, Washington, DC.

Webb DP (1976) Roles of arthropod feces in deciduous litter decomposition processes. Ph.D. Dissertation, University of Georgia, Athens.

Webb DP (1977) Regulation of deciduous forest litter decomposition by soil arthropod feces. pp. 56–69. In Mattson WJ (editor), The Role of Arthropods in Forest Ecosystems. Springer-Verlag, New York.

Webb LJ (1968) Environmental relationships of the structural types of Australian rain forest vegetation. Ecology 49:296–311.

Webster JR (1979) Hierarchical organization and ecosystems. pp. 119–131. In Halfon E (editor), Theoretical Systems Ecology: Advances and Case Studies. Academic Press, New York.

Webster JR (1983) The role of benthic macroinvertebrates in detritus dynamics of streams: a computer simulation. Ecol. Monogr. 53:383–404.

Webster JR and Golladay SW (1984) Seston transport in streams at Coweeta Hydrologic Laboratory, North Carolina, USA. Verh. Int. Ver. Limnol. 22:1911–1919.

Webster JR, Gurtz ME, Hains JJ, Meyer JL, Swank WT, Waide JB, and Wallace JB (1983) Stability of stream ecosystems. pp. 355–395. In Barnes JR and Minshall GW (editors), Stream Ecology. Plenum Press, New York.

Webster JR and Patten BC (1979) Effects of watershed perturbation on stream potassium and calcium dynamics. Ecol. Monogr. 49:51–72.

Webster JR and Waide JB (1982) Effects of forest clearcutting on leaf breakdown in a southern Appalachian stream. Freshwater Biol. 12:331–334.

Webster JR, Waide JB, and Patten BC (1975) Nutrient recycling and the stability of ecosystems. pp. 1–27. In Howell FG, Gentry JB, and Smith MH (editors), Mineral Cycling in Southeastern Ecosystems. ERDA Symposium Series. (CONF-740513).

Weiss H, Bertine K, Koide M, and Goldberg ED (1975) Chemical composition of Greenland Glacier. Geochim Cosmochim. Acta 39:1–10.

Wells CG and Metz LJ (1963) Variation in nutrient content of loblolly pine needles with season, size, soil and position on crown. Soil Sci. Soc. Am. Proc. 27:90–93.

Wetzel RG and Manny BA (1977) Seasonal changes in particulate and dissolved organic carbon and nitrogen in a hardwater stream. Arch. Hydrobiol. 80:20–39.

Wetzel RG and Rich PH (1973) Carbon in freshwater ecosystems. pp. 241–263. In Woodwell GM and Pecan EV (editors), Carbon and the Biosphere. ERDA Symposium Series. (CONF-720510).

Weyman DR (1970) Throughflow on hillslopes and its relation to stream hydrographs. Int. Assoc. Sci. Hydrol. XV Annee, 25–33.

Whetstone BH (1982) Techniques for estimating magnitude and frequency of floods in South Carolina. United States Geological Survey, Water Resources Investigations 82-1, Columbia, South Carolina.

Whipkey RZ (1965) Subsurface storm flow from forested slopes. Bull. Int. Assoc. Sci. Hydrol. 10:74–85.

White DL (1986) Litter production, decomposition and nitrogen dynamics in black locust and pine-hardwood stands of the southern Appalachians. M.S. Thesis, University of Georgia, Athens.

White PS (1979) Pattern, process, and natural disturbance in vegetation. Bot. Rev. 45:229–299.

Whitford WG, Meentemeyer V, Seastedt TR, Cromack K Jr, Crossley DA Jr, Santos P, Todd RL, and Waide JB (1981) Exceptions to the AET Model: Deserts and clear-cut forest. Ecology 62:275–277.

Whitney HT, Sowers GF, and Carter BR (1971) Slides in residual soil from shale and limestone. Proc. 4th Pan-American Conference on soil mechanics and foundation engineering. San Juan, P.R.

Whittaker RH (1956) Vegetation of the Great Smoky Mountains. Ecol. Monogr. 26:1–80.

Whittaker RH (1966) Forest dimensions and production in the Great Smoky Mountains. Ecology 47:103–121.

Whittaker RH (1975) Communities and ecosystems. 2nd Edition. Macmillan, New York.

Whittaker RH, Borman FH, Likens GE, and Siccama TG (1974) The Hubbard Brook ecosystem study: forest biomass and production. Ecol. Monogr. 44:233–252.

Whittaker RH and Likens GE (1973a) Carbon in the biota. pp. 281–302. In Woodwell GM and Pecan EV (editors), Carbon and the Biosphere. ERDA Symposium Series. (CONF-720510).

Whittaker RH and Likens GE (1973b) The primary production of the biosphere. Hum. Ecol. 1:299–369.

Whittaker RH and Woodwell GM (1968) Dimension and production relations of trees and shrubs in the Brookhaven Forest, New York. J. Ecol. 56:1–25.

Whittaker RH and Woodwell GM (1972) The evolution of natural communities. pp. 137–160. In Weins JA (editor), Ecosystem Structure and Function. Oregon State University Press, Corvallis.

Wiederholm T (1984) Responses of aquatic insects to environmental pollution. pp. 508–557. In Resh VH and Rosenberg DM (editors), The Ecology of Aquatic Insects. Praeger Scientific, New York.

Wiklander L (1973/1974) The acidification of soil by acid precipitation. Grundforbattring 26:155–164.

Williams GP and Guy HP (1973) Erosional and depositional aspects of hurricane Camille in Virginia, 1969. U.S. Geological Survey Professional Paper 804,80.

Williams JG (1954) A study of the effect of grazing upon changes in vegetation on a watershed in the southern Appalachian Mountains. M.S. Thesis, Michigan State University, East Lansing.

Winner MD Jr (1977) Ground-water resources along the Blue Ridge Parkway, North Carolina. U.S. Geol. Surv. Water Resour. Invest. No. 77:65,166.

Winterbourn MJ, Rounick JS, and Cowie B (1981) Are New Zealand stream ecosystems really different? N.Z. J. Mar. Freshwater Res. 15:321–328.

Wischmeier WH and Smith DD (1978) Predicting rainfall erosion losses—a guide to conservation planning. Agriculture Handbook 537, United States Department of Agriculture, Washington, DC.

Witherspoon JP (1964) Cycling of cesium-134 in white oak trees. Ecol. Monogr. 34:403–420.

Witherspoon JP, Auerbach SI, and Olson JS (1962) Cycling of cesium-134 in white oak trees on sites of contrasting soil type and moisture. Oak Ridge National Laboratory Publication Number ORNL-3328, 143.

Witherspoon JP and Brown GN (1965) Translocation of cesium-137 from parent trees to seedlings of *Liriodendron tulipifera*. Bot. Gaz. 126:181–185.

Witkamp M (1969) Environmental effects on microbial turnover of some mineral elements. Part I—abiotic factors. Soil Biol. Biochem. 1:167–176.

Witkamp M and Frank ML (1970) Effects of temperature, rainfall and fauna on transfer of ^{137}Cs, K, Mg and mass in consumer-decomposer microcosms. Ecology 51:465–474.

Wood T and Bormann FH (1975) Increases in foliar leaching caused by acidification of an artificial mist. Ambio 4:169–171.

Woodall WR Jr and Wallace JB (1972) The benthic fauna of four small southern Appalachian streams. Am. Midl. Nat. 88:393–407.

Woodmansee RG (1978) Additions and losses of nitrogen in grassland ecosystems. Bioscience 28:448–453.

Woodruff JF and Hewlett JD (1971) Predicting and mapping the average hydrologic response for the eastern United States. Water Resour. Res. 6:1312–1326.

Woods DB and Turner NC (1971) Stomatal response to changing light by four tree species of varying shade tolerance. New Phytol 70:77–84.

Woods FW and Shanks RE (1959) Natural replacement of chestnut by other species. Ecology 40:349–361.

Woodwell GM (1970) Effects of pollution on the structure and physiology of ecosystems. Science 168:429–433.

Woodwell GM, Whittaker RH, Reiners WA, Likens GE, Delwiche CC, and Bodkin DB (1978) The biota and the world carbon budget. Science 199:141–146.

Yandle DO and Harms J (1970) Modeling of optimum multiresource allocation. Unpublished cooperative agreement final report, Duke University, on file at Coweeta Hydrologic Laboratory, United States Department of Agriculture, Forest Service, Otto, North Carolina.

Yamagata N and Shigematsu I (1970) Bull. Inst. Public Health (Tokyo) 19:1–27.

Yarimanoghu M and Ayers JE (1979) Development of a finite element watershed simulation model with a subsurface flow component. pp. 172–183. In Proceedings Hydrologic Transport Modelling Symposium.

Yodis P (1982) The compartmentation of real and assembled ecosystems. Am. Nat. 120:551–570.

Yount JD (1975) Forest-floor nutrient dynamics in southern Appalachian hardwood and white pine plantation ecosystems. pp. 598–608. In Howell FG and Smith MH (editors), Mineral Cycling in Southeastern Ecosystems. ERDA Symposium Series. (CONF-740513).

Zeller J (1982) Einige Hinweise zur Geschiebefuhrung von Wildbachen. Mitt. Forstl. Bundesversuchsanst. Wien 144:125–138.

Zimmerman MH and Brown CL (1974) Trees: structure and function. Springer-Verlag, New York.

Zinke PJ (1967) Forest interception studies in the United States. pp. 137–161. In Sopper WE and Lull HW (editors), International Symposium on Forest Hydrology. Pergamon Press, Oxford.

Zlotin RI and Khodashova KS (1980) The role of animals in biological cycling of forest-steppe ecosystems. English language edition edited by Norman R. French; translated by William Lewis and WE Grant. Dowden, Hutchinson and Ross, Stroudsburg.

Index